LANDSCAPES OF INEQUITY

Landscapes of Inequity

ENVIRONMENTAL JUSTICE IN THE ANDES-AMAZON REGION

Edited and with an introduction by
Nicholas A. Robins and Barbara J. Fraser

UNIVERSITY OF NEBRASKA PRESS | LINCOLN

♾

Library of Congress Control Number: 2019053414

Set in Questa by Laura Buis.
Designed by N. Putens.

CONTENTS

MAPS

TABLES

INTRODUCTION

BARBARA J. FRASER AND NICHOLAS A. ROBINS

More than three decades after the murder of Chico Mendes, the Brazilian rubber tapper whose death in 1988 galvanized global support for rainforest conservation, defense of the environment remains a hazardous commitment. Between 2017 and 2018 more than sixty-five people were killed because of their defense of the environment and land rights in Brazil, the highest number in Latin America, according to the nonprofit organization Global Witness, which tracks such murders worldwide. Colombia placed second in the Amazon region, with fifty-five deaths in those two years.[1] Such murders, in which the perpetrators often go unpunished or are only lightly sanctioned, highlight the threats Amazonian forest dwellers must confront as they continue to face encroaching economic interests backed by powerful politicians.

The killings signal the high stakes at play in the quest for environmental justice in the region that stretches from the eastern slopes of the Andes Mountains, where Amazonian rivers originate, to the lowland plain with its flooded forests. The Amazon basin has come to the fore in discussions of climate change because of its role in regulating global processes related to hydrological cycles and greenhouse gases.[2] The western Amazon is an area of particularly high biological diversity. A popular view of the Amazon as a vast, uninhabited wilderness, however, ignores the existence of indigenous groups that have occupied the region for millennia and have evolved together with the ecosystems they inhabit.

As a result of this "co-evolution of culture with nature," as Porto-Gonçalves and Leff describe it, "these peoples hold an enormous legacy of knowledge welded in their traditional practices for living sustainably within their ecological conditions."[3] Nevertheless the Andes-Amazon region remains a landscape of inequity for many of the communities that depend on its forests, rivers, lakes, and wetlands for their physical, cultural, and spiritual sustenance. Both the land and the people have been exploited for centuries in a cycle of commodity booms and busts. But while much has been written about the biology and ecology of the region, and although the struggle for territorial rights by indigenous Amazonians has become more visible in recent years, there has been little exploration of environmental justice issues focusing specifically on this region.

This book examines environmental justice in the Andes-Amazon region from diverse perspectives, reflecting legal, practical, historical, and spiritual approaches. All, however, are united by a field-based perspective of the ongoing quest of victims of environmental injustice, and their advocates, to secure equitable treatment for their grievances, especially those related to extractive industries and large infrastructure projects. The struggles documented here underscore the challenges indigenous and marginalized groups in the region face in achieving environmental justice and protecting their cultural identity, health, and livelihoods. Those challenges are rooted in historical social, economic, and political exclusion and reflect contemporary power asymmetries, the nexus of patrimonial and neocolonial states and international capital, and a secular religion of development and consumerism.

The concept of environmental justice emerged in the United States with the recognition that people who experience greater social and economic marginalization also suffer disproportionately from environmental hazards. It has since expanded geographically to consider the different forms those injustices take in different parts of the world. Carruthers notes that although "environmental justice in Latin America is not anchored in the hazardous siting inequities that fueled its rise in the United States, environmental concerns are deeply woven into the fabric of Latin American popular mobilization for social justice and equity."[4]

The quest for environmental justice in Latin America, especially in the Andes-Amazon region, is multifaceted and evolving. As Newell points out, there is an environmental justice dimension to many conflicts in Latin America in which indigenous people and small farmers confront landholders or corporations seeking access to resource-rich rural areas being opened for investment.[5] Latin America's growing environmental justice movement encompasses a wide range of issues, described by Martínez-Alier and colleagues, including legacy contamination and ongoing pollution, territorial rights, "land grabbing," seed preservation, food sovereignty, the co-opting of traditional knowledge of plants through biopiracy, water justice, climate justice, citizen science, corporate accountability, and the criminalization of environmental and indigenous rights activists.[6]

The movement also encompasses a range of actors, including rubber tappers, small farmers, descendants of African slaves, and indigenous people, who sometimes form alliances around common interests. Porto-Gonçalves and Leff trace the rise of new movements of indigenous people and small farmers, including rubber tappers in Brazil, Mexico's Zapatistas, and the Colombians of African descent who inhabit tropical forests on that country's Pacific coast.[7]

The movement has also expanded to embrace nonhuman beings. As Schlosberg notes, "When we interrupt, corrupt, or defile the potential functioning of ecological support systems, we do an injustice not only to human beings, but also to all of those non-humans that depend on the integrity of the system for their own functioning."[8] That understanding of the place of humans within nature, rather than apart from it or superior to it, is characteristic of indigenous worldviews and is fundamental to indigenous people's relationship with the territory they inhabit.

Such values are gradually being incorporated into legislation and jurisprudence in some countries. For example, Bolivia and Ecuador have enshrined the rights of nature in their constitutions, explicitly striving toward *sumak kawsay*, a more harmonious relationship between humans and nature. *Sumak kawsay* is a Quechua term that is usually translated into Spanish as *buen vivir*. The closest English translation, "living well," does not reflect the underlying idea that to live well, people do not require the fierce consumerism that drives resource extraction and destruction but rather a relationship with nature based on mutual benefit between humans and their surroundings.

In Colombia, courts have recognized the Amazon River ecosystem as having legal rights, and the Inter-American Court of Human Rights has upheld the right to a healthy environment, with the environmental protections that implies.[9] These developments have been accompanied by increased recognition of indigenous people's territorial rights, especially in international treaties and declarations, although they often are ignored by the very governments that have pledged to enforce them.[10]

Indigenous people, meanwhile, are taking action based on a "process of emancipation and decolonization" that revolves around territory as the space where cultural identities are rooted and reinvented.[11] The indigenous worldview, which draws no sharp distinction between the human and nonhuman beings that coexist and interrelate in Amazonian ecosystems, presents a challenge to Western conservationists, whose approach is rooted in Western science. Increasingly, however, scientists are recognizing the importance of indigenous people's territories and traditional knowledge for the protection of forests, and indigenous leaders are making their views known in international venues such as the United Nations climate summits and are taking their demands for territorial rights to international bodies such as the Inter-American Commission on Human Rights. Although these efforts have yielded important legal precedents, full respect for those rights remains elusive.

A History of Plunder

From the time the first Europeans ventured across the Andes Mountains and into the Amazon lowlands, the region has been viewed by outsiders as little more than a source of raw materials and humans to be exploited. Over the centuries, outside markets eagerly consumed the region's gold and silver, wild game meat and animal skins, timber, oil and gas, rubber, and other commodities. Modern traffickers have added to the list drug crops and illegal trade in exotic pets and ornamental fish, while the illegal harvesting of timber and unregulated gold mining continue, with links to organized crime, money laundering, and even the governments to which they are putatively subject.[12]

Although the products have varied depending on market demand, this exploitation has often been characterized by coerced labor or bonded servitude. During colonial times, thousands of Andean natives provided forced labor in the mines of Potosí, in what is now Bolivia, and the Santa Bárbara cinnabar mine in Huancavelica, Peru, through a system that caused profound upheaval in indigenous communities and fed an ongoing cycle of colonial corruption.[13] Even today the Amazon's colonial legacy of coerced labor and debt peonage persists in remote logging and mining camps and on farms and ranches. Because of threats of violence, withheld wages, or the distance they would have to travel to return home, workers are sometimes unable to escape. Others may remain in the system for years simply because there are no other jobs in the remote areas where they live.

The cruelest episode of this nature was the rubber boom, a relatively short but exceptionally brutal period in the late nineteenth century and early twentieth that led to the deaths of tens of thousands of indigenous people, especially along what was then a disputed border between Peru and Colombia.[14] Although European experiences show that the recognition and repercussions of genocide echo down the generations, the impact of the violence of the rubber era in the Putumayo River basin has received scant attention.

Shamans, healers, and elders died from disease, overwork, or persecution, taking with them knowledge that had been accumulated over generations. Entire communities were relocated to distant lands, sometimes displacing other indigenous populations. Some groups fled to more remote headwaters, where their seminomadic descendants still live, shunning contact with national society. Because these groups, which inhabit remote regions of seven South American countries, lack natural resistance to even the most common diseases, "initial contact" is often devastating.[15] Meanwhile they are under increasing pressure from encroaching development and economic activities.[16] With no direct representation and few ways to express and defend their own views except by their avoidance of contact, their future is largely shaped by debate among policymakers, indigenous rights activists, and anthropologists.

The rubber boom was followed in the twentieth century by greater extraction of the region's underground wealth, as oil exploration and production and mining expanded throughout the region, accompanied by the construction of roads and pipelines for transport of commodities and dams to produce electricity for mines and processing facilities. These activities, along with farming and ranching, have driven relentless deforestation and triggered conflicts with local communities whose lands are affected by "development" projects whose benefits often bypass them.

Amid pressure to extract resources from the Amazon's forests and rivers, the region's indigenous and nonindigenous inhabitants often are caught in the middle. Although they want to defend the ecosystems on which their livelihoods have historically depended, they need cash to purchase goods and pay for schooling for their children. As a result they often find themselves working for the same companies or intermediaries that are damaging or destroying their ecosystems. Meanwhile the revenues generated by the extraction of natural resources from the region rarely trickle down to the local communities, which often lack adequate schools and health care, as well as basic services such as clean water and sanitation.

There remains a historical debt to Amazonian and Andean peoples who have suffered the injustices of slavery or forced labor and the degradation of their environment and who continue to face discrimination from societies that claim to be pluricultural but fail to value a worldview in which humans are set within, and not above, the rest of the natural world. The countries with the greatest biological diversity are also emerging from centuries of poverty. Irrespective of putative ideologies, most have chosen a development path fueled by resource extraction and raw materials exports. These have lowered poverty rates in the region and improved other indicators of well-being, but those gains have come, in part, at the expense of Amazonian ecosystems that are important not only for local people's livelihoods but for regulating the water cycle and climate of the planet.[17] The coming years will be decisive in determining whether those countries will find a value in Andes-Amazon ecosystems that goes beyond immediate economic revenues.

The Quest for Environmental Justice: Making the "Invisible" Visible

Speaking at a conference of environmental journalists in 2008, Robert Bullard, who first drew attention to the "environmental racism" affecting communities of color in the United States, used the term "invisible communities" to describe neighborhoods that suffer from pollution and inadequate living conditions in the urban industrial landscape and that are at a disadvantage because of their lack of economic and political power.[18] It can be argued, however, that the communities suffering the greatest environmental injustice in places such as the Andes-Amazon region are not invisible but lie in plain sight for anyone who makes the effort to look. Rather, they are *invisibilizadas*, or made invisible, by official neglect and by policies that value industrial-scale extraction of natural resources over forests and the people whose livelihoods depend on them. Such policies explicitly or implicitly aim to absorb and deculturate those communities, whose traditions and lifestyles are often incomprehensible to residents of distant capital cities.

The conventional view of the region as a vast forest devoid of people also "invisibilizes" the indigenous people who have lived for millennia in a region where biological and cultural diversity are closely intertwined, as humans have helped shape the ecosystems that sustain them. The myth of an empty Amazonia has a political corollary: a region of such great natural riches should not be left in the hands of relatively few people but should be exploited by those who can turn it into economic wealth. President Alan García of Peru made that argument bluntly in 2007 in an essay in the country's largest-circulation newspaper.[19] In García's view, indigenous communities opposed to the exploitation of natural resources not only failed to take advantage of resources for themselves but also impeded economic development for the entire country. As a result, he said, all of Peru suffered because of the whims of a small percentage of the population.

Two years later García's government attempted to fast-track approval of a package of laws that indigenous groups feared would open up their lands to private economic interests. A protest in the northern Peruvian Amazon led to a violent crackdown in which thirty-four people died and more than two hundred were injured.[20] President Jair Bolsonaro of Brazil, who took office in January 2019, echoed García's view that indigenous people held rights to excessive amounts of resource-rich land. Backed by a congressional bloc aligned with large agricultural interests, Bolsonaro threatened to halt the constitutionally mandated demarcation of indigenous people's territories and open up their lands to mining.[21]

One consequence of the expansion of oil and gas production, mining, dams, and other development projects in the Andes-Amazon region has been the rise of indigenous movements throughout the region, including the creation in 1984 of a pan-Amazonian umbrella group, the Coordinating Committee of Indigenous Organizations of the Amazon Basin (Coordinadora de las Organizaciones Indígenas de la Cuenca Amazónica, COICA). These organizations have been increasingly visible in international climate negotiations, promoting the view that protecting indigenous territories provides a strong defense against the deforestation that is the main source of greenhouse gas emissions in Amazonian countries.[22] They also have played a crucial role in promoting the idea of *sumak kawsay* or *buen vivir*.

That concept was incorporated into Ecuador's Constitution in 2008, in which Article 275 defines the country's development regime as "the organized, sustainable and dynamic set of economic, political, sociocultural and environmental systems that guarantee the fulfillment of *Buen Vivir*, of *sumak kawsay*." That article goes on to explain that "*Buen Vivir* will require that individuals, communities, peoples and nationalities effectively enjoy their rights and exercise responsibilities within the framework of intercultural relations, respect for their diversities, and harmonious coexistence with Nature."[23] Article 71 of Ecuador's Constitution recognizes the right of nature to exist and evolve and served as the basis of a court decision for the protection of the country's Vilcabamba River.[24]

Recognition of nature as a legal entity with rights is also gradually coalescing into law in Latin America. Bolivia gave Mother Earth legal standing as a "collective subject of public interest" in 2010 and in 2012 outlined principles for environmental governance in a framework law on Mother Earth and Integral Development for Living Well. Both Bolivia and Ecuador have recognized the right of nature to have its waters protected, the former in 2010 and the latter in 2014.[25]

Colombia's Supreme Court has explicitly recognized the Atrato River and the country's Amazon region as having legal rights, making the state and its territorial jurisdictions responsible for their "protection, conservation, maintenance and restoration."[26] And in November 2017, in its decision on a case brought by Colombia regarding cross-border environmental impacts of infrastructure projects, the Inter-American Court of Human Rights noted, "[The] right to a healthy environment . . . is a matter of protecting nature and the environment not only in connection with its usefulness to human beings or because of the effects its degradation could have on other rights of persons, such as health, life or personal integrity, but because of its importance for the other living organisms with whom the planet is shared."[27]

In recent years legal protection has been increasing for both the natural world and indigenous people, who are often the ones who are most dependent on forests and rivers and most likely to suffer the negative effects of resource extraction and environmental degradation. The inclusion of terms such as "pluricultural," "intercultural," "pluriethnic," and "multiethnic" in national constitutions and legislation supposedly formalizes legal recognition of the cultural and territorial rights of indigenous peoples. Ironically, and hypocritically, the governments of those same countries continue to promote the exploitation of natural resources on indigenous lands and development projects that threaten indigenous people's territories, livelihoods, and cultures, and the environment that sustains them.

This has led some analysts to question whether such legislation and mechanisms such as free, prior, and informed consent are being used to create the illusion of defending indigenous people's rights while in fact enabling governments and corporations to undermine those rights. Seen from that perspective, "multiculturalism—both in its pre- and 'postneoliberal' variants—may be simply a new means of managing the 'indigenous issue,' as the region's rulers have done since the onset of colonial rule."[28]

At the same time, scientists warn that the Amazon basin is moving toward a "tipping point" at which it would become a net source of climate-warming greenhouse gas emissions, with potentially dire consequences not only for the people living in the region but for the planet as a whole.[29] The next decades will be crucial for determining the fate of the Amazon's forests and rivers and of the people who live in the basin and depend on its resources for their physical, cultural, social, and spiritual sustenance.

The scale of the impacts of land-use decisions in the Amazon has increased exponentially in recent decades. Extractive industries—particularly mining, oil and gas, and logging—overlap or invade indigenous territories and protected areas. Industrial agriculture and land speculation drive deforestation in various parts of the basin, while infrastructure projects—dams, transmission lines, ports, and roads—also expand, often accompanied by corruption at the highest levels of government. The consequences reverberate around the planet, as deforestation changes local and regional rainfall patterns and affects the climate on a global scale.

Amid these looming threats, the inhabitants of the Amazon basin are seeking justice in new and different ways. Indigenous people have organized and forged alliances with nongovernmental organizations, which help them advocate for their rights, and with scientists, who share their sense of urgency about preserving indigenous territories, which often are better buffered against deforestation than government-designated protected areas, according to studies from several countries. Those allies have helped indigenous people gain access to international forums, such as the climate summits where government negotiators set targets for greenhouse gas emissions from various sources, including deforestation.

When justice proves elusive in national courts, communities increasingly turn to international bodies. Decisions by the Inter-American Commission on Human Rights and the Inter-American Court of Human Rights and renewed emphasis on International Labor Organization Convention 169 on the rights of indigenous and tribal peoples are opening new avenues for indigenous communities to demand their rights. Communities and indigenous organizations are drawing on information technologies and a generation of young people trained in law and technical areas to help them evaluate new proposals for compensating them for conserving their lands. And community, indigenous, environmental, and human rights groups are joining forces in efforts to gain recognition for the rights of nature. These efforts have not been entirely successful, but they are gaining traction in ways that seemed unimaginable a decade ago. By exploring these issues from both conceptual and empirical standpoints, this book seeks to illuminate the diverse challenges marginalized groups face in the Andes-Amazon region and highlight the various forms that the quest for justice is taking, thereby contributing to the debate on these complex and interrelated issues.

Landscapes of Inequity: An Overview

In the deep history of the Andes-Amazon region, humans coexisted and interacted with their environment in a way that does not reflect the duality that arrived with the Europeans. The following chapters highlight the evolving nature of people's relationship with their environment in this region, the differences between Eurocentric and indigenous Amazonian views of nature and culture, and the role of economic and political forces in this process. These dynamics are approached from three field-based perspectives.

Part 1, "Extracting Resources, Imposing Inequity," traces the colonial and neocolonial roots of some modern injustices, examining the ways in which extractive policies have created and perpetuated environmental, social, economic, and political inequities. The extraction of raw materials for export is often accompanied by infrastructure to provide energy and facilitate transport. The environmental impacts of these projects disproportionately threaten the material, cultural, and spiritual survival of forest-dwelling peoples. In part 2, "Macro-Development and Marginalization," the focus shifts to the myriad and interlocking effects of macro-development projects on indigenous communities and the responses they provoke. Among those are a renewed recognition and reassertion of territorial claims. Part 3, "Territorial Rights, Ecocosmology, and the Quest for Environmental Justice," explores facets of this trend, including the use of Western legal mechanisms to advance territorial rights, even though they reflect the Eurocentric nature/culture dichotomy.

In part 1 the historical roots of environmental injustices in the region are explored through the lens of legacy contamination from colonial-era mining and toxic discharges from oil operations that began in the mid-twentieth century. Both occurred before environmental regulations were in force and have resulted in multigenerational degradation of the land, water, and food supplies.

In “A Toxic Reckoning: Legacy Contamination in Huancavelica, Peru,” Nicholas A. Robins examines the continuing environmental effects of colonial mercury refining in Huancavelica, Peru. Mercury refined from the Santa Bárbara cinnabar mines was transported throughout the Andes, where it was used to extract silver from ore. The expanded silver production made possible by this process spurred an unprecedented expansion and integration of international commerce, which fueled the Industrial Revolution and, ultimately, modern globalism.

The city of Huancavelica, however, was left with soil contaminated with mercury, lead, and arsenic as a result of cinnabar-refining operations, making it one of the world’s most mercury-contaminated urban areas. Approximately nineteen thousand people are exposed to those toxics daily in homes built of adobe bricks made from contaminated soil. Robins examines the nature and origins of polymetallic contamination in the city, highlights the associated public health risks, charts a course for remedial action, and explores accountability for such legacy contamination.

Contamination by commodity is also a central theme of “When the Rivers Run Black: Oil and Inequity in the Western Amazon,” in which Barbara J. Fraser traces the trajectory of oil exploration in the Amazonian regions of Peru and Ecuador, highlighting the ironic and dialectical nature of development in the region, where the few options for paid employment are in industries that tend to damage the environment on which local communities have traditionally depended for their livelihood.

When oil exploration and production began in the region in the 1930s, it was carried out without regard for environmental concerns, resulting in pollution that contaminated both the land and the rivers, streams, and lakes where people bathed and obtained their drinking water and fish. After decades of neglect by policymakers, indigenous communities began to organize to fight for the right to a healthy environment and pursue justice in national and international courts. And despite the lingering legacy of a polluted past, indigenous peoples have successfully blocked new operations in some areas.

The disproportionate and multifaceted impact of macro-development projects on invisibilized communities in the Andes-Amazon region is the focus of part 2. In "Environmental Justice and Brazil's Amazonian Dams," Philip M. Fearnside analyzes historical and current processes and conflicts concerning hydroelectric projects in Brazil, underscoring the legal, political, and structural obstacles to justice for affected residents. Indigenous and river-dwelling peoples not only are the most severely affected by dam construction, but they also receive the fewest benefits. The electricity generated by Brazil's hydroelectric dams is largely destined for export industries, especially aluminum production, and urban consumers in distant cities.

Fearnside details the challenges that both affected people and environmentalists face in influencing policy, noting that development decisions are commonly made without input from stakeholders or the preparation of adequate impact studies. Even when civil society actors succeed in halting a project, their victory is often short-lived. Besides advocating the development of wind and solar energy, Fearnside calls for substantive reforms, including the rigorous evaluation of technical data, along with meaningful consultation with stakeholders and the integration of their interests, prior to project approval.

The interrelationships among infrastructure, development, and conflict are the focus of "When 'Plurinational' States Undermine Indigenous Territories: TIPNIS in Bolivia." In this chapter Carwil Bjork-James highlights differing concepts of Bolivian indigenous identity and competing visions of the role of and relationship between natural resources and development. Bjork-James explores how such tensions are expressed through an analysis of the conflict over construction of a highway through Bolivia's Isiboro Sécure National Park and Indigenous Territory (TIPNIS in Spanish).

While the government of President Evo Morales has consistently supported the project, it has encountered unrelenting opposition from lowland indigenous people and organizations, who fear massive deforestation, increased production of drug crops, illegal logging, and other illicit activities. The TIPNIS conflict is a microcosm of the larger debate over differing concepts of development and relationships among indigenous peoples, natural resources, and local and global markets.

Juan Pablo Sarmiento Barletti and Anne M. Larson explore other contradictions in development policy in "Environmental Justice in the REDD+ Frontier: Experiences from the Amazon and Beyond." Programs for reducing emissions from deforestation and forest degradation (REDD+) aim to decrease greenhouse gas emissions by providing incentives for the conservation, sustainable management, and expansion of forests and their carbon stocks.

Although REDD+ appears to be a win-win way of compensating communities while helping countries control land-use change, which is the greatest source of greenhouse gas emissions in the Amazon basin, implementation has been accompanied by a number of unintended, and sometimes undesirable, consequences. These include the exclusion of affected groups from effective consultation, the implementation of policies without their consent, and even the removal of indigenous groups from ancestral lands in order to protect the forests on which they have relied for millennia. At the same time, however, it has created channels for affected groups to express their grievances on an international level. In examining how such policies play out on the "frontier," Sarmiento Barletti and Larson highlight the contradictions in REDD+, while emphasizing the potential for integrating environmental justice issues.

While megaprojects and extractive and agro-industries have brought upheaval to many Amazonian communities, their effects have also led to a reassertion of cultural identities that are closely tied to place. Historically, Amazonian indigenous groups were mobile, but the Western concept of property rights, combined with incentives to settle in communities that offer public services such as education and health care, have limited the space they can occupy. This has created social problems, while also spurring demands for territory. The dynamic and evolving relationships between indigenous peoples, their ancestral territories, and their ongoing quest for environmental justice is the focus of part 3.

The situation is especially complex in the case of seminomadic, isolated indigenous groups that shun contact with wider national society. Who represents them if they choose isolation? And how can a state that promotes the very development that threatens their way of life also safeguard their interests? Barbara Arisi and Felipe Milanez examine those issues in "Indigenism, Isolation, and Socioenvironmental Conflicts in the Javari River Valley" by weaving together the divergent and conflicting strands that condition relations between the Brazilian state and the Korubo and Matis people in the Javari Valley Indigenous Land in northeastern Brazil.

A violent encounter between an isolated group of Korubo people and Matis villagers who are considered to be in "recent contact," analyzed in this chapter, highlights contradictions that are often overlooked in studies of state-indigenous relations. Arisi and Milanez introduce the concept of "political isolation," noting that although they shun contact with outsiders, isolated groups are still deeply affected by government policies and are neither completely uncontacted nor isolated from the consequences of wider contact.

Government policies toward indigenous peoples are also the topic of "We Are Here: The State of Community-Based Landscapes in Peru," in which Richard Chase Smith explores the contentious issue of indigenous land rights. In many ways the chapter is a case study of one of the defining themes of Latin American history: the forced integration of native peoples into the global economic system. Within a historical framework, Smith explores how these dynamics and trends have continued in the late twentieth century and early twenty-first, through laws that have weakened the ability of indigenous communities and groups to confront the expansion of foreign and domestic capital.

In the final section of the chapter, Smith details the immense labor of several nongovernmental organizations that have successfully developed a register of indigenous communities and their lands, using georeferencing not only to catalogue these communities but also to show how mining, hydrocarbon, and timber concessions occupy the same space. The result helps to explain many of Peru's current social conflicts, while foreshadowing future disputes.

As they struggle for stronger legal standing and for the right to make decisions about their territories, indigenous peoples call for governments to respect their right to be consulted about projects affecting them. The controversies over this right, and impediments facing peoples in Bolivia,

Colombia, Ecuador, and Peru, are the focus of "In Search of Justice and Power: Contentious Experiences of Free, Prior, and Informed Consent in Latin America" by Roger Merino. Communities increasingly turn for backing to supranational instruments, particularly International Labor Organization Convention 169 on the rights of indigenous and tribal peoples and the UN Declaration on the Rights of Indigenous Peoples. Those treaties and legislation stemming from them should be a major step in protecting territorial rights, but, as the author explains, implementation falls short.

Relations between citizens and government and divergent concepts of development are also the focus of the final chapter of this volume, "Indigenous Amazonian Peoples and the Struggle for Environmental Justice in Lowland South America," in which the anthropologist Jonathan D. Hill probes the origins and nature of the conflicts stemming from government efforts to integrate lowland indigenous peoples into the global economy. Policies designed to protect these peoples' cultures and lifeways often were poorly implemented or ignored as demands from powerful economic interests dominated policy debates.

Lowland indigenous groups also grappled with a new threat, as their ancestral lands were swallowed up by infrastructure projects. This was a double affront, for their lands had profound ecocosmological importance to them. Hill details these connections, which are based on the concept of the intrinsic integration of people with natural elements, and the integration of such elements into humans and their history. He also examines the varied forms that indigenous resistance can take, including street protests, multisectoral mobilizations, and even academic activism.

The chapters that follow span a diversity of topics and time frames, but they all pose similar questions: How can new paths to environmental justice and development be nurtured? How can historical injustices be redressed so that those who have been historically excluded from political and economic power can participate as equals and on their own terms in decisions that affect them, their territory, and their children's future? It is our hope that this volume contributes to this rich, complex, and multifaceted debate.

NOTES

1. Watts, "Almost 4 Environmental Defenders." For 2017 figures see Global Witness, "2017 Defenders Annual Report"; for 2018 figures see Global Witness, "Enemies of the State?"
2. Marengo, "On the Hydrological Cycle," 1.
3. Porto-Gonçalves and Leff, "Political Ecology in Latin America," 74.
4. Carruthers, *Environmental Justice*, 7.
5. Newell, "The Politics of Environmental Justice," 51.
6. Martínez-Alier et al., "Between Activism and Science," 27–33, 35–38, 46.
7. Porto-Gonçalves and Leff, "Political Ecology in Latin America," 73.
8. Schlosberg, "Theorising Environmental Justice," 44.
9. On the rights of nature in Bolivia and Ecuador, see Lalander, "Rights of Nature," 150; on Colombia's recognition of the Amazon as being subject to rights, see Gómez Rojas, "Corte Suprema"; on the inter-American system's recognition of the right to a safe environment, see Inter-American Court of Human Rights, *Opinión Consultiva OC-23/17*.
10. See United Nations Department of Economic and Social Affairs, "United Nations Declaration on the Rights of Indigenous Peoples."
11. Porto-Gonçalves and Leff, "Political Ecology in Latin America," 86.
12. See Dizard, "Gold Is the New Cocaine." See also Kishor and Damania, "Crime and Justice."
13. See Quiroz, *Corrupt Circles*, 31.
14. Hemming, *Tree of Rivers*, 211–19.
15. Shelton et al. *Pueblos Indígenas*, 7–8.
16. Fraser, "Maps Reveal."
17. See United Nations, "We Can End Poverty."
18. Remarks by Robert Bullard during a panel discussion on environmental justice and the poor on October 18, 2008, at the annual conference of the Society of Environmental Journalists in Roanoke VA. An audio recording of the full panel discussion is available at http://sej2008.typepad.com/sej2008/2008/12/saturday-october-18-robert-bullard-rep-nick-rahall-james-burger-biofuels-much-more-.html.
19. García, "El síndrome del perro del hortelano."
20. Cavero, "Después del Baguazo," 7.
21. Borges and Branford, "Bolsonaro Draws Battle Lines."
22. Walker et al., "Forest Carbon in Amazonia."
23. Acosta, *El Buen Vivir*, 6.
24. See Global Alliance for the Rights of Nature, "Celebran el primer caso."
25. D'Andrea, "Can the River Spirit."
26. Gómez Rojas, "Corte Suprema."
27. Inter-American Court of Human Rights, *Opinión Consultiva OC-23/17*
28. Sieder and Barrera Vivero, "Legalizing Indigenous Self-Determination," 21.
29. Lovejoy and Nobre, "Amazon Tipping Point."

BIBLIOGRAPHY

Acosta, A. *El Buen Vivir en el camino del post-desarrollo: Una lectura desde la Constitucion de Montecristi*. Quito: Fundación Friedrich Ebert-ILDIS, 2010.

Adger, W. N., J. Paavola, S. Huq, and M. J. Mace. *Fairness in Adaptation to Climate Change*. Cambridge MA: MIT Press, 2006.

Agyeman, J., R. D. Bullard, and B. Evans, eds. *Just Sustainabilities: Development in an Unequal World*. Cambridge MA: MIT Press, 2003.

Alamgir, M., et al. "Economic, Socio-Political and Environmental Risks of Road Development in the Tropics." *Current Biology* 27, no. 20 (October 23, 2017): 1130–40.

Borges, Thais, and Sue Branford. "Bolsonaro Draws Battle Lines in Fight over Amazon Indigenous Lands." *Mongabay*, April 24, 2019. https://news.mongabay.com/2019/04/bolsonaro-draws-battle-lines-in-fight-over-amazon-indigenous-lands/.

Carruthers, D., ed. *Environmental Justice in Latin America: Problems, Promise and Practice*. Cambridge MA: MIT Press, 2008.

Cavero, O. "Después del Baguazo: Informes, diálogo y debates." *Cuaderno de Trabajo No. 13, Serie Justicia y Conflictos No. 1*. Lima: Pontifical Catholic University of Peru Department of Social Sciences, September 2011.

Cecchini, S. *Reducing Poverty amidst High Levels of Inequality: Lessons from Latin America and the Caribbean*. Santiago, Chile: Economic Commission for Latin America and the Caribbean, May 2017.

Chavez, A., et al. *Superposición espacial en la zonificación de bosques en Madre de Dios: Implicaciones para la sostenibilidad del recurso castañero*. CIFOR Infobrief No. 58. Bogor, Indonesia: Center for International Forestry Research, 2012.

D'Andrea, Ariella. "Can the River Spirit Be a Person in the Eye of the Law?" *International Water Law Project Blog*. Accessed April 6, 2018. https://www.internationalwaterlaw.org/blog/2018/03/26/can-the-river-spirit-be-a-person-in-the-eye-of-the-law/.

da Souza, Jonas Gregorio, et al. "Pre-Columbian Earth-Builders Settled along the Entire Southern Rim of the Amazon." *Nature Communications* 9 (2018). https://www.nature.com/articles/s41467-018-03510-7.

De la Cadena, Marisol. *Earth-Beings: Ecologies of Practice Across Andean Worlds*. Durham NC: Duke University Press, 2015.

Diringer, S. E., et al. "River Transport of Mercury from Artisanal and Small-Scale Gold Mining and Risks for Dietary Mercury Exposure in Madre de Dios, Peru." *Environmental. Science: Processes & Impacts* 17 (2015): 478–87.

Dizard, John. "Gold Is the New Cocaine." *Financial Times*, November 10, 2017. https://www.ft.com/content/bfba8cdc-c5fe-11e7-b2bb-322b2cb39656.

Espinoza, J. C., J. A. Marengo, J. Ronchail, J. Molina Carpio, L. Noriega Flores, and J. L. Guyot. "The Extreme 2014 Flood in South-western Amazon Basin: The Role of Tropical-Subtropical South Atlantic SST Gradient." *Environmental Research Letters* 9, no. 12 (December 2014): 1–9.

Farber, D. "The Case for Climate Compensation: Justice for Climate Change Victims in a Complex World." *Utah Law Review* 377 (2008): 377–413.

Finer, M., et al. "Future of Oil and Gas Development in the Western Amazon." *Environmental Research Letters* 10, no. 1 (January 2015): 1–6.

Fraser, B. "Maps Reveal How Amazon Development Is Closing In on Isolated Tribes." *Science*, August 17, 2017. http://www.sciencemag.org/news/2017/08/maps-reveal-how-amazon-development-closing-isolated-tribes.

García, A. "El síndrome del perro del hortelano." *El Comercio*, October 28, 2007. https://www.scribd.com/document/211691557/ALAN-GARCIA-EL-PERRO-DEL-HORTELANO-pdf.

Global Alliance for the Rights of Nature. "Celebran el primer caso exitoso de exigibilidad de los derechos de la naturaleza en Ecuador." Accessed July 8, 2019. https://therightsofnature.org/celebran-el-primer-caso-exitoso-de-exigibilidad-de-los-derechos-de-la-naturaleza-en-ecuador/.

Global Witness. "Enemies of the State?" July 30, 2019. https://www.globalwitness.org/en/campaigns/environmental-activists/enemies-state/.

———. "2017 Defenders Annual Report." 2017. https://www.globalwitness.org/en/campaigns/environmental-activists/defenders-annual-report/.

Gómez Rojas, Germán. "Corte Suprema ordena protección inmediata de la Amazonía Colombiana." Colombia Corte Suprema. Accessed April 9, 2018. http://www.cortesuprema.gov.co/corte/index.php/2018/04/05/corte-suprema-ordena-proteccion-inmediata-de-la-amazonia-colombiana/.

Goulding, M., R. Barthem, and E. Ferreira. *The Smithsonian Atlas of the Amazon*. Washington DC: Smithsonian Books, 2003.

Hecht, S. *The Scramble for the Amazon and the "Lost Paradise" of Euclides da Cunha*. Chicago: University of Chicago Press, 2013.

Hemming, J. *Tree of Rivers: The Story of the Amazon*. New York: Thames & Hudson, 2009.

Inter-American Court of Human Rights. *Opinión Consultiva OC-23/17 de 15 Noviembre de 2017, solicitada por la República de Colombia*. Accessed April 2, 2018. http://www.corteidh.or.cr/docs/opiniones/seriea_23_esp.pdf.

Kishor, N., and R. Damania. "Crime and Justice in the Garden of Eden: Improving Governance and Reducing Corruption in the Forestry Sector." In *The Many Faces of Corruption: Tracking Vulnerabilities at the Sector Level*, edited by J. E. Campos and S. Pradhan. Washington DC: World Bank, 2007.

Lalander, R. "Rights of Nature and the Indigenous Peoples in Bolivia and Ecuador: A Straitjacket for Progressive Development Politics?" *Iberoamerican Journal of Development Studies* 3, no. 2 (October 22, 2014): 148–72.

Lovejoy, T., and C. Nobre. "Amazon Tipping Point." *Science Advances* 4, no. 2 (February 21, 2018).

Marengo, J. A. "On the Hydrological Cycle of the Amazon Basin: A Historical Review and Current State-of-the-Art." *Revista Brasileira de Meteorologia* 21, no. 3, (2006): 1–19.

Martínez-Alier, J., et al. "Between Activism and Science: Grassroots Concepts for Sustainability Coined by Environmental Justice Organizations." *Journal of Political Ecology* 21 (2014): 19–60.

Munilla, I. *People, Power and Pipelines: Lessons from Peru in the Governance of Gas Production Revenues*. Lima: Oxfam America, 2010.

Negret, P. J., et al. "Need for Conservation Planning in Postconflict Colombia." *Conservation Biology* 31, no. 3 (June 2017): 499–500.

Neves, E. G., J. B. Petersen, R. N. Bartone, and C. A. Da Silva. "Historical and Socio-Cultural Origins of Amazonian Dark Earths." In *Amazonian Dark Earths: Origin, Properties, Management*, edited by J. Lehmann et al. Dordrecht: Kluwer Academic, 2003.

Newell, P. "The Politics of Environmental Justice." In *Environmental Justice in Latin America: Problems, Promise, and Practice*, edited by D. Carruthers. Cambridge MA: MIT Press, 2008.

Porto-Gonçalves, C. W., and E. Leff. "Political Ecology in Latin America: The Social Re-Appropriation of Nature, the Reinvention of Territories and the Construction of an Environmental Rationality." *Desenvolvimento e Meio Ambiente* 35 (December 2015): 65–88.

Potter, L. *Managing Oil Palm Landscapes: A Seven-Country Survey of the Modern Palm Oil Industry in Southeast Asia, Latin America and West Africa*. Occasional Paper 122. Bogor, Indonesia: Center for International Forestry Research, 2015.

Quiroz, Alfonso W. *Corrupt Circles: A History of Unbounded Graft in Peru*. Washington DC: Woodrow Wilson Center Press, 2008.

Ray, R., K. P. Gallagher, C. and Sanborn, C. *Standardizing Sustainable Development? Development Banks in the Andean Amazon*. Lima: University of the Pacific Center for China and Asia-Pacific Studies, 2018.

Revkin, A. *The Burning Season: The Murder of Chico Mendes and the Fight for the Amazon Rain Forest*. Washington DC: Island Press, 2004.

Saunaluoma, S. "Geometric Earthworks in the State of Acre, Brazil: Excavations at the Fazenda Atlântica and Quinauá Sites." *Latin American Antiquity* 23, no. 4 (December 2012): 565–83.

Scheffers, B. R., L. N. Joppa, S. L. Pimm, and W. F. Laurance. "What We Know and Don't Know about Earth's Missing Biodiversity." *Trends in Ecology and Evolution* 27, no. 9 (September 2012): 1–10.

Schlosberg, D. "Theorising Environmental Justice: The Expanding Sphere of a Discourse." *Environmental Politics* 22, no. 1 (2013): 37–55.

Shelton, D., et al. *Pueblos Indígenas en Aislamiento Voluntario y Contacto Inicial*. Copenhagen: IWGIA-IPES, 2012.

Sieder, R., and A. Barrera Vivero. "Legalizing Indigenous Self-Determination: Autonomy and *Buen Vivir* in Latin America." *Journal of Latin American and Caribbean Anthropology* 22, no. 1 (2017): 9–26.

United Nations. "We Can End Poverty." Accessed July 8, 2019. http://www.un.org/millenniumgoals/.

United Nations Department of Economic and Social Affairs. "United Nations Declaration on the Rights of Indigenous Peoples." Accessed July 8, 2019. https://www.un.org/development/desa/indigenouspeoples/declaration-on-the-rights-of-indigenous-peoples.html.

Vadjunec, J. M., C. V. A. Gomes, and T. Ludewigs. "Land-Use/Land-Cover Change among Rubber Tappers in the Chico Mendes Extractive Reserve, Acre, Brazil." *Journal of Land Use Science* 4, no. 4 (2009): 249–74.

Walker, W., et al. "Forest Carbon in Amazonia: The Unrecognized Contribution of Indigenous Territories and Protected Natural Areas." *Carbon Management* 5, nos. 5–6 (2014): 479–85.

Watts, J. "Almost 4 Environmental Defenders a Week Killed in 2017." *The Guardian*, February 2, 2018. https://www.theguardian.com/environment/2018/feb/02/almost-four-environmental-defenders-a-week-killed-in-2017?CMP=share_btn_tw/.

Werth, D., and R. Avissar. "The Local and Global Effects of Amazon Deforestation." *Journal of Geophysical Research: Atmospheres* 107, no. D20 (October 24, 2002): 55-1–55-8.

West, P., J. Igoe, D. and Brockington. "Parks and Peoples: The Social Impacts of Protected Areas." *Annual Review of Anthropology* 35 (June 2006): 251–77.

LANDSCAPES OF INEQUITY

Part 1

Extracting Resources, Imposing Inequity

This section provides a historical context for the entire work, providing an overview of the relationship between the state and the region's indigenous inhabitants since the arrival of Europeans in the Andes-Amazon area. It then explores specific issues in greater detail through the examination of legacy heavy-metal contamination in highland Peru and hydrocarbon extraction in lowland Peru.

1 A Toxic Reckoning

Legacy Contamination in Huancavelica, Peru

NICHOLAS A. ROBINS

Ensconced at just over 12,000 feet in the Peruvian Andes, the remote city of Huancavelica, Peru, is home to about fifty thousand residents and, nearby, the fourth largest cinnabar deposit in the world.[1] The region has long been settled, albeit sparsely, and pre-Hispanic groups such as the Chavín (900 BC–AD 200), Paracas (800–100 BC), Huari (AD 800–1200), and Inca (1425–1532) societies extracted cinnabar from the Chacllatacana mount and processed it in local springs and the Ichu River to produce *llimpi*, or vermilion.[2] The rich red powder had both ritualistic and practical uses among many indigenous societies and was employed in burial rites, as facial war paint for men and cosmetics for high-status women, and as a colorant for dyes and paint. Such was the regional demand for llimpi that isotopic analysis has revealed Huancavelica's cinnabar was exchanged with the Moche (AD 100–700) and Sicán (AD 700–1200) societies in what is now northern Peru.[3]

The extensive ritual use of vermilion underscores its divine associations among indigenous groups, and it imbued the region of Huancavelica with transcendental significance. Although there are numerous theories concerning the origin of the name of the city, most agree that it derives from *huaca*, which is an object of worship, and *velica*, which connotes sacredness and is often associated with prominent geological formations. Thus the word could be translated as "sacred shrine." Alternatively it could be interpreted as "sacred place of the Huancas," referring to the

Huanca people who resided there prior to the Spanish conquest. Despite their differences, both theories highlight the spiritual importance of the region that derived from its seven-million-year-old cinnabar deposits.[4]

Although the Chacllatacana and neighboring Santa Bárbara hills contain the richest deposits, they are a component of a mineralization belt oriented along a southeast-to-northwest axis. Spanning an area approximately sixty kilometers long by three wide, the zone contains upwards of forty other cinnabar deposits.[5] The city of Huancavelica sits atop this belt and is built upon the alluvial deposits of the Ichu River that runs through the city. These soils contain native mercury, or elemental quicksilver in the form of droplets that do not require distillation from ore. Huancavelica's history is replete with accounts of people encountering such deposits, and such events are not the stuff of legend. Even today workers engaged in home and public works construction commonly encounter native mercury during excavation.[6]

Servitude and Sublimation: Unlocking Santa Bárbara's Secrets

While pre-Columbian groups had limited uses for cinnabar, there is no archaeological or documentary evidence that they refined mercury in the region of Huancavelica. Under Spanish domination, however, mercury rendered from Huancavelica's cinnabar deposits would play a vital role in enabling and maintaining Andean silver production and Spain's economy. By utilizing the mercury amalgamation process, refiners were able to process lesser grade argentiferous ores that could not be efficiently smelted. Although there are differing accounts of how the Spanish encountered the massive deposits of the Santa Bárbara mount, all involve an Indian revealing their location to the Spaniard Amador de Cabrera Ulloa y Larrinaga. Based in the nearby village of Acoria, Cabrera was an *encomendero*, a local official who was granted an allotment of Indians whom he was to Christianize and otherwise exploit. On November 3, 1563, Amador de Cabrera and his native informant stood before what would soon be called the Santa Bárbara mine, and soon after the Spaniard would formally claim the deposits in the town of Guamanga, present-day Ayacucho.[7]

Despite the tens of thousands of tons of mercury hidden within the rock, there was very little demand for it at that time in the Andes. However, in New Spain, now Mexico, things were different. In the early 1550s in Pachuca, Bartolomé de Medina tried different approaches to refining silver through mercury amalgamation, a system he had learned from a German in his native Spain. By 1554 he had successfully refined silver using this method, and Mexican miners increasingly embraced it. Previously they had produced silver through smelting, but the process had two drawbacks. One is that it is most effective with high-grade ores, which are usually closer to the surface and soon depleted. The other challenge concerned the cost and logistics of fuel supply, as mineral deposits often lay in treeless areas. Unlike smelting, mercury amalgamation can extract silver from lesser quality ores without the use of heat. Such ores are often deeper underground and hence more costly to extract and transport to the surface. Despite this and the greater amount of ore needed to produce a given amount of silver, the mercury amalgamation process breathed new life into mines that otherwise would not have been viable.[8]

The amalgamation process involves three phases: ore preparation, amalgamation, and firing. To prepare the ore, it was crushed into as fine a powder as possible, and then sieved to ensure consistency. Once the ore had been sifted, it was mixed with water, salt, mercury, iron, and other elements, depending on the quality of the ore. This pasty mixture, known as *torta*, would then be placed in a walled, stone patio, where the amalgamation process would take place. It is for this reason that the technique is also known as the "patio process." Ideally this patio would be built above a furnace, which when charged would expedite the process, reducing it from about a month to about a week. Usually, however, the "cold process," without fire, was employed, the longer processing time being offset by the cost savings of not having to bring in wood. During processing, workers would "tread," or mix, the torta by walking in it knee or thigh deep to promote the amalgamation.[9]

When the reaction was deemed to have finished, the mix was carried to a series of interconnected troughs where it was washed, or rather dumped, allowing the amalgam to fall to the bottom and the impurities to wash away. Once collected, the slippery amalgam was strained to remove as much mercury as possible, placed into a conical ceramic cone, and fired. This would volatize the mercury and yield high-grade silver, known as a *piña* (pineapple) due to the shape of the form. The piña would then be melted into a form and, in theory, brought to the mint, where the *quinto real* (royal fifth) was to be paid, although in practice the amount of tax varied.[10]

Because the technique had not been introduced in the Andes when Amador Cabrera and subsequent miners registered claims and began to produce mercury, their primary market was New Spain.[11] The process for rendering quicksilver was primitive and toxic but effective. Initially workers extracted cinnabar from an open pit at the Santa Bárbara deposit. This ore contained an extraordinary amount of mercury, up to 30 percent. It was also relatively accessible, although at the cost of workers being poisoned by contact with the ore or breathing the vapors that emanated from it. As time went on, however, and the pit became deeper, there were increasing issues with flooding and declining ore quality, which in turn spurred the use subterranean mining.[12] Although this allowed continued production, because extraction took place in an unventilated environment it was much more toxic than open-pit mining. To this was added the ever-present menace of cave-ins and accidents carrying the ore to the surface.[13] Once the ore had been extracted, it was broken into smaller pieces and smelted. Although different types of smelters were employed over the years in Huancavelica, they all involved placing ore in a closed container, smelting it to at least 580 degrees Celsius to precipitate the mercury, condensing the product, and storing it in ceramic vessels before transporting it to mining centers.[14]

In 1564, despite strong demand for mercury in New Spain, there were logistical challenges shipping it there and getting paid for it. As a result, there was some cause for optimism in 1571, when a new viceroy, Francisco de Toledo, arrived in Peru. Among his tasks was boosting flagging silver production by inducing silver miners there to adopt the amalgamation technique. Toledo was relentless and focused much of his efforts on Potosí, in present-day Bolivia, which had produced, through smelting, a spectacular amount of silver since the discovery of the deposits and founding of the city in 1545. By the 1560s, however, the rich surface ores had been worked, extraction costs were increasing, yields were declining, and the city and its economy were in decay. To rectify this Toledo urged miners to adopt mercury amalgamation, offering incentives to offset the significant investment needed to construct industrial-scale mills.[15] In addition, in 1574 he ordered the construction of a series of eighteen interlocking reservoirs in the hills above the city, which, after the excavation of a seventeen-kilometer channel through the city, would link and power the water wheels in the mills spread out along it. Although human- and animal-powered mills continued to operate in the region, these hydraulic mills would henceforth be responsible for most of Potosí's silver production.[16]

More important, Toledo adopted and transformed the Incaic practice of the *mita*, a term-based system of male draft labor. Under the Inca, the mita was employed largely for public works and involved a stint of several weeks of service. Under Toledo, around 12 percent of the tributary Indians between the ages of eighteen and fifty from designated provinces were required to labor for one year on projects deemed to be of public benefit, including mining. In Huancavelica the mita term was set at two months, reflecting the acute toxicity of the mines and smelters there.

Although *mitayos*, or those serving the mita, were to work shifts, in practice the system in both Huancavelica and Potosí was based for most of the colonial period on production quotas. Mitayos were to receive a symbolic wage for their service, but it was insufficient to sustain them and was often paid late, in debased silver, or not at all. As a result, many brought foodstuffs with them and journeyed with family members who could also work independently to help support them during their service.[17] In Huancavelica mitayos were to work for forty nonconsecutive days over a sixty-day period and were to have the remaining twenty days free to work for a regular wage. In practice, however, they often worked twice that amount in order to meet the quotas of their overlords. Service in Huancavelica was accompanied by fear, illness, and frequently death, with contemporaries reporting towns devoid of men, mothers maiming their young to make them ineligible for service, and mercury-poisoned mitayos staggering home to die.[18]

Not all of the mining and refining labor in Huancavelica was based on the mita, although most of it relied on some degree of coercion. Before the mita was introduced by Toledo, Huancavelica's miners used forced laborers whom they controlled through the encomendero system or other means to work there. Once the mita system was in place, it experienced a steady decline as disease and flight continued to take their toll on the population. By the end of the colonial period, most subject communities simply paid a fee in lieu of sending people, with the idea that the revenues generated would be used by the mining guild to hire wage laborers.[19]

Wage laborers were an increasingly important component of the mining labor force, especially after 1642, when the opening of the Bethlehem adit dramatically improved ventilation in and access to the mine, making work relatively less onerous. Moreover declining ore quality had also improved the air quality in the mine by reducing volatilization. Despite working for wages, such workers could hardly be considered free. Many came to work off debts that had been imposed upon them by their *corregidores* (Spanish district governors) and priests, while others were dispatched there by their *curaca* (native governor) to help reduce community debts. Others lived on *haciendas* (agricultural estates) and were at the behest of their overlords, who could order them to work in their own mines or rent them out to other miners. Although there was a mixed workforce, the most dangerous refining tasks were generally carried out by mitayos.[20]

For the mine owners in Huancavelica, Toledo's innovations were encouraging, auguring a new era of silver production, robust demand for the mercury they produced, and lower costs in getting it to market. Seeking to better position themselves for the opportunities they anticipated, on August 4 or 5, 1571, the miners officially founded the town, naming it Pueblo Rico de Oropesa, after Oropesa, Spain, Viceroy Toledo's birthplace. Despite this gesture of goodwill, it ran head-on into the last component of Toledo's plan: the expropriation of the cinnabar mines to generate crown revenue and limit illegal silver production. This would transform the former mine owners into crown contractors, who were to produce a predetermined amount of quicksilver that could only be legally sold for a fixed price to the state under what became known as the *asiento* system.[21] Huancavelica's mining guild had been despoiled of their wealth, and the Habsburg and Bourbon monarchies would retain ownership of the cinnabar deposits for the rest of the colonial period, as Huancavelica became the primary source of mercury to Andean silver mines.[22]

When Peru finally gained independence on July 28, 1821, the Santa Bárbara and Chacllatacana deposits had rendered about sixty-eight thousand metric tons of mercury that had been dispatched to and volatized in countless mining towns in the region.[23] With independence, ownership of the mines passed from the Spanish monarchy to the new Peruvian government, as production and recordkeeping slipped to new lows. The limited refining that then occurred was carried out almost exclusively by *humanchis*, small-scale, independent, and often family-based miners, and sold domestically. Between 1836 and 1900 there were numerous government-sanctioned efforts to reinvigorate mercury production in Huancavelica. None were as productive, however, as the humanchis, and all were, to differing degrees, impeded by arsenic in the mines that limited access to deposits, capital shortages, technical challenges, politicized decisions, and, beginning in the 1840s, stiff competition from mercury produced in California.[24]

In 1901 the Peruvian government issued a new mining code that provided for private ownership of subsoil rights.[25] As a result, beginning in 1901 the Peruvian entrepreneur Augusto Benavides began to acquire the Santa Bárbara and other deposits. It was a smart investment due to the increasing demand for mercury with the outbreak of World War I, and in 1915 Benavides sold his interest to the Negociación Minera Eulogio Fernandini, led by its namesake. Although extensive, the mines were in total disrepair, and it would take Fernandini seven years and considerable investment to resume production.[26] After Fernandini's death in 1938 cave-ins in both the Fernandini and Bethlehem adits largely shut down production, and the enterprise was reorganized as the Negociación Minera Fernandini Clotet Hermanos.[27]

In 1956 Fernandini's heirs restructured their enterprise, renaming it the Sociedad Minera Brocal. With fresh capital and new high-capacity smelters, Brocal conducted open-pit mining, largely in the Chacllatacana and San Roque deposits, beginning in 1968. Although the ore generally contained no more than 3 percent mercury, the new smelters could process 250 metric tons of ore in twenty-four hours, compared to the seven tons in the Gould rotary smelters. Declining international demand for mercury, however, led to the definitive cessation of industrial-scale mercury production there in 1975.[28] Despite the efforts to reinvigorate production in the national period, compared with that of the colonial period production was minimal and amounted to approximately 7,824 metric tons, yielding a total production of approximately 76,024 metric tons by 2014.[29]

Colonialism's Curse

The limited and inconsistent production of mercury during the national period highlights the fact that about 90 percent of Huancavelica's total production occurred during the colonial period, much of it refined in what is now the urban area. Not only was production higher during this period, but so was the release of heavy metals due to better ore quality and the inefficiencies of the smelters that were used there until the twentieth century. These furnaces were poorly sealed, and porous ceramic condensation tubes loosely joined by ash and mud allowed considerable quantities of vapor to escape. Estimates of the emissions factor, or the amount of vapor loss relative to total production, range considerably, from 10 to 60 percent.[30] When Fernandini installed Gould furnaces in the 1920s, tests indicated what was considered an acceptable emissions factor of 20 percent or less, an improvement over the colonial-era smelters.[31] Using an emissions factor of 30 percent for the colonial-era furnaces suggests that about 20,400 metric tons were released into the atmosphere between 1564 and 1825. Once airborne, the mercury ultimately deposited onto the ground and waterways, much of it in the immediate vicinity of the city, where in many cases it remains.[32]

It was not only liquid and vaporous mercury that was released as a result of cinnabar mining. Arsenic is often encountered with cinnabar, and Huancavelica's vast deposits are no exception. Arsenic gas in the mines would on occasion limit excavation of otherwise rich ore. Arsenic was also released during smelting, adding to the toxic woes of those working the smelters.[33] The ore also contained considerable amounts of lead, which was either volatized or discarded with the processed ore. Today these heavy metals permeate the soils of the city and sediments of the Ichu River; the extent of this contamination has been revealed by scientific studies conducted by Peru's Environmental Assessment and Enforcement Agency (Organismo de Evaluación y Fiscalización Ambiental, OEFA) and the Environmental Health Council (EHC), a nongovernmental organization.[34]

As much, if not most, mercury production occurred in what is now the urban area of Huancavelica, contaminated soils are of special concern. EHC soil data are based on samples from sixty-five outdoor locations within the city, drawn in triplicate or as composite samples from depths of between one and three inches, as well as limited samples from borrow piles staged for construction. These samples were subsequently analyzed through atomic absorption spectroscopy (AAS), inductively coupled mass spectroscopy, or x-ray fluorescence (XRF). All samples were analyzed for total mercury, and a subset of forty-four samples was analyzed for lead and arsenic. Mercury content in outdoor soils ranged from 0.1 to 1,201 mg/kg, with a mean of 186 mg/kg, while arsenic concentrations ranged from 5.8 to 2268 mg/kg, and those of lead from 12 to 7,876 mg/kg.[35] On the basis of these findings, the OEFA conducted a more extensive study of Huancavelica's soil, air, water, and sediment quality in 2015.[36] Soil samples were drawn from forty-eight points throughout the city and revealed mercury concentrations from approximately 3.5 to 580 mg/kg, arsenic concentrations ranging from 10.7 to 1,908 mg/kg, and those of lead from 100 to 3,100 mg/kg.[37]

To put these figures in perspective, we can compare them to the screening level (SL) for each element, which represents a level below which lifelong exposure would not pose a health risk. Put differently, the risk of ill effects from exposure increase as concentrations exceed the SL. Peruvian law establishes 6.6 mg/kg as the SL for residential exposure to mercury-contaminated soil, while that of the U.S. Environmental Protection Agency (USEPA) is 23 mg/kg and assumes that all of the mercury is bioavailable, or able to be absorbed into the body through ingestion. In the case of Huancavelica, simulated gastric analysis of soil samples collected there suggest that less than 10 percent is bioavailable. As a result, based on USEPA methods, the EHC developed a higher site-specific SL of 75 mg/kg, the calculation of which accounts for limited bioavailable, daily soil ingestion and body weight.

TABLE 1. Outdoor soil sample results

Metal	Percentage of OEFA samples above screening level	Percentage of EHC samples above screening level
Mercury (SL 6.6 mg/kg)	81	88
Mercury (site-specific SL 75 mg/kg)	23	28
Lead (SL 140 mg/kg)	48	68
Arsenic (SL 50 mg/kg)	92	80

As table 1 indicates, 28 percent and approximately 23 percent of the exterior soil samples collected by the EHC and OEFA, respectively, exceeded the higher, site-specific SL for mercury, while 81 percent of the OEFA samples and 88 percent of the EHC samples exceeded Peru's stricter SL of 6.6 mg/kg.[38]

Results concerning lead in both data sets are even more concerning. The Peruvian screening value for residential lead exposure is 140 mg/kg, while that of the USEPA is 400 mg/kg. In the city the background level of lead is approximately 26 mg/kg. As shown in table 1, in the EHC study 68 percent of the samples exceeded the Peruvian permissible value, while 48 percent did so in the OEFA study.[39]

Arsenic SLs are often lower than background levels, which in Huancavelica is about 16 mg/kg. As shown in table 1, in the EHC study 80 percent of exterior soil and residential earthen home samples exceeded the Peruvian SL of 50 mg/kg for arsenic concentrations in residential and recreational areas. The rate was almost the same, at 77 percent, when only the residential samples were considered. In the OEFA study 92 percent exceeded the same standard.[40] Further corroborating the EHC's data, the OEFA report also indicated that the most contaminated neighborhoods were Yananaco, San Cristóbal, and Ascención. Moreover it identifies the mineralogy of the region as well as previous mining activity as causes for the contamination of the soils and waterways. It also indicates that acid mine drainage from the Yanamina mine is a possible source of lead contamination in the Disparate tributary and Ichu River.[41]

The contamination of the Ichu River and its Disparate, Taxanapampa, and Huaylacucho tributaries appears to be largely a result of runoff from local soils. Mercury deposited into low-energy streams and waterways can be transformed into highly toxic and bioavailable methylmercury when it is consumed and excreted by microbes before entering the food chain via fish.[42] Twenty-six sediment samples gathered upstream of the city, within the city, and downstream were collected by the EHC and analyzed by AAS and XRF. Total mercury concentrations varied from 0.88 mg/kg (upstream) to 1,370 mg/kg (urban area) and 19 mg/kg downstream of the city, while arsenic concentrations ranged from 13.7 to 108, and lead concentrations ranged from 15.4 to 107. In the more extensive OEFA study, mercury concentrations in samples from eleven locations on the Ichu River and its tributaries ranged from 0.16 mg/kg to 74.6 mg/kg, with higher concentrations of heavy metals in the urban area.[43]

There is no specific Peruvian SL for heavy metals in river sediment, due in part to myriad factors that can affect methylation, including the rate of water flow and the amount of carbon and dissolved oxygen in the water. Nevertheless high total mercury concentrations in sediments are of concern due to their methylation potential as well as risks associated with direct dermal absorption, for example when washing clothes in the river.[44] In the OEFA study, researchers adopted the Canadian government's Sediment Quality Guidelines for Protection of Aquatic Life—Fresh Water (2001) for screening and probable effect levels (PEL).[45]

TABLE 2. Sediment sample results

Metal	Percentage of OEFA samples above PEL	Percentage of EHC samples above PEL
Mercury (SL 0.49 mg/kg)	72	77
Lead (SL 91.4 mg/kg)	45	13
Arsenic (SL 17 mg/kg)	100	88

As shown in table 2, results indicated that 72 percent of OEFA's sediment samples were above the PEL of 0.486 mg/kg for mercury, the highest being 153 percent above the level. Forty-five percent of the samples had lead levels above the PEL of 91.4 mg/kg, the highest concentration being 637 percent above the PEL near the San Cristóbal neighborhood. In the case of arsenic, 100 percent exceeded the PEL of 17 mg/kg, the highest concentration being 1,188 percent above the limit. In addition, the highest concentrations of heavy metals were just downstream of the Disparate tributary, which is downstream from tailings from Santa Bárbara and joins the Ichu in the urban area.[46] In the more limited EHC sediment samples, 77 percent exceeded the Canadian PEL for mercury, 88 percent exceeded it for arsenic, and 13 percent exceeded it for lead.[47]

In addition to sediments, the OEFA study evaluated water quality from the Ichu River and its three tributaries in August and December 2015. The two collection events encompassed the dry and wet seasons, respectively, and samples were drawn from eleven points. Water quality was primarily assessed on the basis of whether it was appropriate for irrigation and animal consumption; however water at the intake for the municipal water supply was tested for suitability for conventional treatment for drinking water. Water at the intake upriver of the city was found to be within acceptable parameters, with the exception of during the wet season, when samples revealed high levels of coliforms, aluminum, and chemical oxygen demand, which refers to the ability of water to absorb oxygen during the decomposition of organic material. Water quality declined downstream, however, and was found to be unsuitable even for irrigation or animal consumption, especially in the Huaylacucho tributary, due to high levels of total arsenic, cobalt, aluminum, iron, and manganese. In addition, the waters are characterized by elevated levels of coliforms, thermocoliforms, pH, and chemical oxygen demand. The widespread presence of coliforms and thermocoliforms reflects not only runoff from livestock raising but also the fact that Huancavelica lacks a sewerage treatment plant. As a result, the equivalent volume of three and a half Olympic swimming pools of sewerage and associated residues is released daily into the Ichu or its tributaries.[48]

Despite the contamination of the Ichu River, and the risks it presents to those who come into contact with its sediments or otherwise use it without treatment, it does not appear to be a source of mercury entering the food chain via fish raised at local fish farms. Using AAS, the EHC analyzed forty-eight trout from four fish farms that provision the city. As with a small sample of seven other locally produced and commonly consumed foodstuffs such as alpaca, chicken, guinea pig, and tubers, the trout samples did not contain detectable levels of mercury. Similarly twenty-six samples of drinking water drawn from city supplies and urban wells did not reveal detectable levels of mercury.[49]

Based on the foregoing, local soils contaminated with mercury, arsenic, and lead represent the primary source of heavy metal exposure to residents of Huancavelica. The risk of ingestion is exacerbated by the fact that about half of the homes in the city are of earthen construction, formed as either adobe bricks or larger, molded earthen blocks known as *tapial* construction.[50] In 2011 the EHC collected indoor wall, floor, dust, and air samples from sixty adobe or tapial homes in four of Huancavelica's primary neighborhoods (Yananaco, Santa Ana, Ascención, and San Cristóbal). Mercury vapor in air was measured in 169 samples by either a Jerome 405 or Lumex RA-915 mercury vapor analyzer, while solids were analyzed for total mercury by AAS, inductively coupled mass spectroscopy, and XRF. Later sampling events analyzed arsenic and lead concentrations in a subset of seventeen of the houses, primarily through XRF.

TABLE 3. Indoor soil and wall sample results

Metal	Percentage of EHC samples above screening level
Mercury (SL 6.6 mg/kg)	89
Mercury (site-specific SL 75 mg/kg)	44
Lead (SL 140 mg/kg)	46
Arsenic (SL 50 mg/kg)	77

Mercury concentrations in adobe/tapial ranged from 0.7 to 1,072 mg/kg, while that of arsenic varied from 2 to 1,611 mg/kg, and lead ranged from 2.4 to 5,510 mg/kg. Overall 43 percent of the indoor earthen wall and floor samples were above the EHC's site-specific SL for mercury of 75 mg/kg, and 89 percent were above the Peruvian SL of 6.6 mg/kg. Forty-five percent were above the Peruvian lead SL of 140 mg/kg, and 77 percent exceeded the Peruvian SL of arsenic of 50 mg/kg. Exposure to lead and arsenic in floors and walls is via particulate-bound ingestion, which can originate from indoor or outdoor sources.[51]

Indoor mercury vapor readings ranged from 13 ng/m³ to 12,000 ng/m³.[52] Indoor air contamination is a function of exposed earthen walls and floors off-gassing in a closed space, and outdoor air is not as contaminated. The OEFA study encompassed outdoor air quality, and measurements over a period of five days in two points outside of the city did not indicate the presence of heavy metals above SLs in particulate samples. The objective of this monitoring appears to have been to determine if contaminated dust was being released from tailings in the vicinity of the mine. As a result, the siting of the monitoring stations, and their results, are not necessarily indicative of air quality in the city itself.[53]

Within the city, in 2014 and 2017 the EHC gathered outdoor air samples at different points and analyzed them for mercury vapor with a Lumex RA915 analyzer. The data revealed that background vapor concentrations average around 31 ng/m³, which is twenty times the global average.[54] Elevated though they are, the background concentrations are below USEPA and World Health Organization (WHO) SLs for chronic exposure. Whereas the USEPA limit is 310 ng/m³, that of the WHO is 200 ng/m³. Both agencies have established 1,000 ng/m³ as an action level, requiring the relocation of inhabitants and remediation. Thirty percent of indoor residential air samples were above the WHO SL. It is important to note that because vapor is by nature in a state of flux, concentrations can vary considerably over a short period of time depending on temperature and airflow. In addition, even mild scuffing of dirt floors and walls has been demonstrated to release a spike of mercury vapor, indicating that such vapors are released by human and animal movement within the home.[55]

Overall, in 75 percent of the homes sampled by the EHC, mercury concentrations in the walls and floors and/or mercury vapor concentrations exceeded WHO SLs. In some cases more than one media (soil, vapor) was in exceedance. All indoor soil and floor residential samples exceeded Peru's SL. There are approximately 5,090 earthen homes in Huancavelica's urban area (including the adjoining municipality of Ascención), which account for 37 percent of the housing stock and serve as a residence for just over eighteen thousand people.

Extrapolating the results from the EHC's study of residential mercury contamination, it is estimated that all of these residents are currently exposed to mercury levels in their homes that exceed Peruvian SLs, while 13,500 are living in homes that exceed the WHO SL.[56]

While much is known regarding the individual effects of mercury, lead, and arsenic intoxication, the consequences of polymetallic intoxication are less well understood. Each of these heavy metals can produce miscarriages and birth defects and damage the kidneys, and mercury and arsenic can be passed to an infant through breastfeeding. Lead and mercury poisoning may also lead to impaired cognitive and physical development. In addition, arsenic poisoning can affect the circulatory system, resulting in anemia, increased blood pressure, and an irregular heartbeat. It can also reduce sperm count and likely causes cancer.[57]

As with all toxins, the effects of these heavy metals depend on the amount and means by which they are ingested, as well as exposure duration, the species or compound involved, individual sensitivity, and even nutritional status. In the case of mercury, women of childbearing age are especially at risk, as it can affect female fertility. Children can be disproportionately affected because their brains and central nervous systems are developing rapidly, and a relatively small dose can have a large impact given the foregoing and their lesser body weight relative to an adult.[58]

Elemental, or metallic, mercury is most noxious when its vapors are inhaled, in which case about 85 percent will be retained in the body. This is in contrast with dermal exposure, in which around 2 percent will be absorbed, and with oral ingestion, in which around 10 percent will be retained in the body.[59] In the case of organic mercury compounds, such as methylmercury, the gastrointestinal tract will absorb over 90 percent. These compounds can easily cross the blood-brain barrier and affect the central nervous system. Inorganic compounds, which are more likely to produce kidney damage, result from mercury binding with other minerals or being exposed to sunlight, the latter of which produces divalent mercury.[60]

Acute elemental mercury poisoning through ingestion may produce overt symptoms such as uncontrollable shaking, elevated heartrate, polyneuropathy, delirium, and kidney and gastrointestinal tract maladies. When acute intoxication results from inhaling mercury vapors, the victim may salivate uncontrollably, have difficulty breathing, and suffocate due to inflammation of airways.[61] While these effects may be overt, in the case of chronic mercury poisoning, which is more prevalent in Huancavelica, the psychological symptoms are often subclinical, or not readily apparent. Such symptoms include mood alterations such as depression, anxiety, shyness, indecision, and hostility, in addition to the loss of short-term memory, obsessive-compulsive behaviors, and difficulty concentrating.[62] More readily identified physical symptoms include tremors, uncontrolled salivation, inflamed and discolored gums, wobbly teeth, anemia, and dermatologic disorders.[63] The range of effects, their often subclinical nature, and the fact that other maladies have similar symptoms complicates the diagnosis of chronic mercury poisoning.[64]

Fortunately there are solutions to Huancavelica's contamination and the ongoing risks it presents to those who live there. The most viable approaches center on removal and/or encapsulation of contaminated materials. For the Ichu River, this would involve the control of sources of runoff, from both the city and the mines, the latter of which appear to be a source of acid drainage.[65] In addition it would require extensive dredging and appropriate disposal of sediments along several kilometers of the river. Within the city the continued paving of streets and installation of sidewalks will reduce the amount of outdoor soil available to create dust, while contaminated soils would have to be removed, disposed of, and replaced in unpaved public areas.

The primary risk, however, is borne by those who live in homes made from local soils, who are also Huancavelica's poorest residents and are often found washing their clothes in the Ichu River. Such homes commonly have dirt floors and unsealed walls, which are sources of mercury vapor as well as dust contaminated with mercury, lead, and arsenic. The most effective, and costly, solution is the construction of a new home with noncontaminated materials. This involves the temporary relocation of the residents and the demolition of the existing structure. Such demolition entails measures, such as misting, to reduce the release of dust that could further contaminate the area. The rubble must be relocated and deposited in such a manner that it neither leaches into the soil nor emanates vapors; this can be accomplished by placing the rubble in a pit lined with thick-gauge plastic and sealed with a sufficiently thick layer of fresh, uncontaminated soil.

A less costly alternative is the remediation of the existing structure by encapsulation, which in this case involves sealing the interior walls and installing a concrete floor. A plastic vapor barrier can be applied to walls to limit vapor intrusion, although, while relatively inexpensive, the solution is not durable in a residential context. A somewhat more costly and longer term solution is the encapsulation of the walls with gesso or cement stucco, the latter of which is commonly used in the city. The EHC has experimented with this last solution, pouring a three-inch cement floor and stuccoing walls with a minimum of one-half inch of gesso or cement in eleven homes. In all cases mercury vapor levels were reduced to near background levels, and the technique eliminates indoor-generated sources of mercury-, lead-, and arsenic-containing dust. It does not appear that cement is more efficacious than gesso, the former of which is less expensive.[66]

A Responsible Party?

In Peru, as elsewhere, laws regarding remediation of heavy metals and mining contamination involve the identification of a "responsible party," the entity who is legally liable for the costs of cleanup.[67] In the case of Huancavelica, to issues of legal responsibility are added those of causal and humanitarian responsibility. Causal responsibility is defined here as the ethical if not legal obligation to remediate by the entity, person, or group that was the cause of the contamination.

As we have seen, the vast majority of Huancavelica's quicksilver production occurred during the colonial period (1564–1821) and was rendered by smelters that were inherently inefficient. Subsequent production by humanchis, state-sponsored enterprises, the Negociación Minera Fernandini Clotet Hermanos, and the Sociedad Minería Brocal was at most 10 percent of the total production of the mine. In the twentieth century, industrial refining took place close to the Santa Bárbara mines, which are almost three kilometers outside of and 730 meters above the city. They also utilized smelters that were more efficient than the furnaces that were used historically. Although all refining will involve emissions, that which occurred during the national period was, overall, less than during the colonial period as a result of lesser quality ore, lower production levels, and more efficient refining methods.[68] In addition, while emissions from refining at the Santa Bárbara mine site were released into the atmosphere and ultimately deposited, the distance of the emissions source from the city and its higher elevation suggests that such production was, relatively, not a primary contributor to the contamination of the soils and homes of the city. In contrast, colonial production occurred in several points scattered throughout what is now the city, in addition to smelting at Santa Bárbara, and had greater emissions due to the amount and method of refining and the richness of the ore.[69]

The foregoing suggests that much, if not the vast majority, of the mercury, lead, and arsenic contamination of the city and the Ichu River and its tributaries is a result of colonial-era mining. The picture is complicated, however, by the fact that the city and region are naturally rich in cinnabar, and the soils even contain native mercury. Although it is difficult to distinguish between natural and colonial contamination, the depth at which it occurs and the elements involved can be suggestive of origin. Generally, native mercury is found a foot or more below the surface in Huancavelica; however it is possible that in some cases the deposits are colonial residues from smelters or broken or porous storage containers.[70] Soil samples in the EHC study were collected from between one- and three-inch depths, while the OEFA study drew samples from a depth of approximately one foot. Despite the differing depths, the results were quite similar and are suggestive of deposition rather than natural occurrence. In both studies, soil samples were analyzed by XRF for lead and arsenic, both of which are associated with cinnabar ore. The high levels of lead and arsenic found in the samples suggests that the contamination is the result of refining given lower background levels of lead, arsenic, and mercury in regional soils.[71]

While it may not be possible to conclusively distinguish between naturally occurring heavy metals and colonial contamination, colonial and historic maps indicate smelter locations at different points of time. The greatest concentration was in what was known as the Chunca Horno (Ten Ovens) neighborhood, where mitayos were also required to reside and where they breathed toxic fumes. Today known as the San Cristóbal neighborhood, it is the city's second most contaminated area. Only Yananaco is more contaminated, and this may reflect wind-borne deposition from San Cristóbal as well as refining that took place there. Similarly, refining took place in the Ascención neighborhood, and to a much less extent in that of Santa Ana, which is where many of the town's elite resided. Today Santa Ana is the least contaminated part of the city, having been somewhat protected from the refining directly across the river in San Cristóbal by the prevailing winds.[72] The correlation between colonial refining and contemporary contamination is striking, yet at the same time it is not surprising given the presence of smelters in what is now the city. It is also worth noting that contamination is sometimes plainly visible in earthen home construction, with contaminated soils often having a more reddish color due to cinnabar content and often containing sherds of ceramic condensation tubes and storage vessels used at the smelter.

Although the Spanish crown would grant individual ownership to silver mines, they sought to exercise a monopoly on mercury production and sale as a means of both regulating silver production and generating a source of revenue. This was the impetus for Viceroy Toledo's decision in 1573 to expropriate the cinnabar mines, which reversed claims that had been granted by the crown and transformed the miners' guild from an association of owners to one of state contractors. Thus, unlike most silver mines, beginning in 1573 the Santa Bárbara and adjacent deposits were exclusively held by the Habsburg and, beginning in 1700, Bourbon monarchs. It is for this reason that with Peruvian independence in 1821, ownership of the mines passed to the Peruvian state. As a result, most of the contamination in Huancavelica today is the result of colonial cinnabar mining and refining that was conducted under the aegis of the Spanish monarchy.[73]

But while the Spanish monarchy, through their ownership of the mines during the colonial period, unquestionably had a causal role in the lion's share of the city's anthropogenic toxicity, their legal responsibility may be limited. This is because, in Peru as in most countries, ownership of a mine involves assuming liability for any contamination it may have previously caused. Throughout the nineteenth century the Peruvian state owned the deposits, and the various efforts to reinvigorate production were structured as joint ventures or a lease arrangement. The promulgation of the new mining code of 1901, however, allowed private ownership, an opportunity that was quickly and effectively exploited first by Augusto Benevides, then Eulogio Fernandini, and later their descendants in the Sociedad Minería Brocal.[74]

Brocal is currently the owner of the mine and a subsidiary of the publicly traded Compañía de Minas Buenaventura, one of Peru's largest mining companies. According to current interpretations of Peruvian Law 28271 by the Ministry of Energy and Mines, Brocal is responsible for contamination only at the mining site or installation, and responsibility is assigned by the Ministry's General Directorate. This is despite the fact that Article 2 of the law defines a contaminated site as including "installations, effluents, emissions, remains or storage of residues produced by mining operations, which are currently abandoned or inactive and which constitute a permanent and potential risk to the health of the population, the surrounding ecosystem and property."[75] Since the cessation of mercury production in Huancavelica in 1975, Brocal has conducted remediation work in the immediate vicinity of the refining operations at Santa Bárbara.[76] Despite this it is unclear why the contamination in the city and vicinity of Huancavelica is not considered "remains" or "emissions," as they objectively are, and have been demonstrated to be "a permanent and potential risk to the health of the population [and] the surrounding ecosystem."

As the OEFA study suggested, and Brocal has acknowledged, the mine complex is the source not just of previous contamination but likely of ongoing acid mine drainage, especially from the Yanamina mine. This too would come under the "effluents" category of Law 28271 and should trigger remediation. Acid mine drainage occurs as a result of the interaction of sulfur compounds, especially iron sulfide, with water and air, which produce sulfuric acid. Such drainage affects adjacent soils and, based on the OEFA study, appears to have entered the Ichu River via the Disparate tributary. Exposed workings and tailings at the mine site are also a potential source of contaminated dust; however the OEFA study did not detect airborne contaminants above SLs in the two monitoring stations placed outside the city.[77] While cinnabar mining during the 1960s and 1970s at Santa Bárbara does not appear to be a source of significant legacy contamination, Brocal has responsibilities concerning remediation, which have been circumscribed by ministerial resolutions and a narrow interpretation of Law 28271. In addition, they have specific responsibilities in the formal process of closing the mine, which, according to Peruvian Law 28090, includes sufficient investment and actions for the "protection, preservation and recuperation of the environment and with the purpose of mitigating negative impacts to the health of the population, the surrounding ecosystem and property."[78]

Legal responsibility does not end with Brocal, however, but also encompasses the Peruvian state, which has responsibilities at many levels.[79] One is the enforcement of Peruvian laws and the adherence to international treaties to which Peru is a signatory, such as the Minimata Convention on Mercury. In addition, Article 5 of Law 28271 stipulates that the state is responsible for remediation when a responsible party cannot be identified by the Ministry of Energy and Mines and there is a risk to "health and human security." In such cases, according to Article 9 of the law, the National Environment Fund (Fondo Nacional del Ambiente) is to coordinate remediation efforts. Article 6 of Peru's General Law of the Environment calls for remediation of contaminated sites, and Articles 8 and 9 require the responsible parties to assume the costs of remediation.[80] The government has developed an inventory of 8,616 contaminated sites throughout the country, in which almost half, 4,281, are classified as either "very high risk" or "high risk," and 88 percent do not have an identified responsible party.[81] For unknown reasons, the Peruvian state has not included sites contaminated as a result of colonial mining in their inventory. With such an approach, if contamination does not legally "exist," there is no responsible party. The result has been the "invisibilization" of an entire city.[82]

Even excluding the pervasive polymetallic contamination in the city and vicinity, the department of Huancavelica ranked fourth in the number of contaminated sites, with 858 identified in 2015.[83] Of interest would be Huancavelica's ranking in this regard if the ranking was based on the number of people exposed to polymetallic contamination above SLs. Indeed the regulations concerning identification of contaminated sites calls for prioritization based on the "probability or possibility that a contaminant can have adverse effects on human health . . . based on the characteristics and the quantity that enters into contact with potential receptors, including the consideration of the magnitude or intensity of the associated effects and the number of individuals . . . [who] can be affected in the present or future."[84]

While Peruvian law is clear concerning legal responsibilities associated with cleanup of inventoried sites, enforcement is constricted by narrow and exclusionary legal interpretations of Law 28271, funding, structural issues, and a lack of political will. For example, in 2015 the General Directorate of Mining had only thirteen experts assigned to identify contaminated sites.[85] More broadly, OEFA is the entity responsible for environmental enforcement in Peru; it is a unit within the Peruvian Ministry of the Environment, whose authority has declined in recent years. In 2014 the Humala administration stripped it of the ability to set environmental standards on many contaminants as part of a larger effort to stimulate mining in the country. These new regulations also affected OEFA by reorienting actions toward current mining operations and the prevention of contamination as opposed to remediation and limiting the amount it can fine entities for environmental violations. To some extent OEFA was minimally affected by the 2014 changes, as in only 1.7 percent of the cases they investigated was a fine issued, and according to Peru's Ombudsman's Office, OEFA's enforcement is lax.[86]

In this context OEFA must prioritize their actions; this is routinely done on the basis of "conflictivity," or the degree of social conflict over a given issue. Three years after completing the report cited in this chapter, OEFA had not acted upon it, nor did they offer any specific recommendations, except the pro forma suggestion that the report be sent to both the local OEFA office in Huancavelica and the national supervising director for possible action.[87] This is a remarkably tepid suggestion, as has been the response to what the Peruvian government has acknowledged is widespread heavy metal contamination in the soils and waterways of a city of almost fifty thousand people. Nevertheless Huancavelica must compete in a context where in June 2015, of 210 social conflicts, two-thirds concerned mining.[88]

The Peruvian government has not only failed to enforce existing laws that mandate cleanup of legacy mining sites; they have failed to effectively defend the rights of Huancavelicanos to health and shelter, which are enshrined in the Peruvian Constitution. Specifically Article 2 of the Constitution asserts, "Every person has the inalienable right to live in an environment that is healthy, balanced and adequate for the full development of life . . . particularly assuring the health of people individually and collectively." Similar protections are covered in Article 7, which states, "Every person has a right to the protection of their health."[89] There is extensive scientific evidence concerning the levels of heavy metal contamination in the city's soils, earthen homes, and waterways, and the risks this poses to inhabitants based on Peruvian and international standards. Huancavelica requires a major and multisectoral intervention and, given the scope and severity of the contamination, one that requires the leadership of and investment by the state.

Such an intervention is long overdue and reflects the broader neglect of Huancavelica by Peru's central government. Not only is the city of Huancavelica among the most mercury-contaminated urban areas on earth; it is also Peru's poorest departmental capital. The department itself has the nation's highest levels of extreme poverty, malnutrition among children under five years old, and female illiteracy, at about 33 percent. In 2006, among all of the department's inhabitants, the illiteracy rate of 20 percent was nearly 300 percent above the national average, and 56 percent of residents were malnourished.[90]

The Peruvian Ombudsman's Office, known as the Defensoría del Pueblo, has become involved in the effort to address Huancavelica's contamination. The purpose of the office is to defend the rights of Peruvian citizens and urge the state to meet their legal obligations in this regard. While the Defensoría has the power to summon government officials, their powers are otherwise largely limited to the spheres of persuasion and conciliation. In 2017 I filed a formal complaint with the Defensoría as a result of government inaction on issues concerning Huancavelica's contamination. The Defensoría has convoked numerous meetings on the topic, increasing awareness in various government agencies and the wider society of the contamination issues faced by the residents of Huancavelica. The mere recognition of the problem is itself an important step in eliminating Huancavelica's "invisible" status. Moreover the office has facilitated the development of a broader legislative strategy that involves the modification of laws that, through their structure or interpretation, effectively exclude Huancavelica from the provisions that apply to contaminated sites.

Meanwhile the residents of Huancavelica continue to live in the midst of polymetallic toxicity. Although recent years have seen the advance of many public works in the city, including the paving of streets, much work remains to reduce the prevalence of dust generated by urban activities. Similarly home construction in the city increasingly uses brick or cinderblock instead of contaminated soils. Despite these encouraging developments, however, many still live in adobe or tapial homes, and because they are the most economically disadvantaged sector of the population, they are the least able to obtain credit or otherwise build with more costly materials. While Peru's Ministry of Housing has several "social housing" programs, such as the Fondo Mi Vivienda (My House Fund), they are oriented to rural communities, have numerous restrictions concerning eligibility, and often use construction companies from outside of the region. With an estimated eighteen thousand people living in mercury-contaminated adobe homes in the city, a substantive effort would require the construction of approximately 5,090 homes if the Peruvian SL for mercury were used as an action level. Additional investments would be required for remediation work on the Ichu River and its toxic tributaries, all of which could provide a significant and ongoing economic stimulus in the city and region.[91]

Beyond its duties to its own citizens, the government of Peru also has international obligations under the Minimata Convention on Mercury, which seeks to limit the production, use, and release of mercury on a global scale. Peru became a signatory in October 2013, and the commitments under it came into effect in August 2017. Among the provisions of the Convention is remediation of contaminated sites. Thus, while Brocal has legal and financial obligations concerning remediation, the scope of which is a matter of legal interpretation, the Peruvian government has an obligation to see that this occurs where needed, both under Peruvian law and, one would think, the Minimata Convention.[92] Unfortunately, however, since Huancavelica is not legally considered a contaminated site, it is, again, "invisibilized" and not subject to its provisions. Peru is also a signatory, since 1976, of the International Covenant on Economic, Social and Cultural Rights, in which Article 12 recognizes the "right of everyone to the enjoyment of the highest attainable standard of physical and mental health." And Article 24 of the Convention on the Rights of the Child, which was ratified in Peru in 1990, upholds the right of children to the "enjoyment of the highest attainable standard of health."[93]

Beyond causal and legal responsibilities, there is a more broadly shared humanitarian responsibility to act to protect people's health and development. While the Spanish monarchy may not have a legal obligation to remediate, it is clear that activities conducted under their authority are the primary anthropogenic cause for the toxicity of the city's soils, tributaries, river, and earthen homes. Despite an active foreign development program through the Agencia Española de Cooperación Internacional para el Desarrollo, neither the Spanish government nor the monarchy has invested in remedial work in the city. Similarly, in addition to the Peruvian government's legal responsibilities, it has a humanitarian responsibility to intervene at an appropriate scale to protect the health and well-being of some of their most vulnerable citizens. The results of the OEFA and EHC studies, as well as the socioeconomic indicators for the region, show they have failed in this regard.

While residents of Huancavelica are victims of a toxic colonial legacy, it appears that they will have to actively pressure the authorities and other responsible parties if anything substantive is to occur. By its own admission, OEFA's mobilization and enforcement is a function of conflict. As long as Huancavelicanos remain silent on the issue, there is little probability that OEFA or other entities will act. Attitudes on the issue of contamination in the city are diverse, and sometimes perplexing. Those most affected are the poorest, and their days are framed by a struggle for survival. The insidious and subtle effects of odorless, invisible toxins are, understandably, less of a priority than feeding their family. Many of these residents, and much of the wider population, do not realize that they live in one of the world's most toxic cities and instead evince a pride that they are from the "Land of Mercury," without giving much thought to the implications of that statement. Responses to learning of the risks of polymetallic contamination range from acceptance to defensiveness concerning the human capital of the region. The combination of civic pride and limited knowledge of the situation can produce a form of denial. As with Stockholm syndrome, many embrace what they cannot escape. Without coordinated, multisectoral, and concerted action at a local level, it will be difficult if not impossible to improve the conditions there.

Finally, there is the more diffuse humanitarian responsibility of the international community. While such responsibility is justified by the need to reduce a vulnerable population's exposure to heavy metals, it is enhanced as a result of the contribution that Huancavelica's mercury has made to the development of the global economy in which we live. Were it not for quicksilver from Huancavelica, the Americas would not have produced the tens of thousands of tons of silver it did during the colonial period. Global integration, and even the Industrial Revolution, owe a debt to Huancavelica, for it was mercury that enabled the production of such quantities of silver and the capital accumulation that ensued. Today's global society has benefited indirectly from Huancavelica's toxic treasures and the toil of the laborers who became ill or died producing them.[94] In the end there is not a sole responsible party but rather many irresponsible parties that have failed to act according to the law and ethical and humanitarian principles.

Conclusion

The cinnabar deposits of Huancavelica are among the largest in the world, and over four hundred years of exploitation have rendered approximately seventy-two thousand metric tons of mercury. Most of this was produced during the colonial period, with extensive use of Indian draft labor, and was used to produce silver in the Andes through the mercury amalgamation process. With independence, the ownership of the deposits went from the Spanish monarchy to the Peruvian government. Production during the nineteenth century was sporadic and limited, conducted mostly by freelance humanchis, as well as through joint ventures and a lease arrangement with the state. Following the promulgation of a new mining code in 1901, Eulogio Fernandini acquired and made extensive investments in the mines during the 1920s. In 1956 the Sociedad Minera Brocal was formed and would subsequently produce mercury at Santa Bárbara until the mines were closed in 1975.

Among the results of this historic production of quicksilver is widespread mercury, lead, and arsenic contamination in the city's soils and the Ichu River and its tributaries. Indeed Huancavelica is among the world's most mercury-contaminated urban areas, and the associated health risks are exacerbated by the prevalence of earthen homes there. Not only do such homes contain dangerous levels of mercury, lead, and arsenic, but many off-gas mercury vapor from their walls. The estimated eighteen thousand people who live in adobe or tapial homes that are contaminated above the Peruvian SL for mercury are constantly exposed to toxins when they are in their residence.

Despite the body of scientific evidence indicating the degree of contamination and the unremitting exposure of the population to these toxins, the Peruvian government has not intervened in a substantive, concerted, and consistent manner. While the Spanish monarchy, which owned the mines in the colonial period, bears a causal responsibility for the contamination, the legal responsibility under Peruvian law is shared by the current owner of the mine, the Sociedad Minera Brocal, and the Peruvian state. Brocal's responsibility may be circumscribed by narrow interpretations of applicable laws, but the state has a responsibility to protect its citizens and to abide by the provisions of the Minimata Convention on Mercury, of which it is a signatory. In the final analysis, given that enforcement actions are correlated with social conflict, it appears that sustained and coordinated mobilization of the residents of Huancavelica will be necessary if appropriate remedial actions and investments are to be made.

NOTES

I would like to thank my scientific collaborators for their contributions to and comments on this chapter, especially Dr. Enrique Ecos, Bryn Thoms, RG, Rubén Darío Espinoza Gonzales, Dr. Susan Halabi, and Dr. William Brooks. Further thanks go to all of those who have participated in this study, as well as those who have assisted us in the field, especially Isidora Cauchos, Elsa Matamoros, Vilma Cauchos, Javier Quezada, Jorge Moreira Tora, and Yolanda Castellares Aramburú. I would also like to thank Dallas Noelia Gonzales Malca, Omar Bravo Tirado, and Giuliana Patricia Becerra, of Peru's Ministry of the Environment, for technical support. I am grateful to the Department of History at North Carolina State University and Dr. John Vandenberg and Dr. Nicole Hagan for their support of this research. Special thanks go to Dr. Alicia Abanto, Dr. Paulo Vilca

Arpasi, Dr. Nicasio Valencia Mamani, and Dr. Zeida Patrica Hoces la Rosa for their support for this research and their unwavering efforts to assist the people of Huancavelica.

1. Government of Peru, *Estado de la Población Peruana*, 7; McKee, Noble, and Vidal. "Timing of Volcanic and Hydrothermal Activity," 489; Wise and Féraud, "Historic Maps," 16. For a review of the literature on colonial Huancavelica, see Pearce, "Huancavelica." For works that also encompass the national period, see Robins, *Santa Bárbara's Legacy*, 11–19; Robins et al., "Estimations," 152–53.
2. Cantos de Andrade, "Relación," 303; Ruiz Estrada, *Arqueología*, 30, 33–34, 37–38, 41; Government of Peru, *Almanaque*, 16–17; Burger and Matos, "Atalla," 164–67, 172–73; Brooks et al., *Peru Mercury Inventory*, 3; Cooke et al., "Over Three Millennia," 8833; Cooke et al., "Use and Legacy," 4181–84.
3. Rivero y Ustáriz, "Memoria," 85; Cobo, *Historia*, 1:150; Cantos de Andrade, "Relación." 304; Fonseca, "Informe," 347; Ramírez, "Descripción," 321; Bargallo, *La minería y metalurgía*, 39–40; Crosnier, "Geologie du Perou," 36; Singewald, "Huancavelica Mercury Deposits," 518; Lohmann Villena, *Las minas de Huancavelica*, 11, 13; Patiño Paúl Ortíz, *Huancavelica colonial*, 20–21; Hernández, "Antiguedad y actualidad," 942; Cooke et al., "Use and Legacy," 4181–84; Cooke et al., "Over Three Millennia," 8830, 8833; Ruiz Estrada, *Arqueología*, 30, 37; Petersen, *Mining and Metallurgy*, 29; Brooks et al., *Peru Mercury Inventory*, 3. For more on the topics discussed in this section, see Robins, *Santa Bárbara's Legacy*, 4–5, 23–24, 32–35.
4. Ribera and Chaves y Guevara, "Relación," 182; Delgado de Castro, *El Despertar*, 13; Patiño Paúl Ortíz, *Huancavelica colonial*, 20; Government of Peru, *Almanaque*, 18; Wise and Féraud, "Historic Maps," 16.
5. Hawley, "Notes," 6–7; Berry and Singewald, *Geology and Paleontology*, 19–20, 42; Arana, *Las minas*, 2–3; Strauss, "Quicksilver," 562–63; Yates, Kent, and Fernández, *Geology*, 2; Dueñas, "Fisionomia minera," 160; Singewald, "Huancavelica Mercury Deposits," 518; Loveday, "El azogue de Huancavelica," 83; Gastelumendi, *Huancavelica*, 40; Torrico y Mesa, "Excursión," 891; Tamayo, "Mina de cinabrio," 39; Rivero y Ustariz, *Memoria*, 8–9; Arana, *Las minas*, 2, 11, 44; Patiño Paúl Ortíz, *Huancavelica colonial*, 100–101.
6. Author's field research notes, August 2010, July 2012, June–July 2015; Goldwater, *Mercury*, 49; Singewald, "Huancavelica Mercury Deposits," 522; Strauss, "Quicksilver," 562; Crosnier, "Geologie du Perou," 2, 51–52; Berry and Singewald, *Geology and Paleontology*, 43, 51–52; Gastelumendi, *Huancavelica*, 41.
7. Cantos de Andrade, "Relación," 303; Contreras, *La ciudad del mercurio*, 19; Lohmann Villena, *Las minas de Huancavelica*, 20–21, 23–24; Carrasco, *Cronología de Huancavelica*, 20–22; 90–92; Robins, *Mercury, Mining and Empire*, 22.
8. Lohmann Villena, *Las minas de Huancavelica*, 15; Nriagu, "Mercury Pollution," 168–69; Sánchez Gómez, "La ténica," 84; Fisher, *Silver Mines*, 3; Probert, "Bartolomé de Medina," 110–11; Bakewell, "Registered Silver Production," 81; Brading and Cross, "Colonial Silver Mining," 555; Bargallo, *La amalgamación*, 172; Robins, *Mercury, Mining and Empire*, 2. For more on the topics discussed in this section, see Robins, *Santa*

Bárbara's Legacy, 42–65, 73–95, 102–3, 116–23, 125–28, 141–50; Fisher, *Silver Mines*, 2, 10–12, 21–30, 32–38, 47, 52–62, 70–71, 86–90, 110–12.

9. Juan Alcalá y Amurrio, Directorio del Beneficio del Azogue, en los Metales de Plata, Documentos que se dan en sus Reglas, Oruro, 1781, Archivo y Bibliotecas Nacionales de Bolivia (hereafter ABNB), Ruck 80, 12–14, 21, 26, 52–53; Capoche, *Relación*, 123; Cobo, *Historia*, 1:147; Salinas y Córdoba, *Memorial*, 264; Caravantes, *Noticia*, 4:104–5; Cañete y Domínguez, *Guía histórica*, 50, 65; Haenke, *Descripción del Perú*, 126–30; Llano Zapata, *Memorias histórico*, 169; Bargallo, *La minería y metalurgía*, 110–12, 192, 291; Bakewell, *Miners of the Red Mountain*, 21–22, 148–49; Fisher, "Silver Production," 289; Fisher, *Silver Mines*, 54; Cobb, *Potosí y Huancavelica*, 95; Brading and Cross, "Colonial Silver Mining," 554; Purser, *Metal Mining*, 48; Robins, *Mercury, Mining and Empire*, 10–11, 86.

10. Don Rodrigo de Mendoza y Manrique, administrador y arrendatario que fue de las minas y los ingenios de don Pedro Sorez de Ulloa en el cerro y la ribera de Potosí, con el maestro de campo don Rodrigo Campuzano, hermano de doña Francisca Campuzano, viuda y heredera que fue de dicho don Pedro, sobre la liquidación de los pesos impendidos en el avío de las haciendas mencionadas durante el tiempo de su administración y arrendamiento, La Plata, February 5, 1656–69, ABNB, ALP Minas 15/1, 2; Capoche, *Relación*, 124; Caravantes, *Noticia*, 4:105; Cobo, *Historia*, 1:147–49; Cañete y Domínguez, *Guía histórica*, 65; Juan Alcalá y Amurrio, Directorio del Beneficio del Azogue, en los Metales de Plata, Documentos que se dan en sus Reglas, Oruro, 1781, ABNB, Ruck 80, 20, 53; Arzáns de Orsúa y Vela, *Historia*, 1:170–71; Fisher, *Silver Mines*, 54; Pieper, "Innovaciones tecnológicas," 357–58; Bakewell, "Registered Silver Production," 75–77; Purser, *Metal Mining*, 50–51; Cobb, *Potosí y Huancavelica*, 90; Bakewell, *Miners of the Red Mountain*, 22; Bargallo, *La amalgamación*, 213; Robins, *Mercury, Mining and Empire*, 88.

11. Cobb, *Potosí y Huancavelica*, 35, 49; Bakewell, "Technological Change," 78; Robins, *Mercury, Mining and Empire*, 23.

12. Memorial del capitán Don Pedro Gutíerrez Calderón de algunas advertencies considerables al servicio de Dios Nuestro Señor y de su Real Magestad, Lima, May 2, 1623, Archivo General de Indias (hereafter AGI), Lima, 154, 1; Arana, *Las minas*, 3, 26; Strauss, "Quicksilver," 562, 564; Berry and Singewald, *Geology and Paleontology*, 42, 46; Umlauff, "El Cinabrio De Huancavelica," 43–44; Wise and Féraud, "Historic Maps," 22; Purser, *Metal Mining*, 45; Whitaker, *Huancavelica Mercury Mine*, 14; Tamayo, "Mina de cinabrio," 40; Hawley, "Notes," 8; Hadley, "Report," 44; Molina Martínez, *Antonio de Ulloa*, 84; Brown, "Workers' Health," 71; Robins, *Mercury, Mining and Empire*, 52, 57, 110.

13. Memorial del capitán Don Pedro Gutíerrez Calderón, 1; Emetherio Ramírez de Arellano. Huancavelica, April 3, 1649, AGI, Lima, 279-4, 1–2; Agricola, *De Re Metalica*, 428; Lohmann Villena, *Las minas de Huancavelica*, 182, 229; Patiño Paúl Ortíz, *Huancavelica colonial*, 349; Brown, "Worker's Health," 471, 474; Robins, *Mercury, Mining and Empire*, 52–53.

14. Mendoza y Luna, "Relación," 44; Navarra y Rocaful, "Relacion," 2:175; Cantos de Andrade, "Relación," 304; Rivero y Ustáriz, "Memoria," 130; Llano Zapata, *Memorias histórico*, 218; Montesinos, *Anales del Perú*, 2:128; Lohmann Villena, *Las minas de Huancavelica*,

55–56, 117, 137, 139; Lohmann Villena, "La minería," 652; Brown, "Workers' Health," 479; Strauss, "Quicksilver," 563; Purser, *Metal Mining*, 45–48; Bidstrup, *Toxicology and Mercury*, 5; Goldwater, *Mercury*, 49; Fuentes Bajo, "El azogue," 92; Patiño Paúl Ortíz, *Huancavelica colonial*, 184, 186–87; Carrasco, *Cronología de Huancavelica*, 168; Matilla Tascón, *Historia*, 89–96; Whitaker, *Huancavelica Mercury Mine*, 83; Fernández Alonso, "Los mecenas de la plata," 367; Robins, *Mercury, Mining and Empire*, 57–58.

15. Expediente seguido por don Roque de Reinalte, sobre la prisión a que le han reducido por cierta deuda no obstante de su calidad de minero en Potosí, La Plata, August 14–19, 1697, ABNB, ALP Minas 19/10, 1, 5; Toledo, "Memorial," 101; Acosta, *Historia*, 244; "Historia de la mina de Huancavelica," 65; Caravantes, *Noticia*, 4:147; Lizárraga, *Descripción*, 188; Capoche, *Relación*, 135; Llano Zapata, *Memorias histórico*, 165; Zimmerman, *Francisco de Toledo*, 177, 183, 185; Fisher, *Silver Mines*, 3; Bakewell, "Technological Change," 76, 81–82; Patiño Paúl Ortíz, *Huancavelica colonial*, 105, 167; Ballesteros Gaibrois, *Descubrimiento y fundación del Potosi*, 18; Lohmann Villena, *Las Minas de Huancavelica*, 59; Cobb, *Potosí and Huancavelica*, 55; Bargallo, *La amalgamación*, 172; Robins, *Mercury, Mining and Empire*, 26–27.
16. Cañete y Domínguez, *Guía histórica*, 90; Capoche, *Relación*, 117; Arzáns de Orsúa y Vela, *Historia*, 1:157, 169; Buechler, *Mining Society*, 39; Cobb, *Potosí and Huancavelica*, 91; Rudolph, "Lakes of Potosi," 531–32, 535–36, 543; Bakewell, *Miners of the Red Mountain*, 13; Bargallo, *La amalgamación*, 206; Robins, *Mercury, Mining and Empire*, 27.
17. Expediente promovido por don Nicolás de Sarabia y Mollinedo asentista de la Real Mina de azogue de Guancavelica sre que los corregidores y gobernadores de las provincias que tienen obligación de remitir mitas de indios para el trabajo de otra mina lo ejecuten sin dilación alguna y sre lo ocurrido con este motivo en la de Jauja a causa de haberse intentado cumplir con la remisión de mitayos aquel gobierno como se le mandó que avisa la resistencia que hicieron e insultos al cobrador enterador de mitas don Jacinto Maita, Lima, November 10, 1811, AGI, Indiferente 1335, 24; Carta del Virrey del Perú #38, "Da cuenta a V.Md. de todo lo que en materia de azogue sea escrito," February 2, 1630, AGI, Indiferente general 1777, 3; Armendaris, "Relacion," 162; Torres de Portugal, "Memoria," 207; Navarra y Rocaful, "Relacion," 2:164; Velasco, "Relación," 113; Memorial del capitán Don Pedro Gutíerrez Calderón, 2; Audiencia de Lima, "Relación," 294–95; Capoche, *Relación*, 135, 141; Fonseca, "Informe," 351; Crespo Rojas, "El reclutamiento," 469; Wiedner, "Forced Labor," 369; Bakewell, *Miners of the Red Mountain*, 69, 71–73, 94–95; Zimmerman, *Francisco de Toledo*, 183; Arana, *Las minas*, 12–13; Lohmann Villena, *Las minas de Huancavelica*, 76, 81–82, 104, 109, 139, 175, 232, 386–87, 421; Patiño Paúl Ortíz, *Huancavelica colonial*, 39, 220; Bargallo, *La minería y metalurgía*, 257; Vargas Ugarte, *Historia*, 202; O'Phelan Godoy, *Rebellions and Revolts*, 1, 11–12, 14, 18, 23, 27, 101; Robins, *Mercury, Mining and Empire*, 33–34, 60–61.
18. Fernández de Castro y Andrade, "Carta," 155; Carta del Virrey del Perú #38, 3; Armendaris, "Relacion," 153; Navarra y Rocaful, "Relacion," 2:175; Salinas y Córdoba, *Memorial*, 277–78; Brown, "Worker's Health," 470, 474; Lohmann Villena, *Las minas de Huancavelica*, 268–69, 355, 433; Basadre, *El Conde de Lemos*, 156; Patiño Paúl Ortíz, *Huancavelica*

colonial, 242; Cobb, *Potosí y Huancavelica*, 72; Robins, *Mercury, Mining and Empire*, 32, 47. The following districts were subject to the Huancavelica mita: Andahuaylas, Angaraes, Aymares, Castrovirreina, Chumbivilcas, Cotabambas, Huanta, Lucanas, Jauja, Tarma, Vilcashuamán, and Yauyos.

19. Lohmann Villena, *Las minas de Huancavelica*, 156, 189, 354, 378, 385, 427, 431, 460; Brown, "Worker's Health," 469; Povea Moreno, "Entre la retórica y la disuación," 202; Rivero y Ustariz, *Memoria*, 17.

20. Emetherio Ramírez de Arellano, 2–3; Fernández de Castro, "Advertencias," 254; Patiño Paúl Ortíz, *Huancavelica colonial*, 245, 247; Fisher, *Silver Mines*, 10; Lohmann Villena, *Las minas de Huancavelica*, 242, 253, 330–33, 353, 382; Brown, "Worker's Health," 470, 485; Cook, *Demographic Collapse*, 69, 208; Favre, "La industria minera," 85; Robins, *Mercury, Mining and Empire*, 62.

21. Marqués de Casa Concha, "Relación de estado que ha tenido y tiene la real mina de Guancavelica," 1726, AGI, Lima, 469, 47; Fernández de Cabrera, "Relación," 90; Cueva, "Relacion," 186; Cantos de Andrade, "Relación," 305; Llano Zapata, *Memorias histórico*, 217; Caravantes, *Noticia*, 2:69; Patiño Paúl Ortíz, *Huancavelica colonial*, 24, 29; Carrasco, *Cronología de Huancavelica*, 98, 104, 24; Contreras, *La ciudad del mercurio*, 15; Zimmerman, *Francisco de Toledo*, 96, 97–99; Whitaker, *Huancavelica Mercury Mine*, 10–11; Lizárraga, *Descripción*, 183; Molina Martínez, *Antonio de Ulloa*, 108; Lohmann Villena, *Las minas de Huancavelica*, 62, 67; Robins, *Mercury, Mining and Empire*, 23, 29–30, 70.

22. Manso de Velasco, "Relacion," 165–72, 311–12; Montesinos, *Anales del Perú*, 2:21; Arana, *Las minas*, 40, 81; Nriagu, "Legacy of Mercury Pollution," 589; Cole, *Potosí Mita*, 8; Bakewell, "Registered Silver Production," 84; Lohmann Villena, *Las minas de Huancavelica*, 165, 461; Brown, "La crisis financiera," 362; Patiño Paúl Ortíz, *Huancavelica colonial*, 89; Basadre, *El Conde de Lemos*, 158; Robins, *Mercury, Mining and Empire*, 10, 71, 112.

23. Caravantes, *Noticia*, 4:224; Haenke, *Descripción del Perú*, 266; Rivero y Ustariz, *Memoria*, 20, 45–47; Arana, *Las minas*, 82–83; Loveday, "El azogue de Huancavelica," 82; Patiño Paúl Ortíz, *Huancavelica colonial*, 163–64; Brown, "La crisis financiera," 352, 366, 375, Whitaker, *Huancavelica Mercury Mine*, 6–7, 104; Berry and Singewald, *Geology and Paleontology*, 16; Blanco, *Diario*, 37; Robins, *Mercury, Mining and Empire*, 110.

24. Rivero y Ustariz, *Memoria*, 5; Gastelumendi, *Huancavelica*, 23, 46; Arana, *Las minas*, 16–17, 21; Strauss, "Quicksilver," 561; Yates, Kent, and Fernández, *Geology*, 19; Umlauff, "El Cinabrio De Huancavelica," 17, 46; Contreras and Díaz, *Los intentos*, 8, 11–12, 14, 16–18, 23–25; Berry and Singewald, *Geology and Paleontology*, 24–25; Carrasco, *Cronología de Huancavelica*, 276; Contreras, "El reemplazo," 411.

25. Berry and Singewald, *Geology and Paleontology*, 25.

26. Gastelumendi, *Huancavelica*, 47; Berry and Singewald, *Geology and Paleontology*, 26–27; Yates, Kent, and Fernández, *Geology*, 21; Jiménez, "Estadistica Minera," 205.

27. Berry and Singewald, *Geology and Paleontology*, 27; Jiménez, "Estadistica Minera," 206; Wise and Féraud, "Historic Maps," 21–22; Contreras and Díaz, *Los intentos*, 22; Yates, Kent, and Fernández, *Geology*, 1, 21, 26; Sociedad Minera Brocal. "Nuestra historia."

28. Gastelumendi, *Huancavelica*, 64; Wise and Féraud, "Historic Maps," 22–23; Carrasco, *Cronología de Huancavelica*, 534; Cooke et al., "Over Three Millennia," 8832; Sociedad Minera Brocal. "Nuestra historia."

29. Whitaker, *Huancavelica Mercury Mine*, 81; Wise and Féraud, "Historic Maps," 23; Arana, *Las minas*, 82; Dueñas, "Fisionomia minera," 156; Berry and Singewald, *Geology and Paleontology*, 16; Government of Perú, *Inventario y Evaluación*, 181; Government of Peru, *Centros Poblados*, 260;; Meyer and Greenspoon, "Mercury," 784–85; Pennington and Greenspoon, "Mercury," 826; 761; Burgess, Anderson, and Lester, "Peru," 306; Brown, "Peru," 604; Noe, "Peru," 575; Wessel, "Peru" (1972), 640; Wessel, "Peru" (1974), 718; Martino, *Minerals Yearbook*, 824. This estimate revises national period estimates in Robins, *Santa Bárbara's Legacy*, 127–28. For a range of calculations and estimates, see Loveday, "El azogue de Huancavelica," 83, Whitaker, *Huancavelica Mercury Mine*, 83, Crosnier, "Geologie du Perou," 57; Strauss, "Quicksilver," 562; Yates, Kent, and Fernández, *Geology*, 21; Jiménez, "Estadistica Minera," 17, 21; Wise and Féraud, "Historic Maps," 15; Umlauff, "El Cinabrio De Huancavelica," 41; Rivero y Ustariz, *Memoria*, 20, 45–47; Contreras and Díaz, *Los intentos*, 11–12.

30. Rivero y Ustáriz, "Memoria," 136; Gastelumendi, *Huancavelica*, 56; Strauss, "Quicksilver," 565; Umlauff, "El Cinabrio De Huancavelica," 47; Jiménez, "Estadistica Minera," 203; Arana, *Las minas*, 35–36; Fuentes Bajo, "El azogue," 93; Menéndez Navarro, *Un mundo sin sol*, 127; Robins and Hagan, "Mercury Production," 628.

31. Berry and Singewald, *Geology and Paleontology*, 27; Singewald, "Huancavelica Mercury Deposits," 522; Jiménez, "Estadistica Minera," 206.

32. Robins et al., "Estimations," 146, 148–53; Government of Peru, Dirección de Evaluación, Organisimo de Evaluación y Fiscalización Ambiental, *Informe*, 7–12, 54–61, 64–65; Thoms and Robins, "Remedial Investigation," 9, 20, 21; Robins, *Santa Bárbara's Legacy*, 148–49. For tabulated data, see Robins, *Santa Bárbara's Legacy*, see 199–214.

33. Dueñas, "Fisionomia minera," 161; Hawley, "Notes," 7; Contreras and Díaz, *Los intentos*, 16.

34. Government of Peru, Dirección de Evaluación, Organisimo de Evaluación y Fiscalización Ambiental, *Informe*, 37–38, 42–46, 50–52, 55–57, 60–61, 63–65; Thoms et al., *Data Summary*, 1–5.

35. Thoms et al., *Data Summary*, 1–2.

36. Government of Peru, Dirección de Evaluación, Organisimo de Evaluación y Fiscalización Ambiental, Ministerio del Ambiental, *Informe*, 2.

37. Government of Peru, Dirección de Evaluación, Organisimo de Evaluación y Fiscalización Ambiental, *Informe*, 56, 59–60, 138–46.

38. Hagan et al., "Speciation and Bioaccessibility of mercury," 267–71; Government of Peru, Dirección de Evaluación, Organisimo de Evaluación y Fiscalización Ambiental, *Informe*, 59; Thoms et al., *Data Summary*, 3–4.

39. Government of Peru, Dirección de Evaluación, Organisimo de Evaluación y Fiscalización Ambiental, *Informe*, 60; Thoms et al., *Data Summary*, 2, 4.

40. Government of Peru, Dirección de Evaluación, Organisimo de Evaluación y Fiscalización Ambiental, *Informe*, 56–57; Thoms et al., *Data Summary*, 4.

41. Government of Peru, Dirección de Evaluación, Organisimo de Evaluación y Fiscalización Ambiental, *Informe*, 7–12, 54–61, 64–65. For methods and standards, see 16.
42. Thoms and Robins, "Remedial Investigation," 25; Brown, "Worker's Health," 225; D'Itri and D'Itri, *Mercury Contamination*, 2–3, 47–51.
43. Thoms et al., *Data Summary*, 3–4; Government of Peru, Dirección de Evaluación, Organisimo de Evaluación y Fiscalización Ambiental, *Informe*, 38–42, 136.
44. Thoms et al., *Data Summary*, 2, 4.
45. The screening level for arsenic in sediment is 5.9 mg/kg, while the PEL is 17 mg/kg. For mercury the sediment screening level is .17 mg/kg, and the PEL is .486, and for lead the sediment screening level is 35 mg/kg, while the PEL is 91.3 mg/kg. See Government of Peru, Dirección de Evaluación, Organisimo de Evaluación y Fiscalización Ambiental, *Informe*, 38, 41–42.
46. Government of Peru, Dirección de Evaluación, Organisimo de Evaluación y Fiscalización Ambiental, *Informe*, 6–8, 37–54, 63–64, 136. The screening levels for cadmium and copper, respectively, are .6 mg/kg and 35.7 mg/kg. For methods and standards, see 15–16.
47. Robins, *Santa Bárbara's Legacy*, 207; Thoms et al., *Data Summary*, 3.
48. Mendoza, "95 Litros"; Government of Peru, Dirección de Evaluación, Organisimo de Evaluación y Fiscalización Ambiental, *Informe*, 5–6, 22–37, 62–63. For methods and standards, see 14–15.
49. Thoms et al., *Data Summary*, 2, 4.
50. Robins, *Santa Bárbara's Legacy*, 159–60. See also Hagan et al., "Mercury Hair Levels," 507–14.
51. Hagan et al., "Residential Mercury Contamination," 2–8; Thoms et al., *Data Summary*, 1–3.
52. Thoms et al., *Data Summary*, 1–4.
53. Government of Peru, Dirección de Evaluación, Organisimo de Evaluación y Fiscalización Ambiental, *Informe*, 3–5, 62. For sampling methods and standards, see 12–13.
54. Thoms et al., *Data Summary*, 2.
55. Government of Peru, Ministerio de Energía y Minas, "Contaminación minera," 10; Thoms et al., *Data Summary*, 2, 3–4.
56. Government of Peru, *Resultados Definitivos*, 10464–65, 10490; Thoms et al., *Data Summary*, 1, 3–4.
57. U.S. Environmental Protection Agency, "Lead and Compounds," 2, 7–9. See also Tables 3 and 8 in appendix; World Health Organization, "Lead Poisoning and Health."
58. U.S. Department of Health and Human Resources, "Health Effects," 67, 161, 188, 219, 221, 223, 278, 327; U.S. Environmental Protection Agency, *Mercury Study*, 0-3, 2-6; Rowland et al., "Effect of Occupational Exposure," 28–34; Yeates and Mortensen, "Acute and Chronic," 209–10, 218; Alcser et al., "Occupational Mercury Exposure," 517; Barlow et al., "Reproductive Hazards," 729; D'Itri and D'Itri, *Mercury Contamination*, 26, 135; Schutte et al. "Mercury," 553; Bakir et al., "Methylmercury Poisoning," 234, 238–39; Bidstrup, *Toxicology and Mercury*, 8, 34; Waldron and Scott, "Metals," 103; Evans, "Mercury," 998, 1001; Robins, *Mercury, Mining and Empire*, 104–6, 108, 136.

59. Evans, "Mercury," 1000; Waldron and Scott, "Metals," 91, 102–3, 105; Schutte et al., "Mercury," 552; Bidstrup, *Toxicology and Mercury*, 8, 43, 60; U.S. Department of Health and Human Resources, "Health Effects," 31, 33, 47–49, 54–55, 58, 161–62, 163, 185–86, 264–65, 268, 272; U.S. Environmental Protection Agency, *Mercury Study*, 0-3; D'Itri and D'Itri, *Mercury Contamination*, 26; Robins, *Mercury, Mining and Empire*, 104, 106.
60. D'Itri and D'Itri, *Mercury Contamination*, 2, 23, 105, 118; Goldwater, *Mercury*, 127–28; Marsh, "Organic Mercury," 73; U.S. Environmental Protection Agency, *Mercury Study*, 0-2, 2-1, 3-13, 3-23; Evans, "Mercury," 1860, 1865; Schutte et al., "Mercury," 551; Waldron and Scott, "Metals," 91, 101–3; U.S. Department of Health and Human Resources, "Health Effects," 107, 124, 161, 169, 221; Bidstrup, *Toxicology and Mercury*, 77–78; Robins, *Mercury, Mining and Empire*, 77–78, 103–5.
61. D'Itri and D'Itri, *Mercury Contamination*, 15, 24–25,123, 142, 205; Schutte et al., "Mercury," 551–52; U.S. Department of Health and Human Resources, "Health Effects," 31, 34, 43, 219; Evans, "Mercury," 1001; Evans et al., "Behavioral Effects," 1860, 1865; Bidstrup, *Toxicology and Mercury*, 8, 42–43, 60; Neal et al., *Mercurialism*, 122; Waldron and Scott, "Metals," 103; U.S. Department of Health and Human Resources, "Health Effects," 31, 33, 47–49, 51, 54–55, 58, 162, 163, 185–86, 264–65, 268, 272; U.S. Environmental Protection Agency, *Mercury Study*, 0-3; Hanninen, "Behavioral Effects," 170; Cherry, "Neurotoxic Effects," 79; Piikivi et al., "Psychological Performance," 35; Marsh, "Organic Mercury," 74–75; Robins, *Mercury, Mining and Empire*, 106–7, 110.
62. Parés y Franqués, *Catástrophe morboso*, 294; Waldron and Scott, "Metals," 104–5; Piikivi et al., "Psychological Performance," 35; Schutte et al., "Mercury," 553; Evans, "Mercury," 1001; Neal et al., *Mercurialism*, 122; Hanninen, "Behavioral Effects," 170; Bidstrup, *Toxicology and Mercury*, 42; U.S. Department of Health and Human Resources, "Health Effects," 58, 220; Soleo et al., "Effects," 105; Ehrenberg et al., "Effects," 500; Yeates and Mortensen, "Acute and Chronic," 209–11, 214, 218; Kishi et al., "Subjective Symptoms," 293–95; West and Lim, "Mercury Poisoning," 698; Smith et al., "Effects," 413, 417, 419; Williamson et al., "Occupational Mercury Exposure," 273, 284–85; Robins, *Mercury, Mining and Empire*, 107–8.
63. U.S. Environmental Protection Agency, *Review of the Reference Dose*, 4–3; Parés y Franqués, *Catástrophe morboso*, 99, 120–21, 123, 163, 228, 232–33, 261, 291; Waldron and Scott, "Metals," 104–5; Menéndez Navarro, *Un mundo sin sol*, 166–67; Bidstrup, *Toxicology and Mercury*, 42; Neal et al., *Mercurialism*, 122; U.S. Department of Health and Human Resources, "Health Effects," 44, 56; Piikivi et al., "Psychological Performance," 35; Schutte et al., "Mercury," 552–53; Goldwater, *Mercury*, 268–69; Soleo et al., "Effects," 105; Kishi et al., "Residual Neurobehavioural Effects," 35–38, 40; Salgueiro et al., "Visual Field Losses," 124, 129; D'Itri and D'Itri, *Mercury Contamination*, 134–37; Robins, *Mercury, Mining and Empire*, 107, 109.
64. D'Itri and D'Itri, *Mercury Contamination*, 107; Piikivi et al., "Psychological Performance," 25; Bidstrup, *Toxicology and Mercury*, 43; Cherry, "Neurotoxic Effects," 79; Smith et al.,

"Effects," 413, 417; Uzzell and Oler, "Chronic Low-Level Mercury Exposure," 581–82; Robins, *Mercury, Mining and Empire*, 105, 108.

65. Government of Peru, Dirección de Evaluación, Organisimo de Evaluación y Fiscalización Ambiental, *Informe*, 57, 59, 60, 61.
66. Thoms and Robins, "Remedial Investigation," 16–17; Thoms and Robins, "Results," 3–4.
67. In Peru the identification of a responsible party for legacy sites is governed by Peruvian Law 28271. Government of Peru, Ministerio de Energía y Minas, "Contaminación minera," 17.
68. Gastelumendi, *Huancavelica*, 64; Arana, *Las minas*, 82; Dueñas, "Fisionomia minera," 156; Singewald, "Huancavelica Mercury Deposits," 518, Berry and Singewald, *Geology and Paleontology*, 16; Whitaker, *Huancavelica Mercury Mine*, 81; Wise and Féraud, "Historic Maps," 22–23; Government of Perú, *Inventario y Evaluación*, 181; Government of Peru, *Centros Poblados*, 260; Carrasco, *Cronología de Huancavelica*, 534; Cooke et al., "Over Three Millennia," 8832.
69. Mapa de Huancavelica, ca. 1790, AGI, Mapas y Planos, Peru y Chile, 225: Arana, *Las minas*, map of Huancavelica, n.p.
70. Gastelumendi, *Huancavelica*, 41; Goldwater, *Mercury*, 49; Berry and Singewald, *Geology and Paleontology*, 43, 51–52; Singewald, "Huancavelica Mercury Deposits," 522; Strauss, "Quicksilver," 562; Crosnier, "Geologie du Perou," 2, 51–52; author's field research notes, August 2010, July 2012, June–July 2015.
71. Thoms and Robins, "Remedial Investigation," 20; Government of Peru, Ministerio del Ambiente, *Guia para muestreo de suelos*, 16; Government of Peru, Dirección de Evaluación, Organisimo de Evaluación y Fiscalización Ambiental, *Informe*, 56–57, 59–61.
72. Mapa de Huancavelica, ca. 1790; Government of Peru, Dirección de Evaluación, Organisimo de Evaluación y Fiscalización Ambiental, *Informe*, 56–57, 59–61, 65; Thoms and Robins, "Remedial Investigation," 28.
73. Arana, *Las minas*, 16; Bakewell, "Technological Change," 82–84; Zimmerman, *Francisco de Toledo*, 97–99; Cole, *Potosí Mita*, 9; Lohmann Villena, *Las minas de Huancavelica*, 47, 72, 79, 88; Whitaker, *Huancavelica Mercury Mine*, 10–11; Robins, *Mercury, Mining and Empire*, 29–30; Robins, *Santa Bárbara's Legacy*, 117.
74. Rivero y Ustariz, *Memoria*, 5; Gastelumendi, *Huancavelica*, 23, 46–47; Arana, *Las minas*, 16–17, 21; Jiménez, "Estadistica Minera," 205; Strauss, "Quicksilver," 561; Yates, Kent, and Fernández, *Geology*, 19, 21; Umlauff, "El Cinabrio De Huancavelica," 17–18, 46; Contreras and Díaz, *Los intentos*, 8, 11–12, 14, 16–18, 23–25; Berry and Singewald, *Geology and Paleontology*, 24–27; Carrasco, *Cronología de Huancavelica*, 276; Contreras, "El reemplazo," 411.
75. Government of Peru, Defensoría del Pueblo, *Una llamada a la remediación*, 42; Government of Peru, Ley que regula los pasivos ambientales de la actividad minera.
76. Government of Peru, Ministerio de Energía y Minas, "Contaminación minera," 11.
77. Government of Peru, Dirección de Evaluación, Organisimo de Evaluación y Fiscalización Ambiental, *Informe*, 3–4, 62.

78. Government of Peru, Ministerio de Energía y Minas, "Contaminación minera," 14–15.
79. Government of Peru, Defensoría del Pueblo, *Una llamada a la remediación*, 54. For a detailed discussion of the process of identifying contaminated sites and the steps Peruvian authorities are to take, see 37–115.
80. Government of Peru, Defensoría del Pueblo, *Una llamada a la remediación*, 21.
81. Government of Peru, Defensoría del Pueblo, *Una llamada a la remediación*, 54–56; Government of Peru, Ministerio de Energía y Minas, "Contaminación minera," 2, 6–10.
82. Government of Peru, Defensoría del Pueblo, *Una llamada a la remediación*, 5–7, 10, 46, 108.
83. Government of Peru, Defensoría del Pueblo, *Una llamada a la remediación*, 46–47
84. Government of Peru, Defensoría del Pueblo, *Una llamada a la remediación*, 45.
85. Government of Peru, Defensoría del Pueblo, *Una llamada a la remediación*, 57, 192. For the Peru Ombudsman's Office recommendations regarding contaminated sites in Peru, see 193–98.
86. Government of Peru, Defensoría del Pueblo, *Una llamada a la remediación*, 7, 101–8; Shortell, "In Need of Investment"; Cruz, "Medidas ambientales."
87. Government of Peru, Dirección de Evaluación, Organisimo de Evaluación y Fiscalización Ambiental, *Informe*, 65.
88. Government of Peru, Defensoría del Pueblo, *Una llamada a la remediación*, 31.
89. Government of Peru, Defensoría del Pueblo, *Una llamada a la remediación*, 20; Government of Peru, Constitución Política.
90. Government of Peru, *Censos Nacionales 2007*, 2:11, 941; Government of Peru, *Estado de la Población Peruana*, 7; Government of Peru, Instituto Nacional de Estadística, "Mapa de Pobreza"; Government of Peru, *Encuesta Demográfica*, 180–81, 195–98, 203; Government of Peru, *Compendio Estadístico Departamental*, 300; Robins et al., "Estimations," 152–53; Velázquez, *Salud mental*, 68; Robins, *Santa Bárbara's Legacy*, 156–58.
91. Thoms and Robins, "Remedial Investigation," 38.
92. Government of Peru, Ministerio de Energía y Minas, "Contaminación minera," 2, 6–10.
93. United Nations, "International Covenant on Economic, Social and Cultural Rights"; United Nations, "Convention on the Rights of the Child."
94. Marichal, "Spanish-American Silver Peso," 25–27, 35, 38, 40–42; Flynn and Giraldez, "China and the Manila Galleons," 71–72, 85–86; Cipolla, *Conquistadores*, 57, 63–64, 66, 69–72; Hamilton, *American Treasure*, 46, 51, 53; Attman, *American Bullion*, 5–6, 9, 67, 77, 101; Cross, "South American Bullion," 397, 404, 412–13; Stein and Stein, *Silver*, 24, 26, 261–62; Robins, *Mercury, Mining and Empire*, 4–6.

BIBLIOGRAPHY

Archives and Manuscript Materials

Alcalá y Amurrio, Juan. Directorio del Beneficio del Azogue, en los Metales de Plata.-Documentos que se dan en sus Reglas. Oruro, 1781, ABNB, Ruck 80.

Archivo y Bibliotecas Nacionales de Bolivia (ABNB), Sucre, Bolivia
Audiencia de La Plata, Minas (ALP Minas)
Colección Ruck (Ruck)
Archivo General de Indias, Seville, Spain
Indiferente
Indiferente General
Lima
Mapas y Planos

Carta del Virrey del Perú # 38, "Da cuenta a V. Md. de todo lo que en materia de azogue sea escrito." February 2, 1630, AGI, Indiferente General 1777.

Don Rodrigo de Mendoza y Manrique, administrador y arrendatario que fue de las minas y los ingenios de don Pedro Sorez de Ulloa en el cerro y la ribera de Potosí, con el maestro de campo don Rodrigo Campuzano, hermano de doña Francisca Campuzano, viuda y heredera que fue de dicho don Pedro, sobre la liquidación de los pesos impendidos en el avío de las haciendas mencionadas durante el tiempo de su administración y arrendamiento. La Plata, February 5, 1656–1669. ABNB, ALP Minas 15/1.

Emetherio Ramírez de Arellano. Huancavelica, April 3, 1649. AGI, Lima, 279-4.

Expediente promovido por don Nicolás de Sarabia y Mollinedo asentista de la Real Mina de azogue de Guancavelica sre que los corregidores y gobernadores de las provincias que tienen obligación de remitir mitas de indios para el trabajo de otra mina lo ejecuten sin dilación alguna y sre lo ocurrido con este motivo en la de Jauja a causa de haberse intentado cumplir con la remisión de mitayos aquel gobierno como se le mandó que avisa la resistencia que hicieron e insultos al cobrador enterador de mitas don Jacinto Maita. Lima, November 10, 1811. AGI, Indiferente 1335.

Expediente seguido por don Roque de Reinalte, sobre la prisión a que le han reducido por cierta deuda no obstante de su calidad de minero en Potosí. La Plata, August 14–19, 1697. ABNB, ALP Minas 19/10.

Mapa de Huancavelica. Ca. 1790. AGI, Mapas y Planos, Peru y Chile 225.

Marqués de Casa Concha, Relación del estado que ha tenido y tiene la real mina de Guancavelica. Lima, 1726, AGI, Lima

Memorial del capitán Don Pedro Gutíerrez Calderón de algunas advertencias considerables al servicio de Dios Nuestro Señor y de su Real Magestad," Lima, May 2, 1623, AGI, Lima, 154.

Published Works

Acosta, José de. *Historia natural y moral de las Indias*. Edited by José Alcina Franch. Madrid: Historia 16, 1987.

Agricola, Georgius. *De Re Metalica*. Edited and translated by Herbert C. Hoover and Lou H. Hoover. New York: Dover, 1950.

Alcser, Kirsten, et al. "Occupational Mercury Exposure and Male Reproductive Health." *American Journal of Industrial Medicine* 15, no. 5 (1989): 517–29.

Arana, Pedro. *Las minas de azogue del Perú*. Lima: Imprenta de "El Lucero," 1901.

Armendaris, Marques de Castel-Fuerte, José. "Relacion del estado de los reynos del Perú que hace el Excmo: Señor Don José Armendaris, Marqués de Castel-Fuerte, á su successor el Marqués de Villagarcía, en el año de 1736." In *Memorias de los virreyes que han gobernado el Perú durante el tiempo del coloniaje español*, vol. 3, edited by M. A. Fuentes. Lima: Librería Central de Felipe Bailly, 1859.

Arzáns de Orsúa y Vela, Bartolomé. *Historia de la Villa Imperial de Potosí*. 3 vols. Edited by Lewis Hanke and Gunnar Mendoza. Providence RI: Brown University Press, 1965.

Attman, Arthur. *American Bullion in the European World Trade, 1600–1800*. Translated by Eva Green and Allan Green. Goteborg, Sweden: Kungl. Vetenskaps-och Vitterhets-Samhallet i Goteborg, 1986.

Audiencia de Lima. "Relación que la Real Audiencia de Lima hace al excelentísimso Sr. Marqués de Castel-Dosrius, Virey de estos reinos, del estado de ellos, y tiempo que ha gobernado en vacante." In *Relaciones de los Virreyes y Audiencias que han Gobernado el Perú*, vol. 2, edited by Sebastian Lorente. Madrid: Imprenta y Estereotipia de M. Rivadeneyra, 1871.

Bakewell, Peter. *Miners of the Red Mountain: Indian Labor in Potosí, 1545–1650*. Albuquerque: University of New Mexico Press, 1984.

———. "Registered Silver Production in the Potosi District, 1550–1735." In *Jahrbuch fur Geschichte von Staat, Wirtschaft und Gesellschaft Lateinamerikas* 12 (1975): 67–103.

———. "Technological Change in Potosi: The Silver Boom of the 1570s." In *Mines of Silver and Gold in the Americas*, edited by Peter Bakewell. Brookfield VT: Variorum, 1997.

Bakir, F., et al. "Methylmercury Poisoning in Iraq." *Science* 181, no. 4096 (July 20, 1973): 230–41.

Ballesteros Gaibrois, Manuel. *Descubrimiento y fundación del Potosi*. Zaragoza: Delegación de Distrito de Educación Nacional, 1950.

Bargallo, Modesto. *La amalgamación de los minerales de plata en hispanoamerica colonial*. Mexico City: Compañia Fundidor de Fierro y Acero de Monterrey, 1969.

———. *La minería y metalurgía en la América española durante la época colonial*. Mexico City: Fondo de Cultura Económica, 1955.

Barlow, S. M., et al. "Reproductive Hazards at Work." In *Hunter's Diseases of Occupations*, edited by P. A. B. Raffle et al. 8th ed. London: E. Arnold, 1994.

Basadre, Jorge. *El Conde de Lemos y su tiempo*. Lima: Editorial Huascaran, S.A., 1948.

Bidstrup, P. Lesley. *Toxicology and Mercury and Its Compounds*. New York: Elsevier, 1964.

Berry, Edward, and Joseph T. Singewald. *The Geology and Paleontology of the Huancavelica Mercury District*. Baltimore MD: Johns Hopkins University Press, 1922.

Blanco, José María. *Diario de viaje del Presidente Orbegoso al sur del Perú*. 2 vols. Edited by Felix Denegri Luna. Lima: PUCP-IRA 1974.

Brading, D. A., and Harry E. Cross. "Colonial Silver Mining: New Spain and Peru." *Hispanic American Historical Review* 52, no. 4 (November 1972): 545–79.

Brooks, William E., Estevan Sandoval, Miguel Yépez, and Howell Howard, *Peru Mercury Inventory, 2006*. Washington DC: U.S. Geological Service, 2007.

Brown, Kendall. "La crisis financiera peruana al comienzo del siglo XVIII, la minería de plata y la mina de azogue de Huancavelica." *Revista de Indias* 48, nos. 182–83 (1988): 349–81.

———. "Workers' Health and Colonial Mercury Mining at Huancavelica, Peru." *The Americas* 57, no. 4 (April 2001): 467–96.

Brown, Lester, Jr. "Peru." In *Minerals Yearbook Area Reports: International, 1967*, vol. 4. Washington DC: Bureau of Mines, 1969.

Buechler, Rose M. *The Mining Society of Potosi: 1776–1810*. Syracuse NY: Syracuse University, Department of Geography, 1981.

Burgess, John, Sumner Anderson, and R. Lester Jr. "Peru." In *Minerals Yearbook Area Reports: International 1963*. Vol. 4. Washington DC: Bureau of Mines, 1964.

Burger, R., and R. Matos. "Atalla: A Center on the Periphery of the Chauvín Horizon." *Latin American Antiquity* 13, no. 2 (2002): 10–25.

Cantos de Andrade, Rodrigo. "Relación de la Villa Rica de Oropesa y minas de Guancavelica." In *Relaciones geográficas de Indias*, vol. 1, edited by Marcos Jiménez de Espada. Madrid: Ediciones Atlas, 1965.

Cañete y Domínguez, Pedro Vicente. *Guía histórica, geográfica, física, política, civil y legal del gobierno e intendencia de la provincia de Potosí*. Potosí: Editorial Potosi, 1952.

Capoche, Luis. *Relación general de la Villa Imperial de Potosí*. Edited by Lewis Hanke. Madrid: Ediciones Atlas, 1959.

Caravantes, Francisco López. *Noticia general del Perú*. 6 vols. Edited by Marie Helmer. Madrid: Ediciones Atlas, 1989.

Carrasco, Tulio. *Cronología de Huancavelica (Hechos, poblaciones y personas)*. Lima: Compañía de Minas Buenaventura, 2003.

Cherry, N. M. "Neurotoxic Effects of Workplace Exposures." In *Hunter's Diseases of Occupations*, edited by P. A. B. Raffle et al. 8th ed. London: E. Arnold, 1994.

Cipolla, Carlo M. *Conquistadores, piratas, mercaderes: La saga de la plata española*. Translated by Ricardo González. Buenos Aires: Fondo de Cultura Económica de Argentina, 1999.

Cobb, Gwendolyn. *Potosí y Huancavelica, bases económicas, 1545–1640*. La Paz: Banco Minero de Bolivia, 1977.

Cobo, Bernabé. *Historia del Nuevo Mundo*. Edited by Francisco Mateos, SJ. Madrid: Ediciones Atlas, 1956.

Cole, Jeffrey A. *The Potosí Mita, 1573–1700: Compulsory Indian Labor in the Andes*. Stanford CA: Stanford University Press, 1985

Contreras, Carlos. "El reemplazo del beneficio de patio en la minería peruana, 1850–1913." *Revista de Indias* 59, no. 216 (1999): 391–416.

———. *La ciudad del mercurio: Huancavelica, 1570–1700*. Lima: Instituto de Estudios Peruanos, 1982.

Contreras, Carlos, and Ali Díaz. *Los intentos de reflotamiento de la mina de azogue de Huancavelica en el siglo XIX*. Lima: n.p., 2007.

Cook, Noble David. *Demographic Collapse: Indian Peru, 1520–1620*. Cambridge, U.K.: Cambridge University Press, 1981.

Cooke, Colin, Prentiss Balcom, Harald Biester, and Alexander Wolfe. "Over Three Millennia of Mercury Pollution in the Peruvian Andes." *Proceedings of the National Academy of Sciences* 106, no. 22 (June 2009): 8830–34.

Cooke, Colin, Holger Hintelmann, Jay Ague, Richard Burger, Harald Biester, Julian Sachs, and Daniel Engstrom. "Use and Legacy of Mercury in the Andes." *Environmental Science and Technology* 47, no. 9 (May 2013): 4181–88.

Crespo Rojas, Alberto. "El reclutamiento y los viajes en la 'mita' del cerro Potosí." In *Lamineria hispana e iberoamericana*, vol. 1. León, Spain: Catedra de San Isidoro, 1970.

Crosnier, León. "Geologie du Perou: Notice geologique sur les Departments de Huancavelica et d'Ayachcho." *Annales de mines* 2, 5th series (1852): 1–107.

Cross, Harry E. "South American Bullion Production and Export 1550–1750." In *Precious Metals in the Later Medieval and Early Modern Worlds*, edited by J. F. Richards. Durham NC: Carolina Academic Press, 1983.

Cruz, Alejandra. "Medidas ambientales en las manos del Congreso." *La Republica* (Lima), June 25, 2014.

Cueva, Conde de Castellar, Baltasar de la. "Relacion general que el Excmo: Señor Conde de Castellar, Marqués de Malagon, Gentilhombre de la Cámara de su Majestad, de su Consejo, Cámara y Junta de Guerra de Indias, Virey, Gobernador y Capitan General que fué de estos Reinos, hace del tiempo que los gobernó, estado en que los dejó, y lo obrado en las materias principales con toda distinction." In *Memorias de los virreyes que han gobernado el Perú durante el tiempo del coloniaje español* vol. 1, edited by M. A. Fuentes. Lima: Librería Central de Felipe Bailly, 1859.

Delgado de Castro, Raquel. *El Despertar de Huancavelica.* Lima: C. A. Castrillón, 1927.

D'Itri, Patricia, and Frank D'Itri. *Mercury Contamination: A Human Tragedy.* New York: John Wiley and Sons, 1977.

Dueñas, Enrique, "Fisionomia minera de las provincias de Tayacaja, Angaraes y Huancavelica." *Boletín del cuerpo de ingenieros de minas del Peru.* No. 62. Lima: El Lucero, 1908.

Ehrenberg, Richard, et al. "Effects of Elemental Mercury Exposure at a Thermometer Plant." In *American Journal of Industrial Medicine* 19 (1991): 495–507.

Evans, Hugh. "Mercury." In *Environmental and Occupational Medicine.* edited by William Rom. 3rd. ed. New York: Lippincott-Raven, 1998.

Evans, Hugh, et al. "Behavioral Effects of Mercury and Methylmercury." *Federation Proceedings* 34, no. 9 (August 1975): 1858–67.

Favre, Henri. "La industria minera de Huancavelica en la década de 1960." *Boletín de Lima*, no. 161 (2010): 85–89.

Fernández Alonso, Serena. "Los mecenas de la plata: El respaldo de los virreyes a la actividad minera colonial en las primeras decadas del siglo XVIII. El gobierno del Marques de Casa Concha en Huancavelica, 1723–1726." *Revista De Indias* 60, no. 219 (Madrid: 2000): 345–71.

Fernández de Cabrera, Conde de Chinchón, Luis Jerónimo. "Relación del estado en que el Condede Chinchón deja el gobierno del Perú al Señor Virrey Marqués de Mancera."

In *Colección de las memorias o relaciones que escribieron los virreys del Perú*, vol. 2, edited by Ricardo Beltrán y Rózpide. Madrid: Imprenta del Asilo de Huérfanos del S.C. de Jesús, 1921.

Fernández de Castro y Andrade, Conde de Lemos, Pedro. "Advertencias que hace el Conde de Lemos a la relación del estado del reino que le entregó la Real Audiencia de Lima del tiempo que gobernó en vacante de virrey que fue de año y más de ocho meses, dirigida a la reina nuestra señora en el real y supremo consejo de Indias." In *Los virreyes españoles en America durante el gobierno de la casa de Austria. Perú*, vol. 3, edited by Lewis Hanke and Celso Rodríguez. Madrid: IMNASA, 1978.

———. "Carta del Conde de Lemos a S.M. sobre la Mita de Potosí." In *Pareceres juridicos en asuntos de indias (1601–1718)*, edited by Ruben Vargas Ugarte. Lima: CIP, 1951.

Fisher, John A. *Silver Mines and Silver Miners in Colonial Peru, 1776–1824*. Monograph series No. 7. Liverpool: Center for Latin American Studies, University of Liverpool, 1977.

———. "Silver Production in the Viceroyalty of Peru, 1776–1824." In *Mines of Silver and Gold in the Americas*, edited by Peter Bakewell. Brookfield VT: Variorum, 1997.

Flynn, Dennis, and Arturo Giraldez. "China and the Manila Galleons." In *World Silver and Monetary History in the 16th and 17th Centuries*. Brookfield VT: Variorum, 1996.

Fonseca, George de. "Informe de George de Fonseca del 24 de Julio de 1622." In *Huancavelica colonial: Apuntes históricos de la ciudad minera más importante del Virreynato Peruano*. Edited by Mariano Patiño Paúl Ortíz. Lima: Huancavelica 21, 2001.

Fuentes Bajo, María Dolores. "El azogue en las postrimerías del Perú colonial." *Revista de Indias* 46 (January–June 1986): 75–105.

Gastelumendi, A. G. *Huancavelica Como Región Productora de Mercurio*. Lima: Imp. Torres Aguirre, 1920.

Goldwater, Leonard J. *Mercury: A History of Quicksilver*. Baltimore MD: York Press, 1972.

Government of Peru. *Almanaque de Huancavelica*. Lima: Instituto Nacional de Estadística e Informática, 2002.

———. *Censos Nacionales 2007: XI de Población y VI de Vivienda. Principales Indicadores Demográficos, Sociales y Económicos a Nivel Provincial y Distrital Huancavelica*. Lima: Instituto Nacional de Estadística e Información, 2009.

———. *Centros Poblados: Sexto Censo Nacional de Población, Primer Censo Nacional de Vivienda, 2 de Julio de 1961*. Vol. 2. Lima: Dirección Nacional de Estadística y Censos, 1966.

———. *Compendio Estadístico Departamental*. Lima: Instituto Nacional de Estadística e Informática, 1998.

———. Constitución Política del Peru. Lima: Government of Peru, 1993.

———. *Encuesta Demográfica y De Salud Familiar 2000: Huancavelica*. Vol. 8. Lima: Instituto Nacional de Estadística e Informática, 2001.

———. *Estado de la Población Peruana, 2014*. Lima: Instituto Nacional de Estadística e Informática, 2014.

———. Instituto Nacional de Estadística. Sistema de Consulta de Principales Indicadores de Pobreza. "Mapa de Pobreza." http://censos.inei.gob.pe/Censos2007/Pobreza/.

———. *Inventario y Evaluación de Los Recursos Naturales de La Zona Altoandina del Perú: Reconocimiento, Departamento de Huancavelica*. 2 vols. Lima: Oficina Nacional de Evaluación de Recursos Nacionales, 1984.

———. "Ley que regula los pasivos ambientales de la actividad minera." Law 28271. July 2, 2004.

———. *Resultados Definitivos de los censos nacionales, 2017: Huancavelica*. Vol. 10. Lima: Instituto Nacional de Estadísticas, 2018.

Government of Peru, Defensoría del Pueblo. *Una llamada a la remediación: Avances y pendientes en la gestión estatal frente a los pasivos ambientales mineros y hidrocarburíficos*. Serie Informes Defensoriales Informe 171. Lima: Defensoría del Pueblo, 2015.

Government of Peru, Dirección de Evaluación, Organisimo de Evaluación y Fiscalización Ambiental, Ministerio del Ambiental. *Informe de monitoreo ambiental de calidad de aire, agua, sedimento y suelo, en los distritos de Ascención y Huancavelica, provincial y departmento de Huancavelica, realizado del 11 y 17 de Agosto de 2015 y el 4 de deciembre 2015*. Report No. 83–2016 OEFA/DE-SDCA. Lima: Government of Peru, April 29, 2016.

Government of Peru, Ministerio del Ambiente. *Guia para muestreo de suelos*. Lima: Dirección General de Calidad Ambiental, 2014.

Government of Peru, Ministerio de Energía y Minas, Dirección General de Asuntos Ambientales Mineros. "Contaminación minera en Huancavelica." Presentation at the Defensoría de Pueblo by Milanio Estela Silva and Wilson Sanga Yampasi. Lima, July 12, 2017.

Hadley, William. "Report on the Quicksilver Mines of Huancavelica." *Boletín del Ministerio de Fomento* 2, no. 1 (Lima: 1904): 43–44.

Haenke, Tadeo. *Descripción del Perú*. Lima: Imprenta el Lucero, 1901.

Hagan, Nicole, Nicholas Robins, Rubén Darío Espinoza Gonzales, and Heileen Hsu-Kim. "Speciation and Bioaccessibility of Mercury in Adobe Bricks and Dirt Floors in Huancavelica, Peru." *Environmental Geochemistry and Health* 37 (2015): 263–72.

Hagan, Nicole, Nicholas Robins, Heileen Hsu-Kim, Susan Halabi, Rubén Darío Espinoza Gonzales, Enrique Ecos E., Daniel DeB. Richter, and John Vandenberg. "Mercury Hair Levels and Factors That Influence Exposure for Residents of Huancavelica, Peru." *Environmental Geochemistry and Health* 36, no. 6 (December 2014): 507–14.

Hagan, Nicole, Nicholas Robins, Heileen Hsu-Kim, Susan Halabi, Rubén Darío Espinoza Gonzales, Daniel DeB. Richter, and John Vandenberg. "Residential Mercury Contamination in Adobe Brick Homes in Huancavelica, Peru." *PLoS ONE* 8, no. 9 (September 2013): 1–9.

Hamilton, Earl. *American Treasure and the Price Revolution in Spain, 1501–1650*. Cambridge MA: Harvard University Press, 1934.

Hanninen, Helena. "Behavioral Effests of Occupational Exposure to Mercury and Lead." In *Acta Neurologica Scandinavica* 6, supplement 92 (1982): 167–173.

Hawley, C. E. "Notes on the Quicksilver Mine of Santa Barbara, Peru." *American Journal of Science* 45, 2nd series (1868): 5–9.

Hernández, José A. "Antiguedad y actualidad de la Villa Rica de Oropesa." *Peruanidad* 3, no. 12 (Lima: 1943): 942–44.

"Historia de la mina de Huancavelica." *Mercurio Peruano*, January 30, 1791, 65–68.

Jiménez, Carlos P. "Estadistica Minera en 1917." *Boletín del cuerpo de ingenieros de minas del Peru*. No 95. Lima: Imprenta Americana, 1919.

Kishi, Reiko, et al. "Residual Neurobehavioural Effects Associated with Chronic Exposure to Mercury Vapour." *Occupational and Environmental Medicine* 51, no. 1 (January 1994): 35–41.

Kishi, Reiko, et al. "Subjective Symptoms and Neurobehavioral Performances of Ex-Mercury Miners at an Average of 18 Years after the Cessation of Chronic Exposure to Mercury Vapor." In *Environmental Research* 12, no. 2 (1993): 289–302.

Llano Zapata, José Eusebio. *Memorias histórico, fisicas, critico, apologéticas de la América Meridonal*. Edited by Ricardo Ramírez Castañeda et al. Lima: Instituto Francés de Estudios Andinos, 2005.

Lizárraga, Reginaldo de. *Descripción del Perú, Tucumán, Río de la Plata y Chile*. Buenos Aires: Union Académique Internationale/Academia Nacional de la Historia, 1999.

Lohmann Villena, Guillermo. "La minería en el marco del virreinato peruano: Invenciones, sistemas, técnicas y organización industrial." In *La minería hispana e iberoamericana*, vol. 1. Leon, Spain: Catedra de San Isidoro, 1970.

———. *Las minas de Huancavelica en los siglos XVI y XVII*. Lima: Pontífica Universidad Católica del Perú, 1999.

Loveday, Santiago. "El azogue de Huancavelica." *Boletín de minas, industrias y construcciones* 20 (Lima: 1905): 82–84.

Manso de Velasco, Conde de Superunda, José Antonio. "Relacion que escribe el conde de Superunda, Virrey el Perú, de los principales sucesos de su gobierno, de Real Orden de S.M. comuncada por el Excmo. Sr. Marqués de la Ensenada, su secretario del Despacho universal, con fecha 23 de Agosto de 1751, y comprehende los años desde 9 de Julio de 1745 hasta fin del miso mes en el de 1756." In *Memorias de los virreyes que han gobernado el Peru durante el tiempo del coloniaje español*, vol. 4, edited by M. A. Fuentes. Lima: Librería Central de Felipe Bailly, 1859.

Marichal, Carlos. "The Spanish-American Silver Peso: Export Commodity and Global Money of the Ancien Regime, 1550–1800." In *From Silver to Cocaine: Latin American Commodity Chains and the Building of the World Economy, 1500–2000*, edited by Steven Topik, Carlos Marichal, and Zephyr Frank. Durham NC: Duke University Press, 2006.

Marsh, David O. "Organic Mercury: Methylmercury Compounds." In *Handbook of Clinical Neurology*, vol. 36, edited by P. Vinken and G. Bruyn. New York: Elsevier, 1979.

Martino, Orlando. *Minerals Yearbook Area Reports: International 1976*. Vol. 3. Washington DC: Bureau of Mines, 1976.

Matilla Tascón, Antonio. *Historia de las minas de Almadén*. Vol 2. Madrid: Instituto de Estudios Fiscales, Minas de Almadén y Arrayanes, 1987.

McKee, E. H., D. C. Noble, and Cesar Vidal. "Timing of Volcanic and Hydrothermal Activity, Huancavelica Mercury District, Peru." *Economic Geology and the Bulletin of the Society of Economic Geologists* 81, no. 2 (1986): 489–92.

Mendoza, Gracy. "95 litros de desague contamina río Ichu." *Diario Correo*, September 15, 2012.

Mendoza y Luna, Marqués de Montesclaros, Juan de. "Relación del estado del gobierno de estos reinos que hace el excmo. Señor Don Juan de Mendoza y Luna, Marqués de Montesclaros, al excmo. Señor Príncipe de Esquilache su sucesor." In *Memorias de los virreyes que han gobernado el Perú durante el tiempo del coloniaje español*, vol. 1, edited by M. A. Fuentes. Lima: Librería Central de Felipe Bailly, 1859.

Menéndez Navarro, Alfredo. *Un mundo sin sol: La salud de los trabajadores de las minas de Almadén, 1750–1900*. Granada: Universidad de Granada, 1996.

Meyer, Helena M., and Gertrude N. Greenspoon. "Mercury." In *Minerals Yearbook Metals and Minerals (Except Fuels) 1954*. Washington DC: Bureau of Mines, 1958.

Molina Martínez, Miguel. *Antonio de Ulloa en Huancavelica*. Granada: University of Granada, 1995.

Montesinos, Fernando. *Anales del Perú*, vols. 1–2, edited by Victor M. Maurtua. Madrid: Imprenta de Gabriel L. y del Horno, 1906.

Navarra y Rocaful, Duque de la Palata, Melchor. "Relacion del estado del Perú." In *Memorias de los virreyes que han gobernado el Perú durante el tiempo del coloniaje español*, vol. 2, edited by M. A. Fuentes. Lima: Librería Central de Felipe Bailly, 1859.

Neal, Paul, et al. *Mercurialism and Its Control in the Felt-Hat Industry*. Washington DC: Government Printing Office, 1941.

Noe, Frank. "Peru." In *Minerals Yearbook Area Reports: International 1969*, vol. 4. Washington DC: Bureau of Mines, 1969.

Nriagu, Jerome. "Legacy of Mercury Pollution." *Nature* 363, no. 6430 (June 17, 1993): 589.

———. "Mercury Pollution from the Past Mining of Gold and Silver in the Americas." *Science of the Total Environment* 149 (1994): 168–69.

O'Phelan Godoy, Scarlett. *Rebellions and Revolts in Eighteenth Century Peru and Upper Peru*. Cologne: Bohlau Verlag, 1985.

Parés y Franqués, José. *Catástrophe morboso de las minas mercuriales de la villa de Almadén del azogue (1778)*. Edited by Alfredo Menéndez Navarro. Cuenca, Spain: Ediciones de la Universidad de Castilla–La Mancha, 1998.

Patiño Paúl Ortíz, Mariano. *Huancavelica colonial: Apuntes históricos de la ciudad minera más importante del Virreynato Peruano*. Lima: Huancavelica 21, 2001

Pearce, Adrian. "Huancavelica, 1563–1824: History and Historiography." *Colonial Latin American Review* 22, no. 3 (December 2013): 422–40.

Pennington, J. W., and Gertrude N. Greenspoon. "Mercury." In *Minerals Yearbook: Metals and Minerals (Except Fuels), 1958*. Vol. 1. Washington DC: Bureau of Mines, 1959.

Petersen, Georg. *Mining and Metallurgy in Ancient Perú*. Translated by William E. Brooks. Special Paper 467. Boulder CO: Geological Society of America, 2010.

Pieper, Renate. "Innovaciones tecnológicas y problemas del medio ambiente en la minería novohispana (siglos XVI al XVIII)." In *Europa e Iberoamerica: Cinco siglos de Intercambios*, vol. 2, edited by María Justina Sarabia Viejo et al. Seville: Asociación de Historiadores Latinoamericanistas Europeos, Consejeria de Cultura y Medio Ambiente, 1992.

Piikivi, Leena, et al. "Psychological Performance and Long-Term Exposure to Mercury Vapors." *Scandanavian Journal of Work and Environmental Health* 10 (1984): 35–41.

Povea Moreno, Isabel M. "Entre la retórica y la disuación: Defensores e impugnadores del sistema mitayo en Huancavelica y en las Cortés de Cádiz." In *La Constitución Gaditana de 1812 y sus repercusiones en América*, edited by Alberto Gullón Abao and Antonio Gutiérrez Escudero. Cadiz: Universidad de Cadiz, 2012.

Probert, Alan. "Bartolomé de Medina: The Patio Process and the Sixteenth Century Silver Crisis." In *Mines of Silver and Gold in the Americas*, edited by Peter Bakewell. Brookfield VT: Variorum, 1997.

Purser, W. C. F. *Metal Mining in Peru, Past and Present*. New York: Praeger, 1971.

Ramírez, Balthasar. "Descripción del reyno del Pirú del sitio, temple, provincias, obispados y ciudades: De los naturales, de sus lenguas y traje." In *Juicio de límites entre el Perú y Bolivia*, vol. 1, edited by Víctor Maurtua. Barcelona: Imprenta de Heinrich y Compañía, 1906.

Ribera, Pedro de, and Antonio de Chaves y Guevara. "Relación de la Ciudad de Guamanga y sus terminos: Año de 1586." In *Relaciones geográficas de Indias*, vol. 1, edited by Marcos Jiménez de Espada. Madrid: Ediciones Atlas, 1965.

Rivero y Ustariz, Mariano E. de. *Memoria sobre el Rico Mineral de Azogue de Huancavelica*. Lima: J. M. Masías, 1848.

Rivero y Ustáriz, Mariano Eduardo de [attributed]. "Memoria sobre la mina de azogue de Huancavelica." In *Colección de memorias científicas, agrícolas é industriales publicadas en distintas épocas*, vol. 2, edited by Mariano Eduardo de Rivero y Ustáriz. Brussels: Imprenta de H. Goemaere, 1857.

Robins, Nicholas. *Mercury, Mining and Empire: The Human and Ecological Cost of Colonial Silver Mining in the Andes*. Bloomington: Indiana University Press, 2011.

———. *Santa Bárbara's Legacy: An Environmental History of Huancavelica, Peru*. Leiden: Brill, 2017.

Robins, Nicholas, and Nicole Hagan. "Mercury Production and Use in Colonial Andean Silver Production: Emissions and Health Implications." *Environmental Health Perspectives* 120, no. 5 (May 2012): 627–31.

Robins, Nicholas, Nicole Hagan, Susan Halabi, Heileen Hsu-Kim, Rubén Darío Espinoza Gonzales, Mark Morris, George Woodall, Daniel Richter, Paul Heine, T. Zhang, A. Bacon, and John Vandenberg. "Estimations of Historical Atmospheric Mercury Concentrations from Mercury Refining and Present-Day Soil Concentrations of Total Mercury in Huancavelica Peru." *Science of the Total Environment* 426, no. 11 (June 2012): 146–54.

Rowland, Andrew, et al. "The Effect of Occupational Exposure to Mercury Vapour on the Fertility of Female Dental Assistants." *Occupational and Environmental Medicine* 51, no. 1 (January 1994): 28–34.

Rudolph, William E. "The Lakes of Potosi." *Geographical Review* 26, no. 4 (October 1936): 529–54.

Ruiz Estrada, Arturo. *Arqueología de la ciudad de Huancavelica*. Lima: Servicios de Artes Gráficas, 1977.

Salgueiro, Barboni, et al. "Visual Field Losses in Workers Exposed to Mercury Vapor." *Environmental Research* 107 (2008): 124–31.

Salinas y Córdoba, Fray Buenaventura. *Memorial de las Historias del nuevo mundo Pirú.* Lima: Universidad Nacional Mayor de San Marcos, 1957.

Sánchez Gómez, Julio. "La ténica en la producción de metales monedables en España y en América, 1500–1650." In *La savia del imperio: Tres estudios de economia colonial*, edited by Julio Sánchez Gómez, Guillermo Mira Delli-Zotti, and Rafael Dobado. Salamanca: Ediciones Universidad Salamanca, 1997.

Schutte, Norbert, et al. "Mercury and Its Compounds." In *Occupational Medicine*, edited by Carl Zenz et al. 3rd. ed. St. Louis MO: Mosby, 1994.

Shortell, Paul. "In Need of Investment, Peru Rolls Back Environmental Standards." *World Politics Review*, July 28, 2014. https://www.worldpoliticsreview.com/articles/13955/in-need-of-investment-peru-rolls-back-environmental-standards.

Singewald, Joseph T. "The Huancavelica Mercury Deposits, Peru." *Engineering and Mining Journal* 110, no. 11 (1920): 518–22.

Smith, P. J., et al. "Effects of Occupational Exposure to Elemental Mercury on Short Term Memory." *British Journal of Industrial Medicine* 40 (1983): 413–19.

Sociedad Minera Brocal. "Nuestra Historia." Accessed August 17, 2019. http://www.elbrocal.pe/index.php?fp_cont=903.

Soleo, L., et al. "Effects of Low Exposure to Inorganic Mercury on Psychological Performance." *British Journal of Industrial Medicine* 47, no. 2 (1990): 105–9.

Stein, Stanley, and Barbara Stein. *Silver, Trade and War: Spain and America in the Making of Early Modern Europe.* Baltimore MD: Johns Hopkins University Press, 2000.

Strauss, Lester. "Quicksilver at Huancavelica, Peru." *Mining and Science Press* 99 (October 23, 1909): 561–66.

Tamayo, Augusto. "Mina de cinabrio 'Santa Bárbara' en Huancavelica." *Boletín del Ministerio de Fomento* 2, no. 1 (Lima: 1904): 38–43.

Thoms, Bryn, and Nicholas Robins. "Remedial Investigation: Huancavelica Mercury Remediation Project. Huancavelica, Peru." Unpublished manuscript. Environmental Health Council, July 2015.

Thoms, Bryn, and Nicholas Robins. "Results of June/July 2016 Assessment of Soil, Sediment and Fish: Huancavelica Mercury Remediation Project." Unpublished manuscript. Environmental Health Council, January 2017.

Thoms, Bryn, Nicholas Robins, Enrique Ecos, Rubén Darío Espinoza Gonzales, and William Brooks. *Data Summary: Huancavelica Heavy Metals Cleanup Project.* Unpublished report. Environmental Health Council, March 18, 2017.

Toledo, Francisco de. "Memorial que D. Francisco de Toledo dió al Rey nuestro señor, del estado en que dejó las cosas del Perú, después de haber sido en él Virrey y Capitán General trece años, que comenzaron en 1569." In *Colección de las memorias o relaciones que escribieron los virreys del Perú*, vol. 1, edited by Ricardo Beltrán y Rózpide. Madrid: Imprenta del Asilo de Huérfanos del S.C. de Jesús, 1921.

Torres de Portugal, Conde de Villardompardo, Fernando de. "Memoria gubernativa del Conde del Villardompardo." Lima, May 25, 1592 or 1593. In *Los virreyes españoles en America durante el gobierno de la casa de Austria. Perú*, vol. 1, edited by Lewis Hanke and Celso Rodríguez. Madrid: IMNASA, 1978.

Torrico y Mesa, Juan. "Excursión al departamento de Huancavelica." In *Boletín de minas, industrias y construcciones* 5, no. 12 (Lima: 1889): 889–93.

Umlauff, Augusto F. "El Cinabrio de Huancavelica." In *Boletin del Cuerpo de Ingenieros de Minas del Peru*. No. 7. Lima: Librería Escolar é Imprenta de E. Moreno, 1904

United Nations. "Convention on the Rights of the Child." General Assembly Resolution 44/25, annex, 44 U.N. GAOR Supp. (No. 49) at 167, U.N. Doc. A/44/49 (1989). http://hrlibrary.umn.edu/instree/k2crc.htm.

———. "International Covenant on Economic, Social and Cultural Rights." General Assembly Resolution 2200A (XXI), 21. U.N. GAOR Supp. (No. 16) at 49, U.N. Doc. A/6316 (1966), 993 U.N.T.S. 3. January 3, 1976. http://hrlibrary.umn.edu/instree/b2esc.htm.

U.S. Department of Health and Human Resources, Agency for Toxic Substances and Disease Registry. "Health Effects." In *Toxicological Profile for Mercury*. Atlanta GA: U.S. Department of Health and Human Resources, 1999.

U.S. Environmental Protection Agency. "Lead and Compounds." Integrated Risk Assessment System, Chemical Assessment Summary. Washington DC: National Center for Environmental Assessment, 2004. http://cfpub.epa.gov/ncea/iris/iris_documents/documents/subst/0277_summary.pdf.

———. *Mercury Study Report to Congress*. Vol. 1. Executive Summary. Washington DC: U.S. Environmental Protection Agency, 1997.

———. *A Review of the Reference Dose and Reference Concentration Processes*. EPA/630/P--02/002F. Risk Assessment Forum. Washington DC: EPA, 2002.

Uzzell, Barbara, and Jacqueline Oler. "Chronic Low-Level Mercury Exposure and Neuropsychological Functioning." *Journal of Clinical and Experimental Neuropsychology* 8, no. 5 (1986): 581–93.

Vargas Ugarte, Rubén, SJ. *Historia general del Perú: Virreinato (1551–1596)*. Vol 2. Lima: Carlos Milla Batres, 1966.

Velasco, Luis de. "Relación del Sr. Virrey, D. Luis de Velasco, al Sr. Conde de Monterrey sobre el estado del Perú." In *Colección de las memorias o relaciones que escribieron los virreys del Perú*, vol. 1, edited by Ricardo Beltrán y Rózpide. Madrid: Imprenta del Asilo de Huérfanos del S.C. de Jesús, 1921.

Velázquez, Tesania. *Salud mental en el Perú: Dolor y propuesta: La experiencia de Huancavelica*. Lima: Consorcio de Investigación Económica y Social, 2007.

Waldron, H. A., and A. Scott. "Metals." In *Hunter's Diseases of Occupations*, edited by P. A. B. Raffle et al. 8th ed. London: E. Arnold, 1994.

Wessel, F. W. "Peru." In *Minerals Yearbook Area Reports: International 1972*. Vol. 3. Washington DC: Bureau of Mines, 1972.

———. "Peru." In *Minerals Yearbook Area Reports: International 1974*. Vol. 3. Washington DC: Bureau of Mines, 1974.

West, Irma, and James Lim. "Mercury Poisoning among Workers in California's Mercury Mills." *Journal of Occupational Medicine* 10, no. 12 (December 1968): 697–701.

Whitaker, Arthur. *The Huancavelica Mercury Mine*. Boston: Harvard University Press, 1941.

Wiedner, Donald L. "Forced Labor in Colonial Peru." *The Americas* 16, no. 4 (April 1960): 357–83.

Williamson, A. M., et al. "Occupational Mercury Exposure and Its Consequences for Behaviour." In *International Archives of Occupational and Environmental Health* 50 (1982): 273–86.

Wise, James M., and Jean Féraud. "Historic Maps Used in New Geological and Engineering Evaluation of the Santa Bárbara Mine, Huancavelica Mercury District, Peru." *De Re Metallica* 4 (2005): 15–24.

World Health Organization. "Lead Poisoning and Health." August 23, 2108. http://www.who.int/mediacentre/factsheets/fs379/en/.

Yates, Robert G., Dean F. Kent, and Concha J. Fernández. *Geology of the Huancavelica Quicksilver District, Peru*. Washington DC: U.S. Government Printing Office, 1951.

Yeates, Keith, and Mary Ellen Mortensen. "Acute and Chronic Neuropsychological Consequences of Mercury Vapor Poisoning in Two Early Adolescents." *Journal of Clinical and Experimental Neuropsychology* 16, no. 2 (1994): 209–22.

Zimmerman, Authur F. *Francisco de Toledo, Fifth Viceroy of Peru, 1569–81*. Caldwell ID: Caxton Printers, 1938.

2 When the Rivers Run Black

Oil and Inequity in the Western Amazon

BARBARA J. FRASER

The Amazon basin has seen a series of boom-and-bust commodity cycles: timber, fish, animal pelts, rubber, gold, and more recently oil. They often take advantage of local indigenous populations, exploiting their labor and despoiling the natural resources on which they depend for survival. The rubber boom of the late nineteenth century and early twentieth had particularly far-reaching consequences, as it resulted in the deaths of thousands of indigenous people through overwork, abuse, and disease, and the displacement of thousands more, tearing apart communities and disrupting communal ways of life.

The violence of the rubber boom was still fresh in the experience and memory of Amazonian peoples when foreign companies struck oil in Amazonian Peru and Ecuador. Oil field workers introduced hepatitis and other diseases and fathered children whom they would abandon. Spills fouled water supplies, killed fish, and caused sickness. Local people became dependent for income on the companies that fouled their lands. Out of sight of politicians and the media, the contamination went largely unnoticed for decades.

In the 1990s indigenous organizations, along with nongovernmental organizations that provided technical and legal assistance, began to focus on the impacts of oil pollution. Communities in Ecuador sued Texaco, Achuar communities in Peru sued Occidental, and the community of

Sarayaku in Ecuador took its case to the Inter-American Court of Human Rights. Concern over the environmental impacts of oil drilling led to a campaign to block further drilling in Ecuador's Yasuní National Park; President Rafael Correa pledged to keep the oil in the ground if industrialized countries would provide compensation. That campaign failed, however, and drilling continued in Yasuní. Although the world's largest oil companies no longer have interests in Peru and Ecuador, both countries continue to promote their oil reserves, and conflicts with indigenous communities persist.

In Peru, one group of indigenous federations in the watersheds overlapped by the country's largest Amazonian oil fields has protested repeatedly, winning a battle to require the company operating the wells to reinject underground the hot, salty, metals-laden water that accompanies oil production, which for decades was simply dumped into rivers and streams. Nevertheless, attention to health problems, remediation of environmental damages, compensation for damage, and territorial rights, however, have been more difficult to obtain.

Meanwhile several oil spills in 2014 from the state-owned and -operated pipeline linking the oil fields to the coast drew more public attention to the pollution problem. They were followed in 2016 by a series of pipeline spills, most of which Petroperú, the state-owned oil company, blamed on vandals. The spills and the accusations led to tensions between indigenous communities and the company and other government officials. In 2016 a group of indigenous organizations, mainly in areas downstream from the oil wells, staged a protest lasting nearly four months, blocking the Marañón River, a major transportation route, to press for a series of demands. These included remediation of damages, an independent assessment of the pipeline infrastructure, and public debate about changing the country's energy matrix to other sources.

This chapter describes key events in the history of Amazonian oil production in Peru and Ecuador, highlighting impacts on the residents of local indigenous communities, impacts that are part of a continuum of resource exploitation in the region and that persist even when the immediate effects of an oil spill are no longer visible.

When the Waters Ran Black

Julia Chuje remembers a time, some four decades ago, when the water in streams and rivers near her community of Vista Alegre, on Peru's Tigre River, ran black. To wash clothes, bathe, or draw water for drinking or cooking, she would first clear a small area by pushing away the oily scum on the surface. No one told the residents of Vista Alegre or the other Kichwa villages that polluted water could be hazardous to their health. She learned that later, after she and other mothers watched their children die, vomiting blood.[1]

In the span of two years, twenty-one children and one adult died of a never-identified illness, leaving rows of handmade grave markers in the community's cemetery as silent witnesses. Villagers blamed "some evil sorcerer who wanted to do away with the community out of revenge or envy," wrote the biologist José Alvarez, a former Augustinian missionary who later worked at the Institute for Research in the Peruvian Amazon in Iquitos and in the Ministry of the Environment. "For a small community like Vista Alegre, that death toll meant the loss of almost an entire generation."[2]

Some people thought the disease might have been yellow fever, Alvarez recalled, but no outbreaks had been reported in that area. Another possibility was hepatitis B, perhaps exacerbated by toxic metals from oil operations. When villagers told him the lake where they fished was near an oil well, he warned them against eating the fish.[3]

Several decades later Chuje stood face to face with government officials and representatives of Petroperú, which had operated wells along the Tigre River in the 1970s. Her son—one of her only surviving children—was among the leaders of a protest that had blockaded the Marañón River off and on for more than three months, disrupting the transport of passengers and goods. The protesters demanded an international audit of aging pipelines that had spilled thousands of barrels of oil into rivers and streams, as well as health studies, compensation for environmental damage, and a national debate about weaning the country off fossil fuels.

The river blockade, which began on September 1, 2016, forced government negotiators to travel to Saramurillo, a village beside the pumping station at the start of the pipeline that transports crude oil from the northern Amazonian oil fields to the coast. Three rounds of talks ended with no agreements, and tempers frayed. A new group of government officials arrived on December 14 for what would be the two-day final round of negotiations to end the protest. On the morning of second day, Chuje stood up and faced a cabinet minister. The room full of government representatives, community leaders, and protest participants fell silent as Chuje, speaking sharply in her native language, recounted the history of the arrival of oil companies and the deaths of her children.

At midnight on December 15 the government officials and indigenous leaders finally signed a document containing more than thirty agreements. There was applause, then someone cranked up a stereo and broke out cases of beer. Chuje slipped away before the party began. The next morning, while the revelers slept in, Chuje prepared a kettle of *masato*, a fermented cassava drink. Over a bowl of the beverage, she repeated in Spanish what she had told the government officials emphatically in her native language the day before. "We didn't know what it was," she said of the oil. "The company didn't tell us anything."[4]

Over the years, she said, she had seen various protests that ended in agreements like the ones signed the night before. Most of the pledges had never materialized. She hoped this time would be different, but she was cautious; that was why she had skipped the festivities. "How many months have we waited for these agreements to be signed?" she asked. "When I see things change, that's when I'll celebrate."[5]

Booms, Busts, and Black Gold in the Jungle

The Peruvian government created its Ministry of the Environment, the first cabinet-level environmental institution, in 2008. That year it also invited bids for exploration of twenty-two new oil and gas blocks in Peru's Amazon region.[6] Although government officials speak publicly about the risks of climate change, which is due to greenhouse gas emissions, including those related to fossil fuel use, the country continues to promote exploration for and production of oil and natural gas offshore, on the coast, and in the Amazon basin.

The oil boom in western Amazonia is one of the latest cycles in a boom-and-bust economy based on natural resource extraction that dates back to the arrival of Spanish missionaries in that region in the seventeenth century. Since then, medicinal plants, animal hides, timber, and gold have been collected by indigenous Amazonians for sale to buyers in domestic cities or distant countries, often under a crushing system of debt peonage that still exists in parts of the Amazon.[7]

The most brutal episode was the rubber boom of the late nineteenth century and early twentieth, when possibly hundreds of thousands of indigenous people were enslaved and tens of thousands murdered during the frenzied rush to export this earlier form of "black gold." Many indigenous people were displaced during the rubber boom and the 1932–33 border war between Colombia and Peru. Others fled deeper into the forest, settling in remote areas, especially along the border between Peru and Brazil, where their seminomadic descendants continue to shun most contact with outsiders: "The battles, massacres, persecution, slavery and epidemics caused by the rubber boom of 1880 to 1920 severely impacted . . . these peoples, creating a great mistrust that has been passed down through the generations and can be seen in their current attitude to outsiders and external society. This attitude was further consolidated with the establishment of large estates based on cheap labour and, further still, with the violent invasion of logging and fossil fuel companies onto their territories."[8] As Hvalkof writes, "The demographic patterns and locations of indigenous settlements and populations we find today, therefore, are a direct result of these social disarticulations, dislocations, and forced migrations."[9]

A growing body of literature explores intergenerational trauma among indigenous peoples of the Americas, although there is little directly related to the rubber boom. According to Sotero, "Historical trauma originates with the subjugation of a population by a dominant group. Successful subjugation requires at least four elements: (1) overwhelming physical and psychological violence, (2) segregation and/or displacement, (3) economic deprivation, and (4) cultural dispossession." All of those were characteristic of both the Spanish conquest and the rubber era, at least in certain parts of the Amazon. Although "primary generations are the direct victims of subjugation and loss, which threaten their population and economic and cultural survival," subsequent generations are also affected:

Traumatic events become embedded in the collective, social memories of the population. Offspring are taught to share in the ancestral pain of their people and may have strong feelings of unresolved grief, persecution and distrust. They may also experience original trauma through loss of culture and language, as well as through proximate, first-hand experiences of discrimination, injustice, poverty, and social inequality. Such experiences validate their ancestral knowledge of historical trauma and reinforce the historical trauma experience and response.[10]

Although the rubber boom did not reach all corners of the Amazon, its impacts reverberate throughout the region, shaping relationships between indigenous people and the merchants and landowners who still depend on their labor. Although most of the generation that lived through the worst of the rubber-boom atrocities is gone, the memory remains alive in stories told by their children and grandchildren and in myths in which river dolphins turn into men dressed like overseers and ghostly ships bearing lavish parties appear briefly, then vanish into the depths of the river.[11] A darker part of that memory involves the ways in which some indigenous groups formed alliances with *patrones* to extract revenge on their enemies.[12] Such ambiguous alliances between local indigenous communities and outside economic interests also occurred during the next resource boom, which began less than two decades later, when foreign oil companies arrived in the western Amazon.

The first foray into Ecuador in the 1930s left its mark with a road, an airport, and a town named Shell. The boom began in earnest in the 1960s, when Texaco began operating in Colombia in 1963 and in Ecuador in 1967, building pipelines and roads that opened routes into indigenous lands for settlers from the Andes.[13] In Peru workers struck oil in the Ucayali River basin in the 1950s, but it was the discovery of oil in neighboring Ecuador, combined with Brazil's construction of the Trans-Amazonian Highway, that spurred the Peruvian government's interest in developing its own Amazonian region to firmly stake out its borders.[14] Development of the petroleum industry was one of its goals for the Loreto region.[15]

The military government that took power in 1968 established Petroperú, the state-run oil company, a year later. In the 1970s it gave foreign oil companies tax breaks in exchange for turning over half of the oil produced to the Peruvian state.[16] Petroperú's biggest oil strike, at Trompeteros in the Corrientes River basin in 1971, followed by an Occidental Petroleum discovery nearby, set off a boom. Santos Granero and Barclay note that in 1972 and 1973, eleven foreign companies or consortia received oil concessions in Loreto.[17] Further discoveries failed to materialize, however, and most pulled out of the country by the mid-1970s, leaving Petroperú and Occidental as the largest Amazonian operators.

Occidental began operating the lots that became known as Block 1AB, in the Pastaza, Tigre, and Corrientes watersheds, in 1971, while Petroperú operated Block 8, in part of the Corrientes basin and in the huge wetland and flooded forest area that became the Pacaya Samiria National Reserve. In 1996 Petroperú turned over Block 8 to Pluspetrol, an Argentinian company now domiciled in The Netherlands. Pluspetrol also acquired Block 1AB from Occidental, operating the concession from 2001 until 2015.

As Santos Granero and Barclay observe, "Oil production activated the economy of Loreto and favored the merchant elite of the region by stimulating external trade and river transportation. It also had an important impact in the rural area by creating new job opportunities, introducing internal migration flows, and increasing consumption."[18] In 1976 the national government agreed to allocate a proportion of oil revenues to the Loreto regional government, a system known as the *canon petrolero*. Production increased from 606 barrels in 1963 to more than 47,000 in 1979, making oil one of Loreto's main sources of revenue.[19]

In the 1980s the Peruvian government recognized Block 1AB as the country's most polluted region. Contamination came from oil spills that were never cleaned up, waste dumps that were not properly built and lined, and the hot, salty, metals-laden water called "produced water" pumped out of the well with the oil that was discharged into rivers and streams near the well sites.[20] Although nearly twenty oil and gas blocks were leased in the Peruvian Amazon by 2018, only a few were in production. The largest producers were the former Block 1AB, renamed Block 192, and Block 8, producing oil in the north, and the blocks around the Camisea gas field in southern Peru, which overlap indigenous communities and an area inhabited by seminomadic groups that shun contact with wider Peruvian society.

Seeking Redress for Harm to Environment and Health

Sabin notes, "The conditions of extraction have typically forced native groups to choose between no economic development and a form of industrial activity that would undermine the foundations of native cultural and economic life."[21] In both Ecuador and Peru, native communities have established a form of coexistence with companies that are often the only source of jobs in areas with largely subsistence economies. In some places, companies supplant the state, providing services that the state would typically provide, such as water, electricity, or evacuation in medical emergencies.

That gives companies a disproportionate amount of power in negotiations with both the government and affected communities. In an effort to level the playing field, indigenous organizations have sought alliances with human rights, indigenous rights, and international cooperation organizations, which provide funding, training, and advocacy assistance.

In the 1990s a convergence of factors began to shift the power dynamics in Peru's oil fields. In 1990 the government enacted the Environmental and Natural Resources Code (Legislative Decree 613) that "constituted the country's first systematic effort to define a general framework for environmental protection."[22] It also passed a Hydrocarbon Act (Law 26221) in 1993, which allowed concessions of up to thirty years for oil exploration and forty years for natural gas. During that time, as Bebbington and Scurrah explain, nongovernmental environmental and indigenous rights organizations began raising awareness among indigenous communities and organizations about the impacts of pollution and possible avenues for legal action. They also helped link organizations between countries so they could learn from other experiences.[23]

In the early 2000s four indigenous federations in parts of the Corrientes, Pastaza, Tigre, and Marañón watersheds joined forces to pressure the government to respect their territorial rights and address long-standing environmental and health problems stemming from oil production. Known as the *cuatro cuencas*, or "four watersheds," the organizations that joined forces initially were the Federation of Native Communities of the Corrientes River (Federación de Comunidades Nativas del Río Corrientes, FECONACO), the Federation of Native Communities of the Tigre River (Federación de Comunidades Nativas del Río Tigre, FECONAT), the Quechua Indigenous Federation of the Pastaza River (Federación Indígena Quechua del Pastaza), and the Cocama Association for the Development and Conservation of San Pablo de Tipishca (Asociación Cocama para el Desarrollo y Conservación de San Pablo de Tipishca). Because of intraorganizational divisions, a new federation, the Organization of Kichua Indigenous Peoples on the Peru-Ecuador Amazonian Border (Organización de Pueblos Indígenas Kichuas, Amazónicos Fronterizos del Perú y Ecuador), replaced FECONAT in the process in 2015, and FECONACOR took the place of FECONACO in 2017.

Although scientific studies of the impact of oil spills in the Amazon are scarce, evidence of environmental damage and possible harm to health has gradually emerged, supporting the cuatro cuencas organizations' insistence on environmental health care. Training of environmental monitors in the watersheds enabled the organizations to gather information that they could compare with official data, placing them in a stronger negotiating position with the government and oil companies.[24]

A meta-analysis of water-quality monitoring data from government agencies and the three companies (Occidental Petroleum, Pluspetrol, and Petroperú) that operated the blocks between 1987 and 2013 found that the largest amounts of produced water were discharged into the Corrientes River and its tributaries, and the smallest into the Pastaza. In 2008 an average of 363,000 barrels a day of produced water in Block 8 and 576,000 barrels a day in Block 1AB were dumped into the environment.[25] The produced water contained high concentrations of chloride, barium, hexavalent chromium, lead, and cadmium.[26] Occidental Petroleum had operated Block 1AB during most of that time, and in 2007 five Achuar communities sued the company in the United States for harm to their environment, health, and livelihood from the pollution. They settled the case out of court in 2013 for an undisclosed sum.[27]

Concerns about the possible impacts of the pollution on the health of people living along rivers affected by oil drilling increased after 2006, when the Peruvian Health Ministry released findings from a study conducted in 2005 in seven Achuar communities on the Corrientes River. The study found high levels of lead and cadmium in blood samples taken from 199 residents. The World Health Organization calls lead a "cumulative toxicant that affects multiple body systems, including the neurologic, hematologic, gastrointestinal, cardiovascular, and renal systems."[28] Children and fetuses are at particular risk because their brains and nervous systems are still developing. In 2012 the U.S. Centers for Disease Control and Prevention lowered the "level of concern" for lead in children's blood from 10 to 5 micrograms per deciliter, but experts say no level is safe for children.[29]

Two-thirds of the seventy-four children and adolescents age seventeen and younger in the Corrientes watershed who were tested had blood lead levels exceeding 10 micrograms per deciliter. The Health Ministry report, published in May 2006, does not indicate levels for the twenty-five children and teenagers who were below that threshold. All of the 125 adults tested had blood lead levels below 20 micrograms per deciliter; four out of five had levels between 10 and 20 micrograms per deciliter. In addition, more than 97 percent of children, adolescents, and adults had blood cadmium levels exceeding the recommended threshold of 0.1 micrograms per deciliter.[30] Cadmium is a known carcinogen and is linked to gastrointestinal irritation and kidney disease.[31]

Five months after the report was made public, on October 10, 2006, members of communities along the Corrientes River seized the oil wells, landing strips, and other facilities in Blocks 1AB and 8, shutting down oil production. The subsequent negotiations led to an eleven-point agreement known as the Dorissa Accord, which called for Pluspetrol to reinject all produced water from wells in the two blocks into the subsoil. The environmental and hydrocarbon legislation enacted in the 1990s required reinjection of produced water for all new wells, but existing wells—including those in Blocks 1AB and 8, which produced most of Peru's Amazonian oil—had been exempted. The Accord also called for a comprehensive health plan, better health services, a development program, food aid and safe drinking water, remediation of damages, and community involvement in future monitoring of water quality and remediation.[32]

Pluspetrol completed reinjection four years later, although within a decade some reinjection wells were reportedly leaking.[33] According to Bebbington and Scurrah, "Though the reinjection of produced waters marked a significant 'victory,' it is also important to recognize that this merely brought some operations up to some sort of minimal good practices."[34] The health accords, development plan, and potable water provision were never fully implemented, however, and the government Ombudsman's Office found mismanagement of funds.[35] Pluspetrol claimed to have completed remediation of polluted sites, but after the company abandoned Block 1AB upon expiration of its lease in 2015, community monitors and government inspectors found that hundreds of sites remained to be cleaned up.

The cuatro cuencas federations continued to press their case over the next four years, in negotiations with government officials that led to a wide-ranging agreement signed on March 10, 2015. This included titling of communities in Block 192, remediation of polluted sites (with creation of a $15 million fund), potable water for communities, and an epidemiological and toxicological study.[36] By 2018 temporary water treatment plants had been installed in more than sixty communities in the four watersheds and the government's occupational and environmental health agency was analyzing results of a study conducted in 2016 in communities in the four affected watersheds. Priority sites for remediation had also been identified jointly by the indigenous organizations and government officials, and studies were under way. The $15 million fund, however, was expected to cover studies of a group of priority sites identified by indigenous organizations and government officials, but not the cleanup and remediation of all of the sites.[37]

Ecuador: From Legal Recourse to "Keep It in the Ground"

The history of oil operations in Ecuador is similar to that of Peru. After workers struck oil near the border with Colombia—in a place they called Lago Agrio, after the oil town of Sour Lake, Texas—exploration and production centered on that part of the Ecuadorian Amazon. The Texaco Petroleum Company and Gulf Oil received a twenty-eight-year concession in 1964 to explore in partnership with the state-run oil company,

Petroecuador. When Gulf dropped out in 1977, Petroecuador purchased its share of the investment, giving the state a 62.5 percent majority stake in the operations, while Texaco maintained 37.5 percent. Texaco's concession period ended in 1992, and Petroecuador assumed full operation. Part of Texaco's contribution was construction of a pipeline from the Amazon over the Andes Mountains to the Pacific coast.[38] Over time the government divided most of the rest of Amazonian Ecuador into concessions, mainly in the northeastern Amazon and along the border with Peru.

In Ecuador, as in Peru, development of environmental legislation and regulations lagged behind the oil boom by several decades. Early practices were particularly lax, and waste was dumped into rivers, streams, and forests or buried in poorly lined pits.

In Ecuador the oil boom that began in earnest in the 1960s has led to three emblematic cases that illustrate the evolution of approaches to oil production in Amazonia: the lawsuit filed against Chevron/Texaco in Ecuador and the United States, the triumph of the community of Sarayaku in the Inter-American Court of Human Rights, and the effort to convince industrialized countries to pay to keep oil in the ground under the highly biodiverse Yasuní National Park.

The multibillion-dollar, class-action lawsuit filed in the United States in 1993, a year after Texaco's contract with the Ecuadorian government expired, was among the earliest human rights cases against a transnational oil company operating in a low-income country. "The plaintiffs accused the company of knowingly conducting negligent environmental practices (such as dumping highly toxic water and crude petroleum into the surrounding ecosystem), wrecking traditional ways of life, and increasing health risks for local peoples."[39] Because of disagreements over jurisdiction, the case moved sluggishly through the courts until 2002, when a U.S. appeals court ruled that it could be heard in Ecuador, but the ruling would be binding on Chevron/Texaco (the companies had merged in the meantime) in the United States.[40]

The grassroots groups that filed suit argued that Chevron/Texaco should be held responsible for discharging more than sixteen billion gallons of toxic waste into rivers and streams between 1964 and 1990 and for leakage of toxics from waste pits. The company countered that it had complied with regulations and that state-run Petroecuador was responsible for the pollution.[41] An $18 billion ruling against Chevron by an Ecuadorian court in 2011 was upheld in 2014 by the country's Supreme Court, but reduced to $9.5 million. So far, however, challenges in U.S. courts have kept the plaintiffs from collecting damages there.[42]

Two years after the Texaco lawsuit began, the Ecuadorian government granted two companies a concession to Block 23, which overlapped indigenous communities near the town of Puyo, in the western Amazon basin. The companies reached agreements with several communities to begin operations on their lands, but the Kichwa community of Sarayaku refused. Seismic testing began despite the community's resistance, the conflict escalated, and in 2004 Sarayaku took its case to the Inter-American Commission on Human Rights. The Commission admitted the case in 2009 and submitted it to the Inter-American Court of Human Rights in 2010 on the grounds that the Ecuadorian state had failed to consult the community and obtain its consent and had violated the right to private property; the right to life, judicial guarantees, and judicial protection; the right to freedom of movement and residence; and the right to personal integrity.[43]

The court ruled in Sarayaku's favor in June 2012, ordering the government to remove all seismic explosives from the community's territory and awarding $1.25 million in nonpecuniary damages and $90,000 in pecuniary damages, plus court costs.[44] The case set an important legal precedent for prior consultation of indigenous communities in cases involving extractive industries.

About one-fourth of Ecuador's oil reserves, nearly 900 million barrels, lies east of Sarayaku, beneath Yasuní National Park, a highly biodiverse area that also includes part of the territory used by two seminomadic, isolated indigenous groups, the Tagaeri and Taromenane.[45] Ecuadorian law does not prohibit oil drilling in national parks, and plans were under way for exploration in a block known as Ishpingo-Tambococha-Tiputini (ITT) in 2007, when President Correa proposed keeping the oil in the ground—to avoid greenhouse gas emissions, as well as environmental impacts and impacts on indigenous groups—if international donors would compensate Ecuador for lost revenue. International donations totaling

$3.96 billion would go into a trust fund. The effort failed to attract the necessary commitments, however, and was suspended in 2013. Meanwhile Correa continued to negotiate concession of ITT and a neighboring block. [46] Drilling in ITT began in 2016, with a second phase beginning in January 2018.[47] Although the fund never materialized, the initiative drew attention to the responsibility of industrialized countries for the harm done by oil drilling in Amazonian countries. A new chapter in Ecuador's oil saga opened in February 2019, however, when Waorani people sued the government for attempting to auction Block 22, which overlaps their territory, without first obtaining the free, prior, and informed consent of the affected communities. On April 26, 2019, a three-judge panel ruled in the Waorani plaintiffs' favor, finding that the outcome of the prior consent process did not constitute free, prior, and informed consent. The lawsuit was supported by the government Ombudsman's Office and other indigenous groups in Ecuador and could set a precedent for other indigenous peoples' claims in similar cases.[48]

Beyond Pollution: Social and Cultural Upheaval

Oil production has had an irreversible, and perhaps incalculable and incomprehensible, impact on the lives of indigenous people in the areas affected by drilling and pollution. As part of a study of perceptions of health and pollution among Kichwa people in Peru's Tigre River basin, the Ecuadorian anthropologist María Antonieta Guzmán-Gallegos recorded people's memories of the history of oil operations and their impacts. The Quechua and Achuar people of the Pastaza watershed, the Achuar and Urarina of the Corrientes basin, and the Kukama people living along the Marañón and in and around the Pacaya Samiria reserve were similarly affected.

Oil companies arrived in the Peruvian Amazon just a few decades after another upheaval, the 1941 border war with Ecuador. Until then, people had moved freely along the rivers between the two countries. After the conflict, however, the border was less permeable, and the political boundary divided families. On the Peruvian side, families began to group around military garrisons or other places where schools were established, further restricting their mobility.[49]

Although the indigenous families in those watersheds had not been enslaved by rubber barons in the early twentieth century, as had those on the Putumayo, they traded smaller amounts of rubber, timber, animal hides, rosewood, and a gum-like resin called *leche caspi* with merchants who plied the rivers, receiving cloth and shotguns and hunting ammunition in return. Traders maintained the upper hand by "advancing" payments. As Guzmán-Gallegos notes, "Because the terms of exchange were very unequal, people lived in a constant state of indebtedness."[50]

When the oil companies arrived, workers set up camp in villagers' gardens. One woman, who was just a girl at the time, recalled fleeing into the forest with other children to escape the strangers, fearing they were *pelacaras*, mythical beings that peel the skin from people's faces.[51] Pelacaras, the Amazonian incarnation of the fat-sucking *pishtacos* of the Andean world, are often characterized as "gringos" and are associated with outsiders. The extraction of the skin represents a loss of both personal identity and social identity, of belonging to a people.[52]

Another woman, who had been a child when oil workers arrived in the village of Marsella, recalled helicopters hovering overhead and men being lowered on ropes. The men claimed the government had given the land, which it considered uninhabited, to the company.[53] Families were displaced as a network of oil wells, pipelines, holding tanks, and refinery facilities spread across their territory. Men found work clearing trails for seismic lines or providing wild game meat and fish to company workers. A cash economy began to replace the former pattern of trade, and families paid off their debts to loggers and merchants.[54]

During the years when Occidental Petroleum operated the field, most workers came from Iquitos or other distant towns. That changed when Pluspetrol took over operation of Block 1AB and began to hire local villagers. In the Tigre basin, Marsella was a watering hole for workers, who went there to relax and to drink. Domestic violence increased, and young women became single mothers, abandoned by oil company workers who fathered their children. Few of those workers formed families and remained in the villages, a problem that has persisted around oil facilities for decades.[55] In October 2016 the president of the native community of Saramuro raised the same grievance with the Peruvian minister of justice, when she arrived as part of a government negotiating team during a three-month protest by indigenous people in the neighboring community of Saramurillo, in the lower Marañón River Valley.[56]

During the first two decades of oil operations, the companies operated with virtually no environmental oversight. The noise and movement of people and machinery, combined with hunting, made game scarce, and people recall seeing animals covered with oil or tar after spills. As elsewhere, produced water was simply dumped into rivers and streams or onto the ground. Those places became new salt licks for animals. Researchers have photographed tapirs licking salt from pipeline valves and eating plants in contaminated areas.[57] Some people told Guzmán-Gallegos that peccaries and deer are attracted to those places where the salt and soil are contaminated with toxic metals. People continue to hunt those animals, and the effects of meat consumption on human health are unknown.

The burning of oil residue sometimes sent dense, black smoke billowing over homes and forests. "While the refinery was in Marsella, there were no flies or mosquitoes. That was the only good thing about that black cloud," one man told Guzmán-Gallegos.[58] Villagers along the Tigre River also remember the salty water burning their skin when they bathed. Even water in streams that appeared clear would burn their mouths if they drank it, and people said the fish looked sick and had an unpleasant taste.[59]

Murphy maintains that chemicals are as much a part of industrial infrastructure as pipelines and smokestacks: "By chemical infrastructures, I mean the spatial and temporal distributions of industrially produced chemicals as they are produced and consumed, and as they become mobile in the atmosphere, settle into landscapes, travel in water ways, leach from commodities, are regulated (or not) by states, monitored by experts, engineered by industries, absorbed by bodies, metabolized physiologically, and as they bioaccumulate in food chains, break down over time, or persist. With the term chemical infrastructures, I am naming and imagining the many varied pathways of industrial chemicals as they permeate and structure life, both human and nonhuman."[60]

Like the chemicals, the memories also persist, becoming part of the historical memory of the indigenous peoples of the northeastern Peruvian Amazon.

Pipeline Spills Make Oil Pollution More Visible

Oil pollution in Blocks 192 and 8 attracted little public attention over the years, partly because the area is remote, communication is difficult, and travel is expensive. The problem became more visible in 2014, with the first of a series of oil spills along the Amazonian portion of the Northern Peruvian Oil Pipeline, which carries crude from Pumping Station No. 1, in the village of Saramuro, on the Marañón River, over the Andes Mountains to the Pacific coast. The 687-mile pipeline includes a spur connecting the east-west portion to facilities in Andoas, on the Pastaza River in Block 192.

On July 30, 2014, the pipeline ruptured in the community of Cuninico, on the Marañón River, about forty kilometers upstream from Saramuro. About 2,300 barrels of oil leaked into the flotation canal, the channel in which the pipeline was laid, which was designed to contain spills. This was the rainy season, however, and the canal overflowed, allowing oil to spread into surrounding vegetation and into the Cuninico River, a tributary of the Marañón and the source of water that many community residents used for drinking, cooking, bathing, and washing.[61]

Petroperú hired about twenty men from Cuninico to locate the place where the pipeline, which was under about 1.5 meters of oily water, had ruptured. For nearly a week, with no protective gear, they immersed themselves up to their chests or necks in the crude-filled canal, seeking the leak. Petroperú announced that the pipeline had been deliberately cut, but community members denied the accusation. The government's Environmental Assessment and Enforcement Agency (Organismo de Evaluación y Fiscalización Ambiental, OEFA) attributed the leak to deterioration of the pipeline and fined Petroperú 10 million nuevos soles (equivalent to about $3.1 million) for various infractions, including failure to maintain the pipeline and to adequately remediate the spill.[62]

Shortly after the spill, a national television channel reported that two of the villagers hired in those first days were under age eighteen.[63] The report led to the resignation of the company's top executives and the contracting of a Finnish company with an office in Ecuador to oversee the cleanup. Subcontractors hired workers and provided them with protective coveralls. The cleanup workers scooped oil into barrels, bagged oil-soaked soil, shredded and bagged oily vegetation, and used a degreasing agent to scrub oil off the trunks of standing trees.

Before the spill, a day laborer could earn about 10 soles a day, slightly more than $3, but jobs were scarce. Most villagers fished and raised crops such as cassava and bananas for their families, selling the surplus. Nearly all the adult men in the community and some women took oil spill cleanup jobs, which paid a daily wage of 75 soles, about $25. Many used their earnings to improve or expand their houses.

In mid-December 2014 Petroperú left the community, asserting that the cleanup was finished. Delivery of food and bottled drinking water, which Cuninico's leaders had negotiated with the company, ended at the same time. But the rainy season had begun, and as the water rose, traces of oil reappeared in the Cuninico River. Despite additional cleanup in September 2015, oil remained in sediment under the pipeline.

The 2014 oil spill resulted in a cascade of impacts that the villagers of Cuninico could never have predicted. Buyers rejected fish from lakes near the community, affecting not just Cuninico but several neighboring communities whose residents also fished in those lakes. Fish had been the main source of protein for villagers, and families that generally supplied their own needs at no cash cost were forced to buy fish from other communities at higher-than-usual prices or switch to other sources of protein, such as eggs, also at a cost.[64]

The unusually high wages offered for cleanup work triggered inflation in village shops. Once the work ended and the jobs vaporized, prices dropped but did not return to their previous levels. And after earning 75 soles a day, people were unwilling to work for less afterward. Villagers who needed to hire day laborers to help in their fields or with other work complained that people would not work for less than 30 soles a day, three times the pre-spill rate.[65]

After the spill, women said their children complained of headaches or eye pain or broke out in rashes if they bathed in the river. Stomach problems were also a common complaint. Personnel at the nearest health center, about an hour away by canoe, and a doctor provided by Petroperú during the six-month cleanup operation generally attributed the skin ailments to fungi common in humid climates and stomach problems to parasites. Villagers began to collect rainwater for drinking, cooking, and bathing infants rather than using river water. However, during the dry season, from July to November, when rains were more sporadic, rainwater was insufficient for an average-size family, and some people resorted to drawing water from the Marañón River, upstream from the confluence with the Cuninico. But the water in the Marañón is also deemed unfit for human consumption, so the problem of poor water quality persisted.[66]

In November 2014 another pipeline leak spilled 7,000 barrels of oil in the neighboring community of San Pedro, about twenty kilometers downstream, in a case Petroperú officials blamed on vandalism. Another spill occurred in January 2016, near the town of Chiriaco, in the northern Amazonas region, where shifting earth caused the underground pipeline to buckle. Unlike Cuninico and San Pedro, Chiriaco is accessible by road, and media coverage was prompt and ongoing. One of Peru's largest-circulation national daily newspapers published photographs of children and adolescents in oil-stained clothing and reported that the young people were involved in the cleanup.[67]

Petroperú denied employing minors, but a supervisor at the company's temporary operations center near the spill site confirmed that besides hiring cleanup workers for 150 soles, about $43, a day, the company had publicly offered to pay others the same amount per barrel of oil recovered. In several communities adults responded by using buckets to scoop oil out of the river. In at least one, Nazareth, at the confluence of the Chiriaco and Marañón rivers, children and adolescents did the same, although they were not formally hired to do so.[68] The Chiriaco cleanup was still under way when another leak was reported in the Wampis community of Mayuriaga, on the Morona River, in February 2016. That spill, which was attributed to external corrosion of the pipe due to abrasion from the sandy soil, dumped more than 1,400 barrels of oil onto the ground and into a twenty-three-kilometer stream that empties into the Morona River. More than twenty communities along the river were believed to be affected by the spill. Oddly, Mayuriaga was less directly, although the spill was in its territory, because the oil did not contaminate the river that serves as the community's main water source. The environmental oversight agency OEFA eventually fined Petroperú 49.7 million nuevos soles (equivalent to about $15.4 million) for violations related to the Chiriaco and Morona spills.[69] OSINERGMIN, the regulatory agency responsible for overseeing oil and gas infrastructure, also fined Petroperú 12.6 million nuevos soles (about $3.9 million) for failure to install supports on the pipeline's northern spur, where the Morona River spill occurred; lack of a risk assessment and safety manual; and lack of detailed maps of places where the pipeline crossed rivers.[70]

After the Chiriaco and Mayuriaga spills, government inspectors ordered Petroperú to suspend pipeline operations until the system could be inspected and any necessary repairs made. Nevertheless another leak, again caused by shifting soil, was reported in June 2016 in the community of Barranca, on the Marañón River. In that case, although workers tried to pressure Petroperú to pay 150 soles a day for the cleanup, as it had in Chiriaco and Mayuriaga, the company held firm at 80 soles, about $23, a day, approximately the same amount it had paid in 2014 in Cuninico and San Pedro.

By February 2018 fifteen more spills had occurred in Amazonian communities, mainly along the Marañón River. Petroperú attributed them to vandalism, although OSINERGMIN had not ruled in all of the cases. Petroperú filed official complaints of vandalism with Peru's Public Ministry, without naming perpetrators. Company executives said the wages paid during cleanup operations created a perverse incentive for vandalism, while indigenous leaders objected to the implication that community members were to blame and said the claims of vandalism concealed the fact that the pipeline had been poorly maintained, as described in official government documents.[71]

In September 2016 a group of federations representing communities along the Pastaza, Corrientes, Tigre, Marañón, and Chambira rivers launched a protest about the pollution. At Saramurillo, the native community on the downstream side of Petroperú's Pumping Station No. 1, where the pipeline begins, several hundred people blockaded the Marañón

River, disrupting transportation on a key tributary of the Amazon River. Petroperú shut down the pumping station and evacuated its personnel. Because the station provides running water and electricity to Saramurillo and the neighboring community of Saramuro, the shutdown cut off public services to the two communities during the 105-day protest.

The federations known as the cuatro cuencas did not participate in the protest, and the protest leaders in Saramurillo drew the distinction by referring to their collective as the *cinco cuencas*, or "five watersheds" (the four overlapped by Lots 192 and 8—the Corrientes, Pastaza, Tigre and Marañón—plus the Chambira, which is part of the Corrientes watershed but inhabited mainly by Urarina people). They presented an eight-point platform of demands, including the declaration of an environmental and health emergency in two affected districts, an independent assessment of the pipeline and repair or replacement of corroded sections, and remediation of polluted sites in the oil fields and along the pipeline. They also called for compensation for damages, in the form of health care, education, and income-generating projects in the communities. The demands also included a law to promote environmental monitoring by communities, a truth commission to investigate the impacts of more than four decades of oil operations, and a national debate about shifting the country away from fossil fuels and toward more renewable energy. Finally, they demanded that the government not prosecute protest leaders for blocking a transportation route.

Between October and December 2016 government delegations traveled to Saramurillo four times to meet with the protesters. Each of the first three sessions ended after two days, with little progress. Most of the agreements were defined at the last session, on December 13 and 14, but time lines and next steps were not clear. Tension persisted into 2019 over unfulfilled commitments.

In November 2017 a special congressional commission released a report on its investigation of the pipeline spills. The report did not identify those responsible for the vandalism but highlighted failures by Petroperú and government oversight and regulatory agencies. The commission questioned the hiring of inexperienced local contractors for the cleanup work and noted that some contractors had profited considerably. One company created shortly before the Cuninico spill with an initial investment of about $6,000 billed Petroperú some $18 million in cleanup costs between 2014 and 2017.[72] A relatively small number of contractors shared the approximately $135 million that Petroperú spent on cleanup and recovery of slightly more than 25,000 barrels of oil between 2014 and 2016.[73]

The investigation revealed shortcomings in upgrading the pipeline, as well as in maintenance, security, and contracting of service providers. It recommended further investigation by the Controller's Office and other agencies into possible corruption or mismanagement, and it called for the government to address risks to health and food security in the communities, as well as studies of impacts of the spills on ecosystems and people and remediation of damage. It also recommended changes in policies to ensure appropriate inspection and maintenance of the country's oil pipeline infrastructure.[74]

Toward Justice in the Oil Fields

Amazonian indigenous communities affected by oil pollution are gradually becoming more visible. Cuninico and San Pedro have won court cases demanding safe drinking water, health care, and compensation for damages, although implementation of the court rulings has lagged. A delegation from the Inter-American Commission on Human Rights visited Cuninico, Chiriaco, and the Morona River in June 2017.[75] Media coverage of spills and protests has increased. The communities have also become more proactive. Training of indigenous monitors has given communities the power to oversee company and government inspectors, but funding for community monitoring is not guaranteed.

A U.S. engineering firm helped the Achuar negotiate the agreement for reinjection of produced water in 2006 and advised the federations of the cuatro cuencas on health studies and remediation plans, but communities generally lack expert technical and legal advice, especially in emergencies.[76] Nevertheless increased access to technology—drones, cell phones, social media, and the internet—enables them to distribute photos of oil spills, forcing companies and government agencies to respond to problems they might have ignored in the past.

Technology, however, is not enough to create a level playing field for communities. In remote regions where jobs are scarce, employment in the oil fields is one of the few ways of obtaining cash for necessities such as a new boat motor, a tin roof, school supplies and uniforms, or tuition for teenagers who must travel to a distant city for postsecondary studies. It is both ironic and tragic that an economic activity that degrades the environment, jeopardizes health, and destroys livelihoods is the one that offers the best—or one of the only—opportunities for a decent income. That reality divides communities and organizations, as some people support companies for the jobs they provide, while others oppose extractive activities. Both companies and governments take advantage of those tensions, creating parallel organizations or playing federations or communities against one another.[77]

In November 2017 the Inter-American Commission on Human Rights issued a decision that recognized people's right to a healthy environment, but communities in the affected watersheds still lack safe water for drinking, cooking, bathing, and washing, and they do not know whether the fish they catch are safe to eat. That uncertainty causes stress in communities where health services are marginal and mental health care is nonexistent. Addressing environmental health issues also assumes recognition of the way indigenous communities understand health and illness, an intercultural approach that is not fully implemented in government health services. Health assessments are under way, but results may be inconclusive because there are no control groups outside the affected area with which to compare the levels of toxics found in blood samples, and it is difficult to prove that health problems were caused by pollution.

Steps toward justice must reinforce communities' rights, including territorial rights—land titling and access to natural resources—and the right to give free, prior, and informed consent in cases involving development projects, including resource extraction, that could affect their collective rights, as required under International Labor Organization Convention 169. None of the communities affected by operations that began in the twentieth century was consulted about the plans, and recent studies have found shortcomings in implementation of prior consultation. Indigenous federations argue that projects should not go ahead unless communities give free, prior, and informed consent, but prior consultation laws do not give communities veto power over plans. Instead the goal is to negotiate an agreement acceptable to all parties.[78] Protection of territories used by seminomadic, isolated tribes is especially important. Oil and gas concessions overlap some of those territories, and contact with outsiders can be deadly for isolated people who lack resistance to common illnesses, and for workers who risk violent contact with isolated groups determined to defend their territory.

As important as the rights of communities are the rights of nature, but they are often trumped by economic necessity. Both Ecuador and Bolivia have enshrined the rights of nature or Mother Earth in their constitutions, but hydrocarbons are an economic cornerstone for both countries. Balancing the need for revenue with environmental demands is leading to conflict and to new forms of environmental governance in both countries, as discussed by Andrade.[79] In those countries, as in Peru and most of Latin America, revenues from resource extraction and commodity exports have lowered poverty rates and allowed governments to expand basic services. Indigenous peoples are still the poorest and most marginalized, however, and the benefits of the resource bonanza of the early twenty-first century are slow to trickle down to the remote reaches of the Amazon where the oil and gas wells are located. Peru has yet to debate the inclusion of the rights of nature in its national legislation. Among South American countries, Colombia has taken the lead in recognizing rivers and the Amazon as having rights.[80] How quickly—or whether—other countries will follow suit remains to be seen.

Only in recent years have issues of climate justice entered public discourse, especially with the participation of indigenous leaders at international climate summits. Commitments made at those meetings may force countries like Peru and Ecuador to face the contradiction of dependence on greenhouse gas–emitting fossil fuels and revenues from oil and gas exports while climate change shrinks the glaciers that store crucial water supplies and the tropical forests on which indigenous communities depend for their livelihoods.[81]

At the heart of the quest for justice is the need for government and the dominant society in oil- and gas-producing countries to ensure that affected communities are treated fairly, are visible, and are recognized as citizens; to respect their culture, traditions, and worldview; and to recognize and make reparation for the trauma of repeated waves of exploitation of the communities and their resources. As Guzmán-Gallegos writes, adults in the communities hope for good health for their children, who dream of getting an education that offers them employment options beyond low-skilled labor: "They dream, like everyone, of equity and justice."[82]

NOTES

1. This section is based on the author's coverage of the Saramurillo negotiations as a journalist reporting for Mongabay.com and *EcoAmericas*.
2. Alvarez Alonso, "Peru."
3. Alvarez Alonso, "Peru."
4. Julia Chuje, personal communication, December 16, 2016.
5. Fraser, "Protest."
6. Andina Agencia Peruana de Noticias, "19 postores."
7. Hvalkof, "Outrage," 90.
8. Huertas, *Indigenous Peoples*, 21.
9. Hvalkof, "Outrage," 93.
10. Sotero, "Conceptual Model," 99–100.
11. Tello Imaina and Fraser, "Rubber Barons' Abuses."
12. Echeverri, "To Heal," 54.
13. Hvalkof, "Outrage," 103.
14. Santos Granero and Barclay, *La Frontera Domesticada*, 319.
15. Santos Granero and Barclay, *La Frontera Domesticada*, 312.
16. Santos Granero and Barclay, *La Frontera Domesticada*, 320.
17. Santos Granero and Barclay, *La Frontera Domesticada*, 320.
18. Santos Granero and Barclay, *La Frontera Domesticada*, 321.
19. Santos Granero and Barclay, *La Frontera Domesticada*, 321.
20. Bebbington and Scurrah, "Hydrocarbon Conflicts," 177.
21. Sabin, "Searching," 146.
22. Bebbington and Scurrah, "Hydrocarbon Conflicts," 177.
23. Bebbington and Scurrah, "Hydrocarbon Conflicts," 179–81.
24. Orta-Martínez and Finer, "Oil Frontiers," 214.
25. Yusta-García et al., "Water Contamination," 371.
26. Yusta-García et al., "Water Contamination," 375.
27. Collyns, "Indigenous Peruvians."
28. World Health Organization, "Lead."

29. Caldwell et al., "Measurement Challenges," 1.
30. Government of Peru, Ministerio de Salud, "Visita." The pages of the report are not numbered, but the blood lead levels are reported on pages 24–25 of the PDF version and the cadmium levels on pages 27–28.
31. U.S. Centers for Disease Control, "Cadmium Factsheet."
32. Bebbington and Scurrah, "Hydrocarbon Conflicts," 183–84.
33. Carlos Sandi, personal communication, August 22, 2017.
34. Bebbington and Scurrah, "Hydrocarbon Conflicts," 184.
35. P. Vásquez, "Extractive Industries and Conflicts in Peru: An Agenda for Action," draft of internal paper prepared as part of the United Nations–European Union Partnership on Natural Resources and Conflict Prevention, based on week-long field mission in October 2010, p. 9.
36. An overview of the accords and the process leading up to them can be found on the website managed by the technical team that supports the cuatro cuencas federations: http://observatoriopetrolero.org/memoria-movilizacion-y-reinvencion-de-derechos-en-las-luchas-contra-el-petroleo/.
37. For a list of the priority sites, see OEFA, "Fondo Nacional."
38. Chevron Corporation, "Texaco Petroleum," 2.
39. Valdivia, "'Amazonian Trial,'" 42.
40. Valdivia, "'Amazonian Trial,'" 42.
41. Finer et al., "Ecuador's Yasuní Biosphere Reserve," 7.
42. Associated Press, "U.S. Court."
43. Inter-American Court of Human Rights, *Case of the Kichwa*, 5.
44. Inter-American Court of Human Rights, *Case of the Kichwa*, 91–92.
45. Finer et al., "Oil and Gas Projects," 4.
46. Martin and Scholz, "Policy Debate."
47. Watts, "New Round."
48. Riederer, "An Uncommon Victory."
49. Guzmán-Gallegos, "*Ñukanchi Ñawpa Timputa Yarisha Tantarishun*," 16.
50. Guzmán-Gallegos, "*Ñukanchi Ñawpa Timputa Yarisha Tantarishun*," 20.
51. Guzmán-Gallegos, "*Ñukanchi Ñawpa Timputa Yarisha Tantarishun*," 24.
52. Cure Valdivieso, "Cuidado te mochan la cabeza," 9.
53. Guzmán-Gallegos, "*Ñukanchi Ñawpa Timputa Yarisha Tantarishun*," 25.
54. Guzmán-Gallegos, "*Ñukanchi Ñawpa Timputa Yarisha Tantarishun*," 26.
55. Guzmán-Gallegos "*Ñukanchi Ñawpa Timputa Yarisha Tantarishun*," 27.
56. Indigenous leaders raised this issue with Peru's minister of justice and human rights Marisol Pérez Tello during a conversation in Saramurillo on November 28, 2016.
57. Orta-Martínez et al., "First Evidences," 516.
58. Guzmán-Gallegos, "*Ñukanchi Ñawpa Timputa Yarisha Tantarishun*," 35.
59. Guzmán-Gallegos, "*Ñukanchi Ñawpa Timputa Yarisha Tantarishun*," 33–34.
60. Murphy, "Distributed Reproduction."

61. The information in this section was gathered by the author and Leonardo Tello and Anja Noake, journalists from Radio Ucamara in Nauta, during a visit to the communities of Cuninico, Santa Rosa, Nueva Alianza, and Monterrico on July 7–11, 2014.
62. See OEFA, Resolución Directorial, for the OEFA ruling on violations. On the fine, see OEFA, "El OEFA sanciona."
63. Panorama, "Manto negro en Iquitos."
64. Interviews with local residents in February and October 2015; D. Delgado Pugley, "A Toxic Development: Inequality, Pollution and Change in the Amazonian Oil Frontiers," unpublished paper, 2016.
65. Interviews with villagers in Cuninico, February and September 2015.
66. At the insistence of indigenous federations, various government agencies analyzed water used for human consumption in villages in the lower Marañón River Valley. Results of the various studies are available at Puinamudt, "Acodecospat."
67. See "Amazonía de negro."
68. Peru's labor inspection agency investigated the case. See "Sunafil."
69. See OEFA, "OEFA multa a Petroperú."
70. See Osinergmin, "Osinergmin multa con s/. 12 640 000 a Petroperú."
71. OEFA, Resolución Directorial.
72. Comisión Investigadora, "Informe Final," 300.
73. Comisión Investigadora, "Informe Final," 361.
74. Comisión Investigadora, "Informe Final," 378.
75. On the IACHR visit to areas affected by oil spills in the Peruvian Amazon, see OAS, "IACHR Carries Out Visit." On its precautionary measures for Cuninico and San Pedro, see OAS, "IACHR Grants Precautionary Measure."
76. Bebbington and Scurrah, "Hydrocarbon Conflicts," 185.
77. Bebbington and Bury, "Political Ecologies," 7.
78. Delgado-Pugley, "Contesting the Limits," 170.
79. Andrade, "The Government of Nature," 125–27.
80. See D'Andrea, "Can the River Spirit"; Gómez Rojas, "Corte Suprema."
81. See Baraer et al., "Glacier Recession," 134–50.
82. Guzmán-Gallegos, "*Ñukanchi Ñawpa Timputa Yarisha Tantarishun*," 45.

BIBLIOGRAPHY

Alvarez Alonso, José. "Peru: Los sobrevivientes." Servindi. Accessed December 28, 2017. https://www.servindi.org/actualidad/cronica/5163.

"Amazonía de negro: Derrame de crudo afecta afluente del Marañón." *El Comercio*, February 13, 2016. https://elcomercio.pe/peru/amazonia-negro-derrame-crudo-afecta-afluente-maranon-272580.

Anaya, J. *Informe del Relator Especial sobre la situación de los derechos humanos y las libertades fundamentales de los indígenas*. UN Doc. A/HRC/15/37. July 19, 2010. http://

unsr.jamesanaya.org/annual-reports/report-to-the-human-rights-council-a-hrc-15-37-19-july-2010.

Andina Agencia Peruana de Noticias. "19 postores presentaron ofertas para obtener buena pro de 17 lotes de hidrocarburos." September 8, 2008. http://andina.pe/agencia/noticia.aspx?id=193063.

Andrade A., Pablo. "The Government of Nature: Post-Neoliberal Environmental Governance in Bolivia and Ecuador." In *Environmental Governance in Latin America*, edited by F. de Castro et al. London: Palgrave Macmillan, 2016.

Associated Press. "U.S. Court Rules for Chevron in Ecuador Rainforest-Damage Case." *Los Angeles Times*, August 8, 2016. http://www.latimes.com/business/la-fi-chevron-ecuador-20160808-snap-story.html.

Baraer, Michel, et al. "Glacier Recession and Water Resources in Peru's Cordillera Blanca." *Journal of Glaciology* 58, no. 207 (2012): 134–50.

Bebbington, A., and J. Bury. "Political Ecologies of the Subsoil." In *Subterranean Struggles: New Dynamics of Mining, Oil, and Gas in Latin America*, edited by A. Bebbington and J. Bury. Austin: University of Texas Press, 2013.

Bebbington, A., and M. Scurrah. "Hydrocarbon Conflicts and Indigenous Peoples." In *Subterranean Struggles: New Dynamics of Mining, Oil, and Gas in Latin America*, edited by A. Bebbington and J. Bury. Austin: University of Texas Press, 2013.

Caldwell, K. L., et al. "Measurement Challenges at Low Blood Lead Levels." *Pediatrics* 140, no. 2 (August 2017). https://pediatrics.aappublications.org/content/140/2/e20170272.

Campanario Baqué, Y., and C. Doyle. *El Daño No Se Olvida: Impactos socioambientales en los pueblos indígenas de la Amazonía Norperuana afectados por las operaciones de la empresa Pluspetrol.* Lima: Centro de Políticas Públicas y Derechos Humanos Perú Equidad, 2017.

Chevron Corporation. "Texaco Petroleum, Ecuador and the Lawsuit against Chevron." Accessed September 10, 2019. https://www.chevron.com/~/media/chevron/ecuador/documents/texacopetroleumecuadorlawsuit.

Cleary, David. "Towards an Environmental History of the Amazon: From Prehistory to the Nineteenth Century." *Latin American Research Review* 36, no. 2 (2001): 65–96.

Collyns, Dan. "Indigenous Peruvians Win Amazon Pollution Payout from U.S. Oil Giant." *The Guardian*, March 5, 2015. https://www.theguardian.com/environment/2015/mar/05/indigenous-peruvians-amazon-pollution-settlement-us-oil-occidental.

Comisión Investigadora Multipartidaria para Determinar las Responsabilidades de los Funcionarios y Personas Naturales e Instituciones Públicas y Privadas que Resulten Responsables por los Derrames de Petróleo Ocurridos en el Oleoducto Norperuano. "Informe Final." November 2, 2017. http://www2.congreso.gob.pe/Sicr/TraDocEstProc/Expvirt_2011.nsf/ExpeNotesComInv/D8893A587ABD8E72052583C3006FD61C.

Cure Valdivieso, Salima. "Cuidado te mochan la cabeza: Circulación y construcción de un rumor en la frontera amazónica de Colombia, Perú y Brasil." Master's thesis. Universidad Nacional de Colombia, Leticia, 2005.

D'Andrea, Ariella. "Can the River Spirit Be a Person in the Eye of the Law?" *International Water Law Project Blog*. Accessed April 6, 2018. https://www.internationalwaterlaw.org/blog/2018/03/26/can-the-river-spirit-be-a-person-in-the-eye-of-the-law/.

de Castro, F., et al., eds. *Environmental Governance in Latin America*. London: Palgrave Macmillan, 2016.

Delgado-Pugley, D. "Contesting the Limits of Consultation in the Amazon Region: On Indigenous Peoples' Demands for Free, Prior and Informed Consent in Bolivia and Peru." *Revue général de droit* 43 (2013): 151–81.

Echeverri, J. A. "To Heal or to Remember: Indian Memory of the Rubber Boom and Roger Casement's 'Basket of Life.'" *ABEI Journal* 12 (2010): 49–64.

Finer, Matt, et al. "Ecuador's Yasuní Biosphere Reserve: A Brief Modern History and Conservation Challenges." *Environmental Research Letters* 4, no. 034005 (August 24, 2009): 1–15. http://stacks.iop.org/erl/4/034005.

Finer, M., et al. "Oil and Gas Projects in the Western Amazon: Threats to Wilderness, Biodiversity, and Indigenous Peoples." *PLoS ONE* 3, no. 8 (August 2008): 1–9.

Finer, M., and Orta-Martínez, M. "A Second Hydrocarbon Boom Threatens the Peruvian Amazon: Trends, Projections, and Policy Implications." *Environmental Research Letters* 5, no. 014012 (2010): 1–10.

Fontaine, Guillaume. *El precio del petróleo: Conflictos socio-ambientales y gobernabilidad en la región amazónica*. Quito: FLACSO Ecuador, 2003.

Fraser, Barbara. "Protest against Oil Pollution Ends with Accords on Pipeline Inspection, Remediation, Compensation." *Mongabay*, December 28, 2016. https://news.mongabay.com/2016/12/protest-against-oil-pollution-ends-with-accords-on-pipeline-inspection-remediation-compensation/.

Gamboa, C., V. Cueto, and J. Dávila. *¿El Estado Peruano cumplió con Camisea? Diagnóstica final sobre el cumplimiento de los Compromisos del Perú en el Ámbito del Proyecto Camisea*. Lima: Derecho, Ambiente y Recursos Naturales, 2008.

Goldman, E. S., L. La Torre López, and M. L. Ramos. *A Legacy of Harm: Occidental Petroleum in Indigenous Territory in the Peruvian Amazon*. Lima: EarthRights International, Racimos de Ungurahui, Amazon Watch, WWF Peru, 2007.

Gómez Rojas, Germán. "Corte Suprema ordena protección inmediata de la Amazonía Colombiana." República de Colombia, Corte Suprema de Justicia, April 5, 2018. http://www.cortesuprema.gov.co/corte/index.php/2018/04/05/corte-suprema-ordena-proteccion-inmediata-de-la-amazonia-colombiana/.

Government of Peru, Ministerio de Salud. Comisión Intersectorial para la Prevención y Mitigación de la Contaminación por Plomo y Otros Metales Pesados. "Visita para el Conocimiento de la Evaluación de la Calidad Sanitaria de los Recursos Hídricos y Muestreo Biológico de Comunidades en la Cuenca del Río Corrientes." Lima, May 2006. http://www.minsa.gob.pe/portalMinsa/destacados/archivos/242/RIO%20CORRIENTES.pdf.

Guzmán-Gallegos, María Antonieta. *"Ñukanchi Ñawpa Timputa Yarisha Tantarishun": Surcando nuestra memoria*. Lima: University of Oslo, Norway-FORMABIAP-AIDESEP, 2017.

Huertas Castillo, B. *Indigenous Peoples in Isolation in the Peruvian Amazon: Their Struggle for Survival and Freedom*. Copenhagen: IWGIA, 2004.

Hvalkof, Soren. "Outrage in Rubber and Oil: Extractivism, Indigenous Peoples, and Justice in the Upper Amazon." In *People, Plants and Justice: The Politics of Nature Conservation*, edited by Charles Zerner. New York: Columbia University Press, 1999.

Inter-American Court of Human Rights. *Case of the Kichwa Indigenous People of Sarayaku v. Ecuador*. Judgment of June 27, 2012 (Merits and Reparations). Accessed August 22, 2019. https://www.escr-net.org/caselaw/2012/pueblo-indigena-kichwa-sarayaku-vs-ecuador.

La Torre López, Lily. *All We Want Is to Live in Peace: Lessons Learned from Oil Operations in Indigenous Territories of the Peruvian Amazon*. Amsterdam: IUCN, Racimos de Ungurahui, 1999.

Martin, Pamela L., and Imme Scholz. "Policy Debate: Ecuador's Yasuní-ITT Initiative: What Can We Learn from Its Failure?" *International Development Policy / Revue internationale de politique de développement*, no. 5.2 (2014). http://journals.openedition.org/poldev/1705.

Murphy, Michelle. "Distributed Reproduction, Chemical Violence, and Latency." *Scholar & Feminist Online*, no. 11.3 (Summer 2013). http://sfonline.barnard.edu/life-un-ltd-feminism-bioscience-race/distributed-reproduction-chemical-violence-and-latency/.

OAS. "IACHR Carries Out Visit to Peruvian Amazon Region." Press release. July 25, 2017. http://www.oas.org/en/iachr/media_center/PReleases/2017/105.asp.

———. "IACHR Grants Precautionary Measure for the Communities of Cuninico and San Pedro in Loreto, Peru." Press release. December 12, 2017. http://www.oas.org/en/iachr/media_center/PReleases/2017/205.asp.

OEFA. "El OEFA sanciona a Petroperú S.A. por derrame de petróleo en Cuninico y ordena remediación efectiva e inmediata de las zonas afectadas." Accessed July 9, 2019. https://www.oefa.gob.pe/noticias-institucionales/el-oefa-sanciona-a-petroperu-s-a-por-derrame-de-petroleo-en-cuninico-y-ordena-remediacion-efectiva-e-inmediata-de-las-zonas-afectadas.

———. "Fondo Nacional del Ambiente del Perú: Lista de Sitios Impactados Prioizados." Accessed July 9, 2019. https://www.oefa.gob.pe/avisos/fondo-nacional-del-ambiente-del-peru-lista-de-sitios-impactados-priorizados.

———. "OEFA multa a Petroperú con 49 millones de soles por derrames de petróleo ocurridos en Amazonas y Loreto." Accessed July 9, 2019. https://www.oefa.gob.pe/noticias-institucionales/oefa-multa-a-petroperu-con-49-millones-de-soles-por-derrames-de-petroleo-ocurridos-en-amazonas-y-loreto.

———. Resolución Directorial No. 844-2015-OEFA/DFSAI Expediente No. 1306-2014-OEFA/DFSAI/PAS. Accessed August 22, 2019. https://www.oefa.gob.pe/noticias-institucionales/el-oefa-ordena-a-petroperu-s-a-remediar-area-impactada-por-el-derrame-de-petroleo-ocurrido-en-el-oleoducto-norperuano.

Oficina Nacional de Evaluación de Recursos Naturales. *Inventario y evaluación de recursos naturales de la microregión Pastaza-Tigre*. Lima: Oficina Nacional de Evaluación de Recursos Naturales, 1984.

O'Rourke, D., and S. Connolly. "Just Oil? The Distribution of Environmental and Social Impacts of Oil Production and Consumption." *Annual Review of Environment and Resources* 28 (2003): 587–617.

Orta-Martínez, Martí, and Matthew Finer. "Oil Frontiers and Indigenous Resistance in the Peruvian Amazon." *Ecological Economics* 70 (2010): 207–18.

Orta-Martínez, Martí, et al. "First Evidences of Amazonian Wildlife Feeding on Petroleum-Contaminated Soils: A New Exposure Route to Petrogenic Compounds?" *Environmental Research* 160 (2018): 514–17.

Orta-Martínez, Martí, et al. "Impacts of Petroleum Activities for the Achuar People of the Peruvian Amazon: Summary of Existing Evidence and Research Gaps." *Environmental Research Letters* 2, no. 045006 (2007): 1–10.

Osinergmin. "Osinergmin multa con s/. 12 640 000 a Petroperú por incumplir programa de adecuación de ductos." Accessed July 9, 2019. http://www.osinergmin.gob.pe/empresas/hidrocarburos/noticia/grifos/Osinergmin-multa-con-12640000-soles-aPetroperu-por-incumplir-programa-de-adecuacion-de-ductos.

Panorama. "Manto negro en Iquitos: Derrame de Petroperú y su insólita medida." July 20, 2014. https://panamericana.pe/panorama/nacionales/160500-manto-negro-iquitos-derrame-petroperu-insolita-medida.

Puinamudt. "Acodecospat: Informes ambientales del Marañón y Reserva Pacaya Samiria." Accessed July 9, 2019. http://observatoriopetrolero.org/acodecospat-informes-ambientales-del-maranon-y-reserva-pacaya-samiria/.

Riederer, Rachel. "An Uncommon Victory for an Indigenous Tribe in the Amazon." *New Yorker*, May 15, 2019. https://www.newyorker.com/news/news-desk/an-uncommon-victory-for-an-indigenous-tribe-in-the-amazon.

Sabin, Paul. "Searching for Middle Ground: Native Communities and Oil Extraction in the Northern and Central Ecuadorian Amazon, 1967–1993." *Environmental History* 3 (1998): 144–68.

Santos Granero, Fernando, and Frederica Barclay. *La Frontera Domesticada: Historia económica y social de Loreto, 1850–2000*. Lima: Pontifical Catholic University of Peru, 2002.

Schauwecker, S., et al. "Climate Trends and Glacier Retreat in the Cordillera Blanca, Peru, Revisited." *Global and Planetary Change* 119 (August 2014): 85–97.

Sotero, M. "A Conceptual Model of Historical Trauma: Implications for Public Health Practice and Research." *Journal of Health Disparities Research and Practice* 1, no. 1 (Fall 2006): 93–108.

Stoll, Aurelién. *Impactos Petroleros en Territorios Indígenas: Experiencias del Programa de Vigilancia Territorial del río Corrientes*. Iquitos, Peru: FECONACO and SHINAI, 2011. https://observatoriopetrolero.org/wp-content/uploads/2013/05/FECONACO-Impactos-petroleros-2011.pdf.

"Sunafil investiga empleo de niños durante derrame de petróleo en la selva." *Gestión*, March 6, 2016. https://gestion.pe/economia/sunafil-investiga-ninos-derrame-petroleo-selva-113517.

Tello Imaina, Leonardo, and Barbara Fraser. "Rubber Barons' Abuses Live On in Memory and Myth." *Sapiens*, July 1, 2016. https://www.sapiens.org/culture/rubber-era-myths/.

U.S. Centers for Disease Control and Prevention. "Blood Lead Levels in Children Fact Sheet." Accessed March 28, 2018. https://www.cdc.gov/nceh/lead/acclpp/lead_levels_in_children_fact_sheet.pdf.

———. "Cadmium Factsheet." Accessed February 25, 2018. https://www.cdc.gov/biomonitoring/Cadmium_FactSheet.html.

Valdivia, G. "The 'Amazonian Trial of the Century': Indigenous Identities, Transnational Networks and Petroleum in Ecuador." *Alternatives: Global, Local, Political* 32, no. 41 (2007): 41–72.

Vásquez, P. I. "Extractive Industries and Conflicts in Peru: An Agenda for Action." Internal paper prepared as part of the UN-EU Partnership on Natural Resources and Conflict Prevention, based on a weeklong field mission conducted in October 2010. https://www.un.org/es/land-natural-resources-conflict/pdfs/Peru%20background%20paper%20UN%20EU%20partnership%20october%202010.pdf.

Watts, Jonathan. "New Round of Oil Drilling Goes Deeper into Ecuador's Yasuní National Park." *The Guardian*, January 10, 2018. https://www.theguardian.com/environment/2018/jan/10/new-round-of-oil-drilling-goes-deeper-into-ecuadors-yasuni-national-park.

World Health Organization, International Programme on Chemical Safety. "Lead." Accessed February 25, 2018. http://www.who.int/ipcs/assessment/public_health/lead/en/.

Yusta-García, Raúl, et al. "Water Contamination from Oil Extraction Activities in Northern Peruvian Amazon Rivers." *Environmental Pollution* 225 (2017): 370–80.

Part 2

Macro-Development and Marginalization

This section focuses on three types of projects that are generally described as bringing development to remote parts of the Amazon basin but that often perpetuate inequities. The displacement, corruption, and other problems associated with hydroelectric power development in Brazil are an example of the difficulties posed by major infrastructure projects. The TIPNIS controversy in Bolivia highlights the environmental and cultural impacts of road construction, as well as conflicts over such projects within a country. This section also examines "green economy" projects that target forest communities, highlighting the policy contradictions and unintended consequences of the United Nations program on Reducing Emissions from Deforestation and Forest Degradation.

3 Environmental Justice and Brazil's Amazonian Dams

PHILIP M. FEARNSIDE

"Environmental justice" refers mainly to unequal environmental impacts on human groups based on differences such as race, ethnicity, and income. Definitions of both the "environmental" and the "justice" portions of the term are varied and continually evolving.[1] Amazonian dams clearly fall within any definition of the concept. These dams particularly affect those who live along and depend upon Amazonian rivers, in other words, traditional residents such as indigenous peoples and *ribeirinhos* (traditional riverside dwellers). Benefits of the dams accrue to urban consumers and especially to industries. The export of electricity in the form of electro-intensive commodities such as aluminum illustrates inequality on an international scale, with Amazonia and its people paying the environmental price for consumption and industrial employment in countries that would themselves no longer accept the impacts of the kind unleashed by these dams.

Amazonian dams (map 1) have a history of impacts causing environmental injustice. Part of this is inherent in this energy option: concentrated impacts on riverside residents and indigenous peoples in the Amazonian interior versus diffuse benefits to distant beneficiaries. This aspect is often written off by dam builders with the shibboleth "You have to break a few eggs to make an omelet." Of course this logic is much easier to apply when the eggs to be broken refer to poor people spread along Amazonian rivers far from the centers of power and political influence.

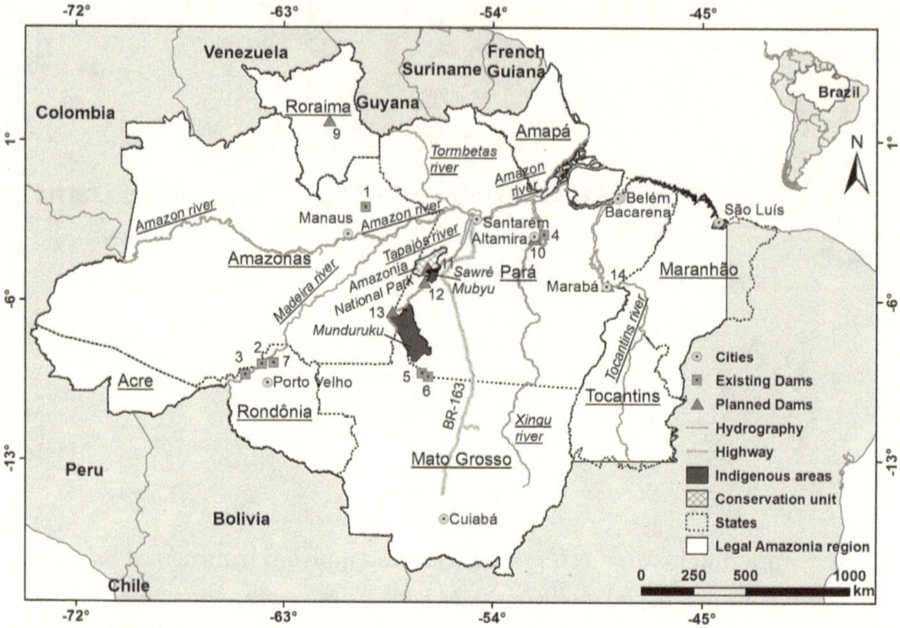

1. Brazil's legal Amazon region and locations mentioned in the text. Existing dams: (1) Balbina, (2) Santo Antônio, (3) Jirau, (4) Belo Monte, (5) São Manoel, (6) Teles Pires, (7) Samuel. Planned dams: (8) Cachoeira Ribeirão, (9) Cachoeira Porteira, (10) Babaquara/Altamira, (11) São Luiz do Tapajós, (12) Jatobá, (13) Chacorão, (14) Marabá. Created by author.

Ironically, one frequently hears discourse from the electrical sector arguing that more dams are needed because millions of Brazilians live without electricity. However, this argument bears little relation to electricity distribution in Brazil. Rural electrification has not been a high priority in government budgets, and recent advance of the Light for Everyone (Luz para Todos) program represents a minuscule fraction of the country's electricity use and an even smaller fraction of the portion of the country's electricity that is connected to the national grid and hence to hydroelectric dams. A poignant example of the traditional low priority for rural electrification is the Tucuruí Dam, completed in 1984, where twenty-nine years later twelve thousand families around the reservoir still had no access to electricity.[2] High-voltage transmission lines carry most of the dam's power directly to aluminum smelters in Barcarena, Pará, and São Luis, Maranhão.[3] Hydroelectric companies advertise dams by trumpeting astronomical figures for the number of homes that can be supplied with power from the dams. However, most electricity in Brazil is not for domestic use, which accounts for 22 to 29 percent of the total depending on the year.[4] The fact that the largest share of electricity from dams goes to industry is not advertised.

This kind of injustice could be reduced somewhat by better measures to resettle the displaced people and replace their livelihoods, but the fundamental structure of the injustice is not altered. In addition to this inherent type of injustice, Amazonian dam projects have shown a consistent pattern of human-rights violations and of actions without consideration for social and environmental impacts in general.

Brazil's existing Amazonian dams can be divided into those that were completed or under construction during the country's 1964–85 military dictatorship and those built in more recent times under the current environmental-licensing system, which began on January 23, 1986. The impacts of the earlier dams are often dismissed by Brazil's electrical sector as irrelevant—mistakes of the past that would not recur today under a democratic government with an environmental licensing system. Unfortunately these cases are still highly relevant, and the basic structure of decision making has changed little. Decisions are no longer made by a small group of military officers; they are instead made by a small group of officials in the Civil House of the presidential office and in the Ministry of Mines and Energy. The real decision to build a dam is made long before the environmental studies are done, therefore with no input on social and environmental impacts even if these considerations were given the importance they deserve. The licensing procedures that follow, such as drafting the environmental impact study and holding public hearings, merely legalize the dam project, with the maximum effect being suggestions for alteration of mitigation programs and minor adjustments in design rather than the existence of the project as such.

Amazon Dams and Decision Making

Brazil's Amazon dams have a notoriously poor record as examples of rational decisions on building infrastructure and of governance before, during, and after the dams are built. Dams initiated during Brazil's 1964–85 military dictatorship, such as Tucuruí, Balbina, and Samuel, provide examples that are still highly relevant today.[5] During the military dictatorship environmental studies were done, although not under the current system of federal licensing that began in 1986. (Works in progress such as Balbina and Samuel were exempted.) These environmental studies were done while the dams were under construction, the decision to build them having been made years earlier.

The record has not improved much since implementation of the current licensing process and since the advent of guarantees for the environment and human rights included in Brazil's 1988 Constitution.[6] The inadequacy of the initial decision making, the environmental licensing, and the subsequent mitigation measures are manifold in the cases of the Santo Antônio Dam (reservoir filled in 2011) and the Jirau Dam (filled in 2013), both on the Madeira River.[7] Licensing was approved over the formal objections of the technical staff of the Brazilian Institute for the Environment and Renewable Natural Resources (Instituto Brasileiro do Meio Ambiente e dos Recursos Naturais Renováveis, IBAMA).[8] South America's richest inland fishery has been destroyed, and conflicts with the thousands of displaced fishermen and women continue. The murder of the outspoken leader of one of the fishing cooperatives remains unsolved: the body of Nilce de Souza Magalhães, known as "Nicinha," was found weighted down with rocks at the bottom of the Jirau Reservoir.[9]

In the case of the Belo Monte Dam, on the Xingu River, licensing was approved over the formal objections of IBAMA's technical staff by replacing the head of the agency.[10] Impacted indigenous people were not consulted, a violation of International Labor Organization Convention 169 and Brazilian Law No. 5051 of April 19, 2004.[11] This resulted in over twenty legal suits still awaiting judgment despite the dam's having been allowed to continue to completion on the strength of a one-person decision by the head of Brazil's Supreme Court.[12] Postconstruction mitigation has been disastrous by any measure.[13] Fishing populations lost their livelihoods in the reservoir area and in a hundred-kilometer "reduced flow" stretch below the main dam, a stretch that includes two indigenous lands plus an area fished by a third indigenous group that resides in a tributary. The 2001 murder of another dam-opposition leader, Ademir Albeu Federicci, known as "Dema," also remains unsolved.[14]

The Tapajós River dams have followed a similar path.[15] The Teles Pires Dam, filled in 2015, flooded the Sete Quedas rapids, which was the most sacred site of the Munduruku indigenous people who live along the Tapajós.[16] The São Manoel Dam, under construction since 2014, is only seven hundred meters from the Kaiabí Indigenous Land and has provoked a series of conflicts.[17] In 2012 Federal Police invaded a Kaiabí village and opened fire on the population, killing Adenilson Kirixi Munduruku.[18]

Planned Dams

Many of the officially planned dams in Brazilian Amazonia have evident impacts. For example, the Marabá Dam, which is considered to be "in execution" by Brazil's Program for the Acceleration of Growth, would displace tens of thousands of people (with estimates as high as forty thousand), mostly traditional riverside dwellers known as *ribeirinhos*.[19]

Some of the greatest impacts would result from "unofficially planned" dams, these being dams that have featured in past official plans but that currently have disappeared from publicly announced plans. One is the Babaquara Dam (renamed the Altamira Dam but best known by its original name), which is one of five dams in the Xingu River basin that were long planned to store water upstream of Belo Monte in order to run the 11,000 megawatts of installed capacity in Belo Monte's main powerhouse during the low-flow season.[20] The natural river flow is insufficient to turn even one of the eighteen turbines in the main powerhouse for three months of the year and is sufficient for only a few turbines for several other months.[21] Climate change and deforestation are projected to reduce Belo Monte's power output substantially.[22] Even without these impacts on the Xingu River's future flow, this dam would be financially unviable without the massive subsidies it has received from Brazilian taxpayers through the National Bank for Social and Economic Development (Banco Nacional de Desenvolvimento Econômico e Social, BNDES). Only these subsidies, combined with political pressure, were sufficient to induce a consortium of companies to invest in the dam.[23] BNDES financed 80 percent of the construction cost with loans at 4 percent annual interest, while the federal government finances itself by selling bonds at 10 percent annual interest.[24] Funds from taxpayers in Europe and North

America have also contributed through World Bank development policy loans to BNDES.[25]

The financial unviability of Belo Monte without the upstream dams is a strong indication that at least the Babaquara/Altamira Dam continues as part of the real plan.[26] The Belo Monte Dam by itself defies basic economic logic.[27] Other indications include the history of plans for the number of turbines to be installed in Belo Monte; in 2003 officials considered installing only half of the total 11,233-megawatt capacity, which would be more compatible with operations assuming an unregulated river flow, but ultimately they reverted to the original design, which had been planned when the intention to build the upstream dams was publicly admitted.[28] Another indication that Babaquara/Altamira continues in undeclared plans was President Dilma Rousseff announcement in 2013 of a switch in policy to favor dams with "big reservoirs" rather than run-of-river dams like Belo Monte, a policy change that was confirmed by the President Michel Temer's administration in 2016.[29] The Xingu River is the obvious place planned for development with large storage dams.

The Chacorão Dam on the Tapajós River is a case with a strong similarity to Babaquara/Altamira as an elephant in the room that is absent from official discussions. Chacorão would flood 11,700 hectares of the Munduruku Indigenous Land. This dam no longer appears in official plans for hydropower, but the Tapajós Waterway (*hidrovia*) that is planned to carry soybeans from northern Mato Grosso to ports with access to the Amazon River continues to be a government priority, and the barges would not be able to cross the Chacorão rapids without the dam.[30]

Obstacles to Environmental Justice

The Current Decision-Making System

The presumption that what is forbidden by Brazil's laws or by the Constitution will simply not happen in real life is very naïve. After all, the Belo Monte Dam was described by the Federal Public Ministry (public prosecutors charged with defending the people's interests) in Belém as "totally illegal," but it now stands on the Xingu River as a concrete fact.[31]

The current decision-making system in Brazil represents a major obstacle to incorporating concerns for environmental justice into government planning and policy. Decisions to build dams and other infrastructure that provoke obvious injustices are common, in part due to the influence of political donations and outright corruption from parties with financial interests in the projects. In 2013 Brazil's Supreme Electoral Court released data for the first time on political donations, indicating that in the previous ten years the top four donors were large construction companies that build dams in Amazonia.[32] The Belo Monte Dam provides a well-documented example where both those on the paying and those on the receiving end have made affidavits affirming both legal and illegal donations.[33] The Santo Antônio Dam on the Madeira River also yielded illegal campaign donations to the ruling political party, according to confessions by the Odebrecht construction firm.[34]

The decision-making system also tends to ignore environmental justice concerns because the real decisions on infrastructure and other major projects are made by a handful of people before information on environmental and social impacts has been collected, and usually before any public discussion of the projects in question. The system needs to be reformed so that information can be collected and discussed publicly before the real decisions are made.

Brazil's environmental licensing system is another impediment. Licensing should not be confused with decision making: in practice, the current licensing system essentially serves to ratify what has already been decided on political grounds. The licensing allows alterations in mitigation and compensation requirements and the addition of small changes in construction plans, but virtually never extends to questioning the wisdom of the project's existence.

Erosion of Environmental Licensing

Federal licenses are required for major infrastructure projects such as large dams. However, various trends are eroding the influence of this requirement. One is a trend to decentralize licensing, transferring progressively more authority from the federal to the state level. Compared to the federal government, the state governments are more directly influenced by local economic and political interests and are less subject to scrutiny by the press or by environmental nongovernmental organizations. An example is the redefinition of a "large" dam in Brazil in 2004: prior to that year all dams with 10 megawatts or more of installed capacity were considered large and were required to have a federal license. This limit was raised to 30 megawatts, creating a class of dams between 10 and 30 megawatts, called "small hydroelectric centers," that require only a state license. Dams of this size have substantial impacts, and they are being built by the hundreds with little public visibility. Because state licenses are much cheaper and easier to obtain than federal ones, there is often a tendency for hydroelectric companies to invest in multiple small hydroelectric centers rather than in a large dam when the choice exists. In 2011 a further weakening occurred when a law was enacted allowing large dams where impacts are considered to be restricted to a single state to be licensed by the government of the state in question.[35] In 2019 the Sinop Dam in Mato Grosso was completed under this new loophole.[36]

Another trend weakening licensing is to accelerate the approval process, often irrespective of the completeness of the information or of the compliance of the project proponent with demands that may have been made by the environmental agency. One example of a measure to accelerate approvals is a June 2017 change in IBAMA's procedure for granting pay bonuses to its technical staff. Previously bonuses were granted on the basis of the number of technical opinions (*pareceres*) produced, a policy that introduces a motive for staff to accelerate the approval process even if corners are cut. The 2017 change creates an even stronger bias, by granting the bonus only if the technical opinions are favorable, not if they recommend against approving a project.[37]

Political pressure on top environmental officials to approve high-priority hydroelectric projects has been common; this happened for the Santo Antônio, Jirau, and Belo Monte dams.[38] In all of these cases government prosecutors threatened the individual technical staff members in the licensing department of IBAMA with prosecution for "bad faith" in recommending against approval of the licenses.[39] To minimize such threats, the signatures of the technical staff no longer appear on the technical opinions, such as the opinion recommending against approval of the São Manoel Dam.[40]

Strategies to Evade Control

One strategy used by dam-building interests to obtain approval of dams with severe environmental-justice impacts is to deny the existence of plans for these dams until approval is obtained for other dams in a complex of projects on the same river. The Babaquara/Altamira Dam on the Xingu River and the Chacorão Dam on the Tapajós River are examples of this, where public denial of plans appears to be disinformation, or a deliberate spreading of false information.[41] Another is the Cachoeira Riberão Dam on the Madeira River (also known as the Guajará Mirim or Binacional Dam). In 2006, while preparations were under way for the Santo Antônio and Jirau Dams, the head of the Oderbrecht construction firm in Porto Velho told this author that his staff were forbidden to discuss Cachoeira Riberão until the other dams were approved.

Another strategy is to simply not divulge plans for controversial dams. In December 1987 Brazilian Electrical Centers (Centrais Elétricas Brasileiras, ELETROBRÁS) released its 2010 Plan (after the plan had leaked to the public). This document presented dams with over 100 megawatts of installed capacity to be built by the year 2010, but also included a list of dams planned for construction in other years. A total of seventy-nine of these dams were listed for the Legal Amazonia region, including five dams in the Xingu basin upstream of Belo Monte (which was then known as Kararaô).[42] A storm of criticism ensued, and ELETROBRÁS has never again released information on planned dams without being limited to a relatively short timeline, such as the ten-year plans and occasional reports for longer periods (twenty or thirty years).

Dams that have vanished from the announced plans can resurge years later. An example is the Cachoeira Porteira Dam, planned for the Trombetas River. Environmental impact statements demonstrated that *quilombos* (communities of descendants of escaped African slaves) would be affected, which may explain why the dam vanished from official plans.[43] *Quilombolas* (members of quilombo communities) have the same rights as indigenous peoples, in accord with Brazil's current Constitution, yet the ease with which the Belo Monte Dam gained congressional approval appears to have encouraged other dam plans that impact quilombolas and indigenous people.[44] In addition to quilombolas, the Cachoeira Porteira

Dam would displace communities of traditional ribeirinhos and Brazil nut gatherers; it would also affect Amazonia's largest turtle breeding beach, located downstream of the dam.[45] The Cachoeira Porteira Dam reappeared in official plans in 2014.[46] In January 2019 President Jair Bolsonaro's administration announced the Barão do Rio Branco program, to which this dam is central and which includes a highway opening the Trombetas River basin and fifteen planned dams, several of them impacting indigenous lands and quilambolas.[47]

Projects such as these are known as "vampire projects" because, like a vampire asleep in his coffin, they can rise up later; also like vampires, which can be killed only with a stake through the heart, these projects are very hard to cancel definitively. A major case that appears to have become a vampire project is the São Luiz do Tapajós Dam, which would flood part of the Sawre Muybu indigenous area.[48] On April 19, 2016, the head of IBAMA "archived" the licensing process, with the approval of the minister of environment.[49] However, heads of IBAMA and ministers of environment change frequently, and it is quite probable that the new occupants of these offices could "de-archive" the licensing process.[50] ELETROBRÁS officials continued to plan for the dam after the licensing was archived, and the company's 2017–26 ten-year plan states, "The process that involves this dam continues to be accompanied by EPE [Energy Research Enterprise, of the Ministry of Mines and Energy] and, once all environmental questions are resolved, it can become part of the basket of candidates for expansion of [the] energy supply."[51] In September 2018 the National Agency for Electrical Energy requested that the National Water Agency grant a "reservation of water availability in the federal domain" for the São Luiz do Tapajós Dam.[52] These developments suggest that those in power expected the environmental licensing system as a whole to be scrapped by the National Congress. The removal of licensing barriers is now even more likely under Bolsonaro, who took office in January 2019.

Legislative Proposals to Gut the Licensing System

"Ruralists," or representatives of large landholders, have quickly risen to control the Brazilian government. Although large landholders represent a minuscule percentage of Brazil's population, of the 513 members of the

Chamber of Deputies (the house of the National Congress where representation is proportional to population), an estimated 210 (41 percent) were ruralists during the Temer administration, and a similar percentage continues in the Bolsonaro administration.[53]

This voting block has been the critical part of the coalition supporting presidents since the mid-2000s. In addition to obtaining support by means of laws and infrastructure projects that benefit agribusiness at the expense of environmental justice and other concerns, the ruralists have been able to place their people in many of the key appointed positions in the government. The power of the ruralist block was dramatically demonstrated during the 2011–12 passage of the reform that gutted Brazil's 1965 Forest Code, a package of regulations governing deforestation. On the day of the vote in the Chamber of Deputies, the environmentalists José Cláudio Ribeiro da Silva and Maria do Espírito Santo da Silva were murdered near Marabá, in southern Pará; when a deputy from the Green Party tried to speak of this during his address from the Chamber podium, he was shouted down by the multitude of deputies opposed to the Code.[54] The initial vote approved gutting the Code by a ratio of seven to one.[55] This is extraordinary, given that 85 percent of Brazil's population is urban and therefore has no financial interest in being allowed to deforest steep hillsides, near rivers, and elsewhere. In fact at the time of the vote, public opinion polls indicated that 80 percent of the population was opposed to any change in the Forest Code.[56] The rise in influence of the ruralists is explained by the fact that soybeans and other agricultural commodities make up a greatly increased percentage of the country's gross national product, impressing government policymakers who are also responsive to the ruralists' generous political campaign contributions, both legal and illegal.[57]

The ruralist block in the National Congress is pushing a series of bills for legislation and proposed constitutional amendments (PECs) that would cause serious environmental and social impacts. Constitutional amendments are much easier to pass in Brazil than in most other countries: the current Constitution, which was enacted in October 1988, had been amended ninety-seven times by December 2017. Since the ruralists' power is sufficient to block a presidential impeachment, this, along with distraction from other congressional business, offered opportunities for the ruralist agenda to make great advances during the impeachment trial of President Rousseff in 2016 and in the weeks preceding two congressional votes on initiating investigations for impeachment of President Temer in 2017.[58]

Proposals include measures to weaken or effectively abolish environmental licensing, such as PEC-65, which would make the mere submission of an environmental impact study an automatic approval of any development project, and proposed laws PLS-654/2015 and PL-3.729/2004, which would eliminate two of the current three required licenses and would establish impossible deadlines for IBAMA to approve the projects, after which the license would be approved automatically.[59]

One major blow to environmental licensing was approval of Law 13,334 of September 13, 2016 (formerly Provisional Measure 727), establishing a program of "public-private partnerships" with a governing board that was given the power to force IBAMA, National Indian Foundation (Fundação Nacional do Índio, FUNAI), or any other agency to approve any project that the board considers to be "strategic."[60] Yet another blow was the approval in December 2016 of PEC-95 (formerly PEC-55), freezing the federal budget for the next twenty years at the current level, a budget that has already been reduced to the point where agencies such as IBAMA and FUNAI are in many aspects inoperable.[61] Only the Ministries of Health and Education are allowed to have increases in their shares of the budget pie, which means that as demands inevitably increase, the Ministry of Environment and other agencies will shrink further, making proper enforcement of environmental laws unviable. In addition to inadequate allocations in the federal budget for less powerful sectors such as the Ministry of Environment, Brazil has a unique system that results in the real amounts spent in these sectors being even less than the budget would suggest. One feature is *contingenciamento*, placing the funds in contingency status when tax revenues are insufficient to pay the promised allocations. The budget amounts are not released at one time at the beginning of the year but rather are released in installments as the year progresses. The ministries are therefore pitted against each other in a continual struggle to get their installments released, and not all are successful. Another feature of the system is that less powerful ministries are likely to have much of their money released in the closing days of the fiscal year, when it is bureaucratically unviable to complete the lengthy bidding process needed to make most expenditures, and the funds are therefore returned to the treasury unspent.

Other proposals that are moving forward include one to open indigenous lands to mining (PL-1.610/1996).[62] There is also the notorious PEC-215, which would remove the authority of the Chico Mendes Institute for the Conservation of Biodiversity (Instituto Chico Mendes de Conservação da Biodiversidade) to create "conservation units" and of FUNAI to create "indigenous lands."[63] This authority would be transferred to the National Congress, which is controlled by ruralists and thus would effectively end the creation or expansion of protected areas. Various moves to reduce existing protected areas are progressing, such as removal of one million hectares from protected areas in the southern part of the state of Amazonas and the removal of protection in 2017 from 346,000 hectares along the Santarém-Cuiabá Highway in Pará.[64]

Removal of Protected Areas as Barriers to Dams

Protected areas, including indigenous lands, represent a barrier to hydroelectric development in many locations, but this barrier is progressively weakening. Parts of reserves can be removed to make way for dams, as was done, initially though presidential provisional measures (*medidas provisórias*), for the planned São Luiz do Tapajós Dam for five protected areas, including Amazonia National Park.[65] The degazetting and downgrading of reserves is an increasing threat throughout Brazilian Amazonia.[66]

Dam plans can also block creation of new protected areas, as has occurred for an extractive reserve for ribeirinhos in the area of the planned Jatobá Dam on the Tapajós River and the Sawre Muybu indigenous area that would be partially flooded by São Luiz do Tapajós, and where, despite the 2016 decree initiating the process of creating an official indigenous land, the area has yet be officially demarcated on the ground and *homologado* (confirmed by a higher authority).[67]

In addition to gaining concessions through the legislature, ruralists also make gains by convincing the president to issue medidas provisórias.

Ruralists are pressing for such a measure to allow "partnerships" between indigenous peoples and nonindigenous actors, thus allowing areas in indigenous lands to be rented out to agribusiness.[68] This would be catastrophic both for the forest and for the indigenous people. The opening of indigenous lands for leasing by agribusiness companies was proposed by President Bolsonaro and reinforced on February 13, 2019, when the ministers of agriculture and environment visited and praised an illegal soybean plantation in an indigenous land in Mato Grosso.[69] Indigenous lands protect more Amazon forest than do the "conservation units" that are administered by the Chico Mendes Institute for the Conservation of Biodiversity in the Ministry of Environment.[70] Indigenous lands have so far been the most effective type of protected area in avoiding deforestation, but this effectiveness cannot be taken for granted.[71]

Security Suspensions

An infallible means for dam proponents to circumvent restrictions on social and environmental impacts, including environmental justice concerns, is the "security suspension." This device stems from a law enacted during Brazil's military dictatorship (Law 4348 of June 26, 1964) that has been maintained and broadened in the intervening years (Law 8437 of June 30, 1992; Law 12,016 of August 7, 2009).[72] It allows any judge to overturn any judicial decision blocking a project if stopping the project would cause "grave damage to the public economy." Since hydroelectric dams are invariably important for the economy, decisions can be overruled regardless of how many laws, constitutional protections, or international conventions have been violated. By 2014 security suspensions had been invoked twelve times in the case of the Belo Monte Dam and fourteen times in the case of the Tapajós dams.[73] They have been used several times since in both cases. The Brazilian public is generally unaware of the existence of security suspensions, thus guaranteeing a complete lack of pressure on lawmakers to revoke the security suspension laws.[74]

Brazil's court system has proved incapable of enforcing relevant laws such as Law 5051 of April 19, 2004, which implements International Labor Organization Convention 169 requiring consultation of indigenous peoples.[75] The legality of Belo Monte is contested in over twenty still-undecided suits from the Public Ministry (a public prosecutor's office created by the 1988 Constitution to defend the interests of the people); in the meantime the dam has been built. In the one case that reached a court decision, the ruling was in favor of the indigenous people.[76] But the presidential administration appealed the ruling to the Supreme Court, and, after receiving four government representatives and none from civil society, Chief Justice Ayres Britto ruled that the dam construction could continue, pending a decision on the merits of the case.[77] The ruling was made without consulting the other members of the court, just fifteen days before Britto's retirement date, and in the midst of trying the "Mensalão" corruption scandal.[78] Since then the Belo Monte case has not even appeared on the radar for inclusion in the Supreme Court's agenda, and the dam has been built in practice.

Alternatives to Dams

Energy alternatives abound in Brazil, making the government's plans for many more dams in Amazonia unnecessary. The plans themselves are based on flawed projections, the costs and construction times of dams being systematically underestimated, and Brazil's assumed future energy demands are based on wildly unrealistic projections of economic growth.[79] One obvious step is for Brazil to stop exporting electricity in the form of electro-intensive commodities like aluminum, which generate very little employment in Brazil while wreaking great damage through the dams built to supply these industries.[80]

A highly visible efficiency option in Brazil is to stop using electric showerheads to heat bathwater. These are officially estimated to consume 5 percent of all of the electricity in Brazil.[81] Brazil has great potential for heating bathwater with solar heaters without use of electricity, and since the 2008 National Program for Climate Change it has been a government objective to phase out electric showerheads, but essentially nothing has been done; instead the priority continues to be building more hydroelectric dams.[82]

Brazil's inefficient electrical transmission system, which wastes 20 percent of the power transmitted, could be greatly improved without the impact of building more dams.[83] The country also has enormous undeveloped wind and solar resources, yet in January 2016 the president vetoed the inclusion of any funding for "non-hydraulic renewable energy" in the next five-year development plan.[84]

Needed Reforms

Environmental justice concerns need to be incorporated into the decision-making and licensing systems such that the current pattern of building dams and other projects regardless of these impacts is ended. Subterfuges need to be eliminated that currently allow projects to go forward even when legal protections are violated, especially the security suspension laws.

A profound reform of the decision-making process is needed such that likely social and environmental impacts are studied, democratically debated, and given real consideration before the critical decisions are made on dam construction.[85] Note again that the real decisions on dam construction are not the same as the officially designated ones in the licensing process, which, in practice, only ratify prior decisions that have been made in the absence of information, public debate, and consideration of social and environmental consequences.

These changes alone are insufficient. Brazil, and many other countries where similar situations apply, must also pursue alternatives to dams and other development modes that inherently provoke environmental injustices and other unacceptable impacts.

The environmental justice concerns provoked by Amazonian dams are of a type that is qualitatively different from many of the other concerns over dams, such as their financial cost or their legality. Environmental justice concerns have been shown to be important in bringing about change both at individual and societal levels.[86] For these changes to take hold in Brazilian decision making on Amazonian dams, it is essential that these concerns not only be formally documented, studied, and explained by academics and others but also that they be understood and internalized by society at large.

Possible Policy Change on Amazon Dams

A development in January 2018 offered hope of a change in policy on energy priorities, although it is improbable that Amazon dams will be foregone.

Two key officials—the executive secretary of the Ministry of Mines and Energy and the head of the Energy Research Enterprise, which is the agency under the Ministry of Mines and Energy responsible for energy planning—stated that the heavy environmental and social impacts of Amazon dams mean that priority should shift to other energy sources, such as wind and solar.[87] However, there was no indication that any of the planned Amazonian dams that are listed for construction by 2026 would be canceled.[88]

The stated motive of heavy impacts may not be the key factor, as economic realities would logically lead to the same policy shift.[89] Contrary to frequent claims by dam proponents, hydropower is not "cheap energy," even ignoring the nonfinancial cost of social and environmental impacts. The financial cost of dealing with social and environmental impacts has increased greatly in Brazil over the past twenty years, making it an important factor in the worsening economics of building new Amazonian dams.[90] A worldwide survey shows the normal pattern to be for dams to have much higher financial costs and to take much longer to begin generating power than is thought at the time construction decisions are made.[91] This is supported by the Madeira River dams and Belo Monte, which cost more than double the amounts officially expected.[92]

Uneconomic dams like Belo Monte have gone forward with massive government subsidies that are connected to the role of construction companies in financing political campaigns (both legally and illegally) and to simple corruption at the level of individuals.[93] This influence is presumably decreasing due to the prohibition of campaign contributions from corporations beginning with the elections of 2016, reaction to the ongoing "Lava Jato" corruption investigations, and the diminished cash reserves of BNDES.[94] The August 2017 announcement of the government's intent to privatize ELETROBRÁS by the end of 2018 could also be a factor in the change; there is uncertainty concerning this privatization since 70 percent of Brazil's population is opposed to it, and there are unresolved legal issues.[95] As a government company, ELETROBRÁS is headed by political appointees and has been subject to political pressure to promote uneconomic dams that would never be considered by a private company.[96] An increase in economic rationality is desirable, but it is important to remember that the invisible hand of the economy is not always kind to issues of environmental and social impacts. Incorporation of environmental justice concerns is also essential to prevent the problems so often seen in Brazil's Amazonian dams. It would be a great advance if the January 2018 statements were transformed into a change in the role of these impacts in decision making.

The chances of these statements being acted upon diminished further in April 2018, when both of the officials who had made the statements were removed from their posts when a new minister of mines and energy was appointed. The new minister, Wellington Moreira Franco, was under investigation in the *lava jato* (car wash) corruption probe, and his appointment was widely viewed as a means of giving him the protection of the "privileged forum," whereby ministers can be tried only by the eternally backlogged Supreme Court and are therefore, in practice, effectively immune from prosecution, along with any other individuals who might be named should the minister turn state's evidence in exchange for a light sentence.[97] When asked to explain why he removed the two technical experts who had questioned the priority for dams, he limited his reply to saying "Positions here are not permanent."[98] Since then the Bolsonaro administration has only decreased further the chances of dams with high environmental and social impacts being foregone.

Setbacks under Bolsonaro

The October 2018 election of Jair Bolsonaro, who became president on January 1, 2019, signals significant downgrading of protections both for the environment and for human rights.[99] Bolsonaro's statement during the campaign that he would withdraw Brazil from the United Nations because "it is a gathering of communists, of people who have no commitment to South America," was later clarified as referring only to the UN Commission on Human Rights.[100] His campaign attacks on environmental restrictions included frequent promises to strip IBAMA of its licensing power and distribute that authority to the ministries in each subject area, such as the Ministry of Mines and Energy in the case of dams.[101] He also promised to abolish the Environment Ministry altogether and transfer its functions to the Agriculture Ministry, but after taking office he was convinced by prominent ruralists to maintain the Environment Ministry in order not to provoke restrictions by countries importing Brazil's agricultural products.[102] However, Bolsonaro achieved the same effect by transferring the deforestation monitoring and enforcement functions to the Agriculture Ministry, which is headed by a ruralist congresswoman known as the "muse of the poison" for her role in championing the removal of restrictions on agrochemicals.[103] The Brazilian Forest Service was also moved from the Environment to the Agriculture Ministry.[104] What remained of the Environment Ministry was neutralized by appointing a ruralist as minister and a head of IBAMA who wants environmental licenses to be granted automatically by simply filling out an online form.[105] Another move to neutralize enforcing environmental laws is a change being drafted in the Environment Ministry's rules to make individual inspectors financially responsible if any infractions they have reported are later struck down, as by a court appeal; the measure is even retroactive for five years.[106] Meanwhile, since Bolsonaro's election the frequency of threats and attacks on IBAMA inspectors and on their vehicles has greatly increased.[107]

NGOs are a particular target of Bolsonaro, who has vowed to expel international environmental organizations such as Greenpeace and the World Wildlife Fund and has called for an end to "activists."[108] Ricardo Salles made one of his first acts as environment minister a ninety-day suspension of all projects the ministry had contracted with NGOs, after which some might be reinstated after a "fine-toothed comb" evaluation.[109]

Bolsonaro's denial of anthropogenic climate change and his campaign promises to abandon the Paris Agreement have important implications for deforestation, dams, and other developments in Amazonia.[110] Honoring Brazil's commitment to reduce its emissions by 43 percent below the 2005 level by 2030 would require reversing the trend to increasing deforestation rates that has prevailed since 2012.[111] The Paris commitment has been an important justification for the Environment Ministry's obtaining funds from the federal budget for its efforts to control forest loss; one study has calculated that deforestation rates could almost triple if all of Bolsonaro's proposals take effect.[112] During the 2018 campaign, with Bolsonaro well ahead in the polls and widespread euphoria among ruralists in anticipation of his victory, deforestation rates increased by 36 percent between June and September compared to the same months in 2017; the explosion of deforestation increased further just before the October 6 election, with a 49 percent increase in the August–October period as compared to 2017.[113]

Bolsonaro and his ministers have made contradictory statements regarding withdrawal from the Paris Agreement. This campaign promise became prominent after Bolsonaro's son Eduardo traveled to New York in August 2018 to meet with Steve Bannon, who is credited with convincing President Donald Trump to withdraw from the agreement.[114] Bolsonaro has appointed climate deniers to head the Environment Ministry and the Foreign Affairs Ministry, and both have abolished the sections within their ministry that dealt with climate change.[115]

A statement by Bolsonaro before the October 2018 runoff election was widely reported as a reversal of his intention of abandoning the Paris Agreement, but there was actually no reversal. What Bolsonaro said was that Brazil would remain in the agreement if "someone" could give him a written guarantee that there would be no "Triple A" project and no "independence of any indigenous area."[116] "Triple A" refers to a proposal by a Colombian NGO for a chain of protected areas linking the Andes to the Atlantic, which Bolsonaro believes is a foreign conspiracy to take Amazonia away from Brazil, and "independence of any indigenous area" refers to his belief that foreign governments are trying to convince indigenous peoples to declare independence from Brazil so that the conspiring governments can recognize the areas and gain control over Amazonia. Since no written guarantee can be expected regarding these imagined threats, Bolsonaro's promise to leave the Paris Agreement remained intact.[117] He later said that Brazil could stay in the Agreement, but only if it were changed to suit his demands.[118] Bolsonaro also characterized the Climate Convention's 2019 Conference of the Parties as a threat because "'Triple A' is at stake in this agreement," and after the election he (successfully) requested the lame-duck Temer administration to revoke Brazil's invitation to host the conference.[119]

After Bolsonaro took office, he was convinced to keep Brazil in the Paris Agreement "for now" as a result of pressure from some parts of his administration and from international sources, including President Emmanuel Macron of France, who required Brazil to remain in the Paris Agreement as a condition for France's support for a trade accord between the European Union and the Common Market of the South (Mercado Comum do Sul, Mercosul).[120] When corporate CEOs questioned Bolsonaro in a closed session at the World Economic Forum in Davos, Switzerland, on January 22, 2019, he said Brazil would remain in the Agreement, with the added qualification "for now."[121] The next day he made clear that there were strings attached, including payment to Brazil for emissions reductions and "respectful" treatment of the country.[122] When the vice president was questioned about Bolsonaro's position on the Paris Agreement he dodged by saying only that the president was aware that Brazil "cannot escape the question . . . of climate."[123] Bolsonaro's statements in Davos on the environment were contested by fact checkers and generally bore little relation to his actual actions in Brazil.[124]

Indigenous peoples and their lands have been important impediments to destruction of Amazonian forests and rivers by deforestation and dams.[125] They are also a target for Bolsonaro, who has characterized indigenous peoples as isolated in their "reserves" like "animals in zoos."[126] During his campaign, Bolsonaro promised not to allow demarcation of "a single centimeter" of additional indigenous lands.[127] He has transferred the responsibility for demarcating indigenous land from FUNAI to the Agriculture Ministry in a sector headed by a ruralist.[128] What remains of FUNAI has been moved from the Justice Ministry to the Ministry of Women, Family and Human Rights, headed by Damares Regina Alves, a preacher who represents the evangelical voting block that was a critical factor in Bolsonaro's election and that wants restrictions removed on proselytizing in indigenous areas.[129] In 2009 Atini-Voz Pela Vida, an evangelical organization cofounded by Alves, was denounced by indigenous leaders to the Human Rights Commission of the Chamber of Deputies for illegal adoption of indigenous children, and three cases against the organization are currently advancing through the courts.[130] Sergio Olímpio Gomes (known as "Major Olímpio"), an important Bolsonaro supporter who was elected senator from the state of São Paulo, made a campaign promise to "sell" the country's indigenous lands.[131] Legislative proposals by Bolsonaro supporters include making indigenous peoples "partners" in hydroelectric projects by offering tribal leaders royalties in exchange for supporting dams in their territories.[132] All of these developments would speed the construction of Brazil's planned Amazonian dams. The announcement of the Barão do Rio Branco plan in January 2019 indicates an acceleration of Amazonian hydroelectric projects, beginning with the Trombetas River.[133]

Conclusions

Brazil's Amazonian dams provoke severe social and environmental impacts and have shown a pattern of human rights violations. Environmental justice concerns should motivate a rethinking of this development strategy, shifting energy development to alternatives such as reduced use of electricity, cessation of energy export in the form of aluminum and other electro-intensive commodities, elimination of waste and inefficiency, and generation from wind and solar resources.

Past injustices provoked by existing Amazonian dams need to be given priority in government programs to reestablish the livelihoods and quality of life of affected populations. Safeguards need to be strengthened in Brazilian development and regulatory agencies and in both Brazilian and international financial institutions in order to avoid the kinds of environmental injustices illustrated by Amazonian dams.

NOTES

The author's research is financed by Conselho Nacional do Desenvolvimento Cientifico e Tecnológico (Proc. 304020/2010–9; 573810/ 2008–7), Fundação de Amparo a Pesquisa do Estado do Amazonas (Proc. 708565) and Instituto Nacional de Pesquisas da Amazônia (PRJ 15.125). The discussion of the Paris Agreement in the section on setbacks under President Bolsonaro is adapted from Fearnside, "Will President Bolsonaro Withdraw?"

1. Schlosberg, *Defining*; Schlosberg, "Theorising"; Schlosberg and Carruthers, "Indigenous Struggles."
2. "As margens da usina de Tucuruí."
3. Fearnside, "Social Impacts."
4. Bermann, "O setor de eletro-intensivos"; Fearnside, "Environmental and Social Impacts."
5. Fearnside, "Social Impacts"; Fearnside, "Environmental Impacts"; Magalhães et al., *Energia*; Fearnside, "Brazil's Balbina Dam"; Rodrigues and Fearnside, "Índios"; Fearnside, "Brazil's Samuel Dam."

6. Government of Brazil, Presidência da Republica, *Constituição.*
7. Fearnside, "Decision-making"; Fearnside, "Impacts"; Fearnside, "As barragens e as inundações"; Fearnside, "As barragens do rio Madeira."
8. Deberdt et al., "Parecer Técnico"; Fearnside, "Brazil's Madeira River Dams."
9. Aranda, "Nicinha"; Toledo, "Após 5 meses."
10. Government of Brazil, Instituto Brasileiro do Meio Ambiente e dos Recursos Naturais Renováveis, "Parecer Técnico No. 114/2009"; Government of Brazil, Instituto Brasileiro do Meio Ambiente e dos Recursos Naturais Renováveis, "Parecer Técnico No. 06/2010"; Fearnside, "Belo Monte Dam: A spearhead."
11. Government of Brazil, Presidência da Republica, "Decreto No 5.051."
12. International Rivers, "Supreme Court Judge."
13. Fearnside, "Belo Monte"; Fearnside, "Brazil's Belo Monte Dam"; Magalhães and da Cunha, *A expulsão de Ribeirinhos*; Villas-Bôas et al., *Dossiê Belo Monte.*
14. Switkes, "Leader."
15. Fearnside, "Brazil's São Luiz do Tapajós Dam"; Fearnside, "Amazon Dams and Waterways"; Alarcon, Millikan, and Torres, *Ocekadi*; Nitta and Naka, *Barragens do rio Tapajós*; de Sousa Júnior, *Tapajós.*
16. Branford and Torres, "End."
17. Branford and Torres, "Brazil's Indigenous Munduruku."
18. Forest Comunicações, "Trailer"; Silva, "Vídeo"; Sposati, "Por que a Polícia Federal matou."
19. Government of Brazil, Ministério do Planejamento, "Hidrelétricos"; Rodrigues and Ribeiro Júnior, "Construção."
20. Sevá Filho, "Works"; Fearnside, "Dams in the Amazon."
21. Fearnside, "Planned Disinformation."
22. Angelo and Feitosa, "País"; Margulis and Untersell, "Shaping Up"; Sorribas et al., "Projections"; Stickler et al., "Dependence."
23. Rojas and Millikan, "El BNDES."
24. Leitão, "Belo Monte's Avatar."
25. Bank Information Center, "World Bank"; Fearnside, "Belo Monte."
26. Fearnside, "Planned Disinformation."
27. de Sousa Júnior and Rei, "Uncertainties"; de Sousa Júnior, Reid, and Leitão, *Custos*; Fearnside, "Dams in the Amazon."
28. Pinto, "Corrigida."
29. Borges, "Dilma defende usinas"; Borges, "Diretor-geral"; Fearnside, "Dams with Big Reservoirs."
30. Government of Brazil, Ministério dos Transportes, "Diretrizes"; Fearnside, "Amazon Dams and Waterways."
31. Miotto, "Norte Energia."
32. Gama, "Maiores doadores."
33. "Delator deverá revelar propina"; do Amaral, "Anexo 07 Belo Monte."
34. Francis, "Delação da Odebrecht."

35. Government of Brazil, Presidência da Republica, "Lei Complementar N° 140."
36. Fearnside, "Brazil's Sinop Dam."
37. Borges, "Governo."
38. Fearnside, "Brazil's Madeira River Dams"; Fearnside, "Belo Monte"; Fearnside, "Brazil's Belo Monte Dam."
39. Advocacia-Geral da União, "Advocacia-Geral pede ao CNMP definição"; Advocacia-Geral da União, "AGU pede ao CNMP que sejam coibidos abusos."
40. Government of Brazil, Instituto Brasileiro do Meio Ambiente e dos Recursos Naturais Renováveis, "Parecer Técnico n° 93/2017"; Fearnside, "Amazon Dam Defeats."
41. Fearnside, "Planned Disinformation."
42. Government of Brazil, Centrais Elétricas Brasileiras S/A, *Plano 2010*; Fearnside, "Hydroelectric Dams in the Brazilian Amazon" Fearnside, "Dams in the Amazon."
43. Government of Brazil, Centrais Elétricas do Norte do Brasil, *Aproveitamento*; Farias, "Hidrelétricas."
44. Fearnside, "Belo Monte"; Fearnside, "Brazil's Belo Monte Dam."
45. Teixeira, *Energy Policy*, 253–317. Eisemberg et al., "Eventos climáticos extremos."
46. Comissão Pró-Índio de São Paulo, "Estudos."
47. Rocha, "Bolsonaro Government"; Comissão Pró-Índio de São Paulo, "Governo."
48. Fearnside, "Brazil's São Luiz do Tapajós Dam."
49. de Araújo, "Despacho 02001.018080/2016-41."
50. Fearnside, "A Hidrelétrica de São Luiz do Tapajós."
51. Nunes and Neder, "Eletrobrás"; Government of Brazil, Empresa de Pesquisa Energética, *Plano Decenal*, chapter 3, 61.
52. Alves, "Despacho."
53. Carvalho et al., "Em véspera."
54. Milanez, "Zé Claudio e Maria."
55. "Veja como os deputados votaram o Código Florestal."
56. Lopes, "Datafolha."
57. Fearnside and Figueiredo, "China's Influence"; Zucco and Lauderdale, "Distinguishing."
58. Fearnside, "Brazilian Politics." Fearnside, "Environmental Nightmare"; Fearnside, "Indigenous Lands."
59. Government of Brazil, Senado Federal, "Proposta"; Government of Brazil, Senado Federal, "Texto"; Government of Brazil, Câmara dos Deputados, "Projeto de Lei N°. . . , de 2004."
60. Government of Brazil, Presidência da Republica, "Lei N° 13.334."
61. Government of Brazil, Câmara dos Deputados, and Senado Federal, "Emenda Constitucional No–95."
62. Government of Brazil, Câmara dos Deputados, "PL-1.610/1996."
63. Government of Brazil, Senado Federal, "Proposta de Emenda à Constituição N°. . . , de 2012"; Instituto Socioambiental, "Impactos."
64. Fearnside and Lovejoy, "Political Attack"; Chagas, "Após veto"; Fearnside, "Environmental Nightmare."

65. Sanson, "Movimentos sociais."
66. Bernard, Penna, and Araújo, "Downgrading."
67. Fearnside, "Brazil's São Luiz do Tapajós Dam."
68. Borges, "Ruralistas."
69. Gonzales, "Brazil."
70. Nogueira et al., "Brazil's Amazonian protected areas"; Nogueira et al., "Carbon Stocks."
71. Nepstad et al., "Inhibition"; Vitel, Fearnside, and Graça, "Análise"; Fearnside, "Indigenous Peoples"; Fearnside and Ferraz, "Conservation"; Vitel et al., "Land-Use Change."
72. Government of Brazil, Presidência da Republica, "Lei nº 4.348"; Government of Brazil, Presidência da Republica, "Lei nº 8.437"; Government of Brazil, Presidência da Republica, "Lei nº12.016."
73. Palmquist, "Usina Teles Pires."
74. Fearnside, "Amazon Dams and Waterways."
75. Government of Brazil, Presidência da Republica, "Decreto No 5.051."
76. Silva and Santos, "O caso Belo Monte."
77. Britto, "Medida."
78. International Rivers, "Supreme Court Judge"; Peres, "Ayres Britto"; Sevá Filho, "Profanação hidrelétrica."
79. Ansar et al., "Should We Build?"; Prado et al., "How Much?"
80. Fearnside, "Environmental and Social Impacts."
81. Government of Brazil, Comitê Interministerial sobre Mudança do Clima, *Plano*, 58.
82. Costa, "Viabilidades."
83. Rey, "Um olhar."
84. Baitelo, "Energias renováveis"; Baitelo et al., *Revolução energética*; Instituto Socioambiental, "Dilma."
85. Fearnside, "Challenges."
86. Reese and Jacob, "Principles."
87. Ventura, "Fase."
88. Government of Brazil, Empresa de Pesquisa Energética, *Plano Decenal*; Fearnside, "Possível mudança"; International Rivers, "Era."
89. Branford, "Brazil Announces."
90. Hirata, "Custo."
91. Ansar et al., "Should We Build?"
92. Fearnside, "Belo Monte."
93. For instance, Branford, "Brazil Announces"; Fearnside, "Brazil's Belo Monte Dam."
94. Falcão, "STF proíbe"; "Lava Jato"; Lima and Vettorazzo, "BNDES."
95. Ventura, "Fase"; Hirata "Sete em cada"; Reuters, "Liminar."
96. Branford, "Brazil Announces."
97. Brandino, "Temer empossa."
98. Dias and Wiziack, "Sem dinheiro."
99. Fearnside, "Why Brazil's New President."

100. BBC-Brasil, "Bolsonaro presidente."
101. Masionnave, "Bolsonaro."
102. Bragança, "Bolsonaro"; Watanabe, "Bolsonaro."
103. Phillips, "Jair Bolsonaro"; Boldrini, "Ruralistas."
104. Albuquerque and Parreira, "Ministra da Agricultura."
105. Guerra and Ribeiro, "Indicado"; Borges, "Novo chefe."
106. Boghossian, "Ministério."
107. Masionnave, "Com cartazes."
108. ClimaInfo, "Mídia internacional"; Seto, "Bolsonaro."
109. Azevedo and Grandelle, "Ministério."
110. Darby, "Brazil"; Darby, "Meet"; Fearnside, "Amazonia"; Gaier, "Bolsonaro."
111. Government of Brazil, "Intended"; Fearnside, "Business as Usual."
112. Soterroni et al., "Future"; Soterroni et al., "Fate."
113. Observatório do Clima, "Desmatamento"; Masionnave, "Desmatamento."
114. Darby, "Brazil."
115. Bilenky, Fernandes, and Watanabe, "Aquecimento"; Di Cunto, Araújo, and Freitas, "Novo chanceler"; Alencastro, "Governo"; Estadão Conteúdo, "Itamaraty"; Tuffani, "Governo."
116. "Discurso de Bolsonaro é o 'mais ameaçador.'"
117. Fearnside, "Why Brazil's New President."
118. Soares and Grandelle, "Bolsonaro."
119. Marques and Fernandes, "Bolsonaro."
120. "Bolsonaro recua mais uma vez"; "Macron isola Bolsonaro."
121. Coelho, Neves, and Frias, "Brasil"; "Bolsonaro diz que 'por ora,' Brasil permanecerá."
122. *Bloomberg News*, "Bolsonaro."
123. Fernandes, "Mourão."
124. Frias, Coelho, and Neves, "Leia"; Rittle, "Discurso."
125. Nepstad et al., "Inhibition."
126. "Índios em reservas são como animais em zoológicos."
127. "Nem um centímetro a mais para terras indígenas."
128. Sassine, "Ruralista."
129. de Moura e Souza, "Igreja."
130. Balloussier and Linhares, "ONG"; Estadão Conteúdo, "ONG."
131. "Senador, Major Olímpio apoiará venda de reservas indígena."
132. Pamplona, "Produtores."
133. Comissão Pró-Índio de São Paulo, "Governo"; Rocha, "Bolsonaro Government."

BIBLIOGRAPHY

Advocacia-Geral da União. "Advocacia-Geral pede ao CNMP definição de limites para atuação de procuradores que ameaçam agentes técnicos do Estado." May 26, 2011. http://www.agu.gov.br/page/content/detail/id_conteudo/158804.

———. "AGU pede ao CNMP que sejam coibidos abusos de procuradores contra agentes técnicos do Estado." August 29, 2012. http://agu.gov.br/page/content/detail/id_conteudo/207524.

Alarcon, D. F., B. Millikan, and M. Torres, eds. *Ocekadi: Hidrelétricas, conflitos socioambientais e resistência na Bacia do Tapajós*. Brasília: International Rivers Brasil, Programa de Antropologia e Arqueologia da Universidade Federal do Oeste do Pará, 2016. https://www.internationalrivers.org/files/attached-files/tapajos_digital_0.pdf.

Albuquerque, L., and M. Parreira. "Ministra da Agricultura anuncia deputado Valdir Colatto como novo chefe do Serviço Florestal." *Globo*, January 16, 2019. https://g1.globo.com/politica/noticia/2019/01/16/ministra-da-agricultura-anuncia-deputado-valdir-colatto-como-novo-chefe-do-servico-florestal.ghtml.

Alencastro, C. "Governo acaba com secretaria dedicada a mudanças climáticas e gera temor entre cientistas." *O Globo*, January 7, 2019. https://oglobo.globo.com/sociedade/sustentabilidade/governo-acaba-com-secretaria-dedicada-mudancas-climaticas-gera-temor-entre-cientistas-23352452.

Alves, R. F. F. "Despacho." *Diário Oficial da União*, no. 186, section 1 (September 26, 2018): 84.

Angelo, C., and C. Feitosa. "País poderá viver drama climático em 2040, indicam estudos da Presidência." *Observatório do Clima*, October 30, 2015. http://www.observatoriodoclima.eco.br/pais-podera-viver-drama-climatico-em-2040/.

Ansar, A., B. Flyvbjerg, A. Budzier, and D. Lunn. "Should We Build More Large Dams? The Actual Costs of Hydropower Megaproject Development." *Energy Policy* 69 (2014): 43–56. https://doi.org/10.1016/j.enpol.2013.10.069.

Aranda, A. "Nicinha: Um corpo à espera uma voz que não se cala." *Amazônia Real*, July 18, 2016. http://amazoniareal.com.br/nicinha-um-corpo-a-espera-do-exame-de-dna-uma-voz-que-nao-se-cala/.

"As margens da usina de Tucuruí, 12 mil famílias vivem sem energia." *Folha de São Paulo*, January 7, 2013.

Azevedo, A. L., and R. Grandelle. "Ministério do Meio Ambiente suspende todos os convênios e parcerias com ONGs." *Extra Globo*, January 15, 2019. https://extra.globo.com/noticias/brasil/ministerio-do-meio-ambiente-suspende-todos-os-convenios-parcerias-com-ongs-23375079.html.

Baitelo, R. "Energias renováveis: Eólica e solar." In *Setor Elétrico Brasileiro e a Sustentabilidade no Século 21: Oportunidades e Desafios*, edited by P. F. Moreira. 2nd ed. Brasília: Rios Internacionais, 2012. http://www.internationalrivers.org/node/7525.

Baitelo, R., M. Yamaoka, R. Nitta, and R. Batista. *Revolução energética: A caminho do desenvolvimento*. São Paulo: Greenpeace Brasil, 2013. http://www.greenpeace.org/brasil/Global/brasil/image/2013/Agosto/Revolucao_Energetica.pdf.

Balloussier, A. V., and C. Linhares. "ONG de ministra é acusada de incitar ódio a indígenas e tirar criança de mãe." *Folha de São Paulo*, December 15, 2018. https://www1.folha.uol.com.br/poder/2018/12/ong-de-ministra-e-acusada-de-incitar-odio-a-indigenas-e-tirar-crianca-de-mae.shtml.

Bank Information Center. "World Bank Environmental Policy Loan to BNDES: Moving Money or Mainstreaming Environmental Sustainability?" In *IFI Infobrief*. Washington DC: BIC, September 2009.

BBC-Brasil. "Bolsonaro presidente: As propostas com as quais Jair Bolsonaro se elegeu presidente do Brasil." October 28, 2018. https://www.bbc.com/portuguese/brasil-46012309.

Bermann, C. "O setor de eletro-intensivos." In *Setor Elétrico Brasileiro e a Sustentabilidade no Século 21: Oportunidades e Desafios*, edited by P. F. Moreira. 2nd ed. Brasília: Rios Internacionais, 2012. http://www.internationalrivers.org/node/7525.

Bernard, E., L. A. O. Penna, and E. Araújo. "Downgrading, Downsizing, Degazettement, and Reclassification of Protected Areas in Brazil." *Conservation Biology* 28 (2014): 939–50. https://doi.org/10.1111/cobi.12298.

Bilenky, T., T. Fernandes, and P. Watanabe. "Aquecimento global é tema secundário, diz futuro ministro." *Folha de São Paulo*, December 10, 2018. https://www1.folha.uol.com.br/ambiente/2018/12/vamos-preservar-o-ambiente-sem-ideologia-diz-futuro-ministro-de-bolsonaro.shtml.

Bloomberg News. "Bolsonaro fala à Bloomberg em Davos; veja a íntegra." *BOL*, January 23, 2019. https://www.bol.uol.com.br/noticias/2019/01/23/bolsonaro-fala-a-bloomberg-em-davos-veja-a-integra.htm.

Boghossian, B. "Ministério quer punir fiscais que apliquem multas consideradas inconsistentes." *Folha de São Paulo*, January 21 2019. https://www1.folha.uol.com.br/ambiente/2019/01/ministerio-do-meio-ambiente-quer-punir-fiscais-que-apliquem-multas-consideradas-inconsistentes.shtml.

Boldrini, A. "Ruralistas festejam 'musa do veneno' em festa após aprovação de relatório sobre agrotóxicos." *Folha de São Paulo*, June 26, 2018. https://www1.folha.uol.com.br/mercado/2018/06/ruralistas-festejam-musa-do-veneno-em-festa-apos-aprovacao-de-relatorio-sobre-agrotoxicos.shtml.

"Bolsonaro diz que 'por ora,' Brasil permanecerá no acordo do clima de Paris." *J10*, January 22, 2019. http://g1.globo.com/globo-news/jornal-das-dez/videos/v/bolsonaro-diz-que-por- ora-brasil-permanecera-no-acordo-do-clima-de-paris/7320409/.

"Bolsonaro recua mais uma vez e mantém Brasil no Acordo de Paris." *Brasil247*, January 14, 2019. https://www.brasil247.com/pt/247/poder/380321/Bolsonaro-recua-mais-uma-vez-e-mantém-Brasil-no-Acordo-de-Paris.htm.

Borges, A. "Dilma defende usinas hidrelétricas com grandes reservatórios." *Valor Econômico*, June 6, 2013. http://www.valor.com.br/imprimir/noticia_impresso/3151684.

———. "Diretor-geral da ANEEL defende retorno de hidrelétricas com grandes reservatórios." *O Estado de São Paulo*, September 30, 2016. http://economia.estadao.com.br/noticias/geral,diretor-geral-da-aneel-defende-retorno-de-hidreletricas-com-grandes-reservatorios,10000078947.

———. "Governo liga pagamento de bônus para servidores a licenças ambientais." *O Estado de São Paulo*, June 29, 2017. http://sustentabilidade.estadao.com.br/noticias

/geral,governo-liga-pagamento-de-bonus-para-servidores-a-licencas-ambientais,70001869763.

———. "Novo chefe do Ibama quer licenciamento ambiental automático." *Terra*, December 21 2018. https://www.terra.com.br/economia/futuro-presidente-do-ibama-quer-licenciamento-ambiental-automatico,a8b3d565bc174aeeb35da55ae8653c3c3oxcjdku.html.

———. "Ruralistas negociam com governo MP para arrendar terras indígenas." *O Estado de São Paulo*, October 4, 2017. http://economia.estadao.com.br/noticias/geral,ruralistas-negociam-com-governo-mp-para-arrendar-terras-indigenas,70002027426.

Bragança, D. "Bolsonaro defende o fim do Ministério do Meio Ambiente." OECO, October 1, 2018. https://www.oeco.org.br/reportagens/bolsonaro-defende-o-fim-do-ministerio-do-meio-ambiente/.

Brandino, G. "Temer empossa 11 novos ministros; veja como fica a Esplanada." *Folha de São Paulo*, April 11, 2018. https://www1.folha.uol.com.br/poder/2018/04/temer-empossa-onze-novos-ministro-veja-como-fica-a-esplanada.shtml.

Branford, S. "Brazil Announces End to Amazon Mega-Dam Building Policy." *Mongabay*, January 3, 2018. https://news.mongabay.com/2018/01/brazil-announces-end-to-amazon-mega-dam-building-policy.

Branford, S., and M. Torres. "Brazil's Indigenous Munduruku Occupy Dam Site, Halt Construction." *Mongabay*, July 19, 2017. https://news.mongabay.com/2017/07/brazils-indigenous-munduruku-occupy-dam-site-halt-construction/.

Branford, S., and M. Torres. "The End of a People: Amazon Dam Destroys Sacred Munduruku 'Heaven.'" *Mongabay*, January 5, 2017. https://news.mongabay.com/2017/01/the-end-of-a-people-amazon-dam-destroys-sacred-munduruku-heaven/.

Britto, A. "Medida Cautelar na Reclamação 14.404 Distrito Federal." Supremo Tribunal Federal, 2012. http://www.stf.jus.br/arquivo/cms/noticiaNoticiaStf/anexo/rcl14404.pdf.

Carvalho, D., A. Boldrini, G. Uribe, B. Boghossian, M. Dias, and M. Prado. 2017. "Em véspera de votação, Temer afaga ruralistas e baixo clero." *Folha de São Paulo*, August 2, 2017. http://amazonia.org.br/2017/08/em-vespera-de-votacao-temer-afaga-ruralistas-e-baixo-clero/.

Chagas, P. V. "Após veto, governo envia ao Congresso novo projeto que reduz floresta no Pará." Amazonia, July 15, 2017. http://amazonia.org.br/2017/07/apos-veto-governo-envia-ao-congresso-novo-projeto-que-reduz-floresta-no-para/.

ClimaInfo. 2018. "Mídia internacional comenta as ameaças de Bolsonaro à floresta amazônica." October 10, 2018. http://climainfo.org.br/2018/10/10/climainfo-10-de-outubro-de-2018.

Coelho, L., L. Neves, and M. C. Frias. "Brasil ficará em acordo sobre o clima, afirma Bolsonaro." *Folha de São Paulo*, January 23, 2019. https://www1.folha.uol.com.br/ambiente/2019/01/bolsonaro-diz-a-executivos-que-brasil-ficara-no-acordo-do-clima.shtml.

Comissão Pró-Índio de São Paulo. "Estudos das hidrelétricas no Rio Trombetas são retomados sem consulta prévia." August 24, 2014. http://comissaoproindio.blogspot.com.br/2014/08/estudos-das-hidroeletricas-no-rio.html.

———. "Governo anuncia nova hidrelétrica na Amazônia que impactará Terras Indígenas e Quilombolas." January 23, 2019. http://cpisp.org.br/governo-anuncia-nova-hidreletrica-que-impactara-terras-indigenas-e-quilombolas/.

Costa, R. N. A. "Viabilidades Térmica, Económica e de Materiais de um Sistema Solar de Aquecimento de Água a Baixo Custo para Fins Residenciais." Master's thesis, Universidade Federal de Rio Grande do Norte, 2007. http://bdtd.bczm.ufrn.br/tedesimplificado//tde_arquivos/10/TDE-2008-02-21T011110Z-1119/Publico/RaimundoNAC.pdf.

Darby, M. "Brazil: Bolsonaro Threatens to Quit Paris Climate Deal." *Climate Exchange News*, August 14, 2018. http://www.climatechangenews.com/2018/08/14/brazils-bolsonaro-threatens-quit-paris-climate-deal/.

———. "Meet the Political Dynasty of Climate Science Deniers Threatening to Withdraw Brazil from the Paris Agreement." *DesmogUK*, August 15, 2018. https://www.desmog.co.uk/2018/08/15/meet-political-dynasty-climate-science-deniers-threatening-withdraw-brazil-paris-agreement.

de Araújo, S. M. V. G. "Despacho 02001.018080/2016-41 Gabinete da Presidência/IBAMA. Assunto: Processo nº 02001.003643/2009-77-AHE São Luiz do Tapajós. August 4, 2016." Brasília: IBAMA, 2016. http://www.mpf.mp.br/pa/sala-de-imprensa/documentos/2016/arquivamento.pdf.

Deberdt, G., I. Teixeira, L. M. M. Lima, M. B. Campos, R. B. Choueri, R. Koblitz, S. R. Franco, and V. L. S. Abreu. "Parecer Técnico No. 014/2007–FCOHID/CGENE/DILIC/IBAMA." Brasília: Instituto Brasileiro do Meio Ambiente e dos Recursos Naturais Renováveis-IBAMA, 2007. http://philip.inpa.gov.br/publ_livres/Dossie/Mad/Documentos%20oficiais/Madeiraparecer.pdf.

"Delator deverá revelar propina em Belo Monte." *Amazonas em Tempo*, March 7, 2015.

de Moura e Souza, M. "Igreja quer liberdade para converter Índios." *Valor Econômico*, December 11, 2018. https://www.valor.com.br/politica/6019979/igreja-quer-liberdade-para-converter-indios.

de Sousa Júnior, W. C., ed. *Tapajós: Hidrelétricas, Infraestrutura e Caos*. São José dos Campos: Instituto Tecnológico de Aeronáutica-ITA, 2014. http://www.riosvivos.org.br/arquivos/site_noticias_2134831519.pdf.

de Sousa Júnior, W. C., and J. Reid. "Uncertainties in Amazon Hydropower Development: Risk Scenarios and Environmental Issues around the Belo Monte Dam." *Water Alternatives* 3, no. 2 (2010): 249–68. www.water-alternatives.org/index.php/volume3/v3issue2/92-a3-2-15/file.

de Sousa Júnior, W. C., J. Reid, and N. C. S. Leitão. *Custos e Benefícios do Complexo Hidrelétrico Belo Monte: Uma Abordagem Econômico-Ambiental*. Lagoa Santa: Conservation Strategy Fund-CSF, 2006. http://conservation-strategy.org/sites/default/files/field-file/4_Belo_Monte_Dam_Report_mar2006.pdf.

Dias, M., and J. Wiziack. "Sem dinheiro na Eletrobrás, apagão vai ser inevitável." *Folha de São Paulo*, April 13, 2018. https://www1.folha.uol.com.br/mercado/2018/04/sem-dinheiro-na-eletrobras-apagao-vai-ser-inevitavel-diz-moreira-franco.shtml.

Di Cunto, R., C. Araújo, and C. Freitas. "Novo chanceler diz que esquerda criou 'ideologia da mudança climática.'" *Valor Econômico*, November 15, 2018. https://www.valor.com.br/politica/5985233/novo-chanceler-diz-que-esquerda-criou-%3Fideologia-da-mudanca-climatica.

"Discurso de Bolsonaro é o 'mais ameaçador' para Amazónia e Acordo de Paris." *Jornal Económico com Lusa*, November 4, 2018. https://jornaleconomico.sapo.pt/noticias/discurso-de-bolsonaro-e-o-mais-ameacador-para-amazonia-e-acordo-de-paris-373733.

do Amaral, D. "Anexo 07 Belo Monte." In *Termo de acordo de colaboração premiada. Petição 5952-22/02/2016*. Brasília: Supremo Tribunal Federal, 2016. https://drive.google.com/file/d/0BzuqMfbpwX4wYVJlak1qdmIyWUE/view.

Eisemberg, C. C., V. C. D. Bernardes, R. A. M. Balestra, M. B. O. Silva, and R. C. Vogt. "Eventos climáticos extremos relacionados ao ENSO e o sucesso reprodutivo da tartaruga-da-amazônia (*Podocnemis expansa*) na Reserva Biológica do Rio Trombetas: Relatório preparado para o Instituto Chico Mendes de Conservação da Biodiversidade (ICMBio)." Manaus: Instituto Nacional de Pesquisas da Amazônia-INPA, 2016. https://doi.org/10.13140/RG.2.2.12142.20806.

Estadão Conteúdo. "Itamaraty também deixa de ter uma divisão sobre mudança do clima." January 10, 2019. https://sustentabilidade.estadao.com.br/blogs/ambiente-se/itamaraty-tambem-deixa-de-ter-uma-divisao-sobre-mudanca-do-clima/.

———. "ONG de futura chefe da Funai foi denunciada por discriminação indígena." *Exame*, December 7, 2018. https://exame.abril.com.br/brasil/ong-de-futura-chefe-da-funai-foi-denunciada-por-discriminacao-indigena/.

Falcão, M. "STF proíbe doações de empresas para campanhas eleitorais e partidos." *Folha de São Paulo*, September 17, 2015. http://www1.folha.uol.com.br/poder/2015/09/1683012-stf-proibe-doacoes-de-empresas-para-campanhas-eleitorais-e-partidos.shtml?loggedpaywall.

Farias, E. "Hidrelétricas no rio Trombetas preocupam quilombolas e indígenas do Pará." *Amazônia Real*, April 21, 2014. http://amazoniareal.com.br/hidreletricas-no-rio-trombetas-preocupam-quilombolas-e-indigenas-do-para/.

Fearnside, P. M. "A Hidrelétrica de São Luiz do Tapajós: 22-Pós-escrito." *Amazônia Real*, December 12, 2016. http://amazoniareal.com.br/hidreletrica-de-sao-luiz-do-tapajos-22-pos-escrito/.

———. "Amazon Dam Defeats Brazil's Environment Agency." *Mongabay*, September 20, 2017. https://news.mongabay.com/2017/09/amazon-dam-defeats-brazils-environment-agency-commentary/.

———. "Amazon Dams and Waterways: Brazil's Tapajós Basin Plans." *Ambio* 44, no. 5 (2015): 426–39. https://doi.org/10.1007/s13280-015-0642-z.

———. "Amazonia and the Setbacks of Brazil's Political Moment." *Mongabay*, October 12, 2018. https://news.mongabay.com/2018/10/amazonia-and-the-setbacks-of-brazils-political-moment-commentary/.

———. "As barragens do rio Madeira como espada de Dâmocles." *Amazônia Real*, December 14 and 21, 2015. http://amazoniareal.com.br/as-barragens-do-rio-madeira-como-espada-de-damocles-1-mudancas-nas-enchentes/; http://amazoniareal.com.br/as-barragens-do-rio-madeira-como-espada-de-damocles-2-o-risco-para-porto-velho/.

———. "As barragens e as inundações no rio Madeira." *Ciência Hoje* 53, no. 314 (2014): 56–57. http://cienciahoje.uol.com.br/revista-ch/2014/314/barragens-e-inundacoes-no-rio-madeira.

———. "Belo Monte: Actors and Arguments in the Struggle over Brazil's Most Controversial Amazonian Dam." *Die Erde* 148, no. 1 (2017): 14–26. https://doi.org/10.12854/erde-148-27.

———. "Belo Monte Dam: A Spearhead for Brazil's Dam Building Attack on Amazonia?" GWF Discussion Paper 1210 (Canberra, Australia: Global Water Forum), 2012. http://www.globalwaterforum.org/wp-content/uploads/2012/04/Belo-Monte-Dam-A-spearhead-for-Brazils-dam-building-attack-on-Amazonia_-GWF-1210.pdf.

———. "Brazilian Politics Threaten Environmental Policies." *Science* 353 (2016): 746–48. https://doi.org/10.1126/science.aag0254.

———. "Brazil's Balbina Dam: Environment versus the Legacy of the Pharaohs in Amazonia." *Environmental Management* 13, no. 4 (1989): 401–23. https://doi.org/10.1007/BF01867675.

———. "Brazil's Belo Monte Dam: Lessons of an Amazonian Resource Struggle." *Die Erde* 148, nos. 2–3 (2017): 167–84. https://doi.org/10.12854/erde-148-26.

———. "Brazil's Madeira River Dams: A Setback for Environmental Policy in Amazonian Development." *Water Alternatives* 7, no.1 (2014): 156–69. http://www.water-alternatives.org/index.php/alldoc/articles/vol7/v7issue1/244-a7-1-15/file.

———. "Brazil's Samuel Dam: Lessons for Hydroelectric Development Policy and the Environment in Amazonia." *Environmental Management* 35, no. 1 (2005): 1–19. https://doi.org/10.1007/s00267-004-0100-3.

———. "Brazil's São Luiz do Tapajós Dam: The Art of Cosmetic Environmental Impact Assessments." *Water Alternatives* 8, no. 3 (2015): 373–96. http://www.water-alternatives.org/index.php/alldoc/articles/vol8/v8issue3/297-a8-3-5/file.

———. "Brazil's Sinop Dam Flouts Environmental Legislation." *Mongabay*, March 1, 2019. https://news.mongabay.com/2019/03/brazils-sinop-dam-flaunts-environmental-legislation-commentary/.

———. "Business as Usual: A Resurgence of Deforestation in the Brazilian Amazon." *Yale Environment 360*, April 18, 2017. http://e360.yale.edu/features/business-as-usual-a-resurgence-of-deforestation-in-the-brazilian-amazon.

———. "Challenges for Sustainable Development in Brazilian Amazonia." *Sustainable Development* 26, no. 2 (2018): 141–49. https://doi.org/10.1002/sd.1725.

———. "Dams in the Amazon: Belo Monte and Brazil's Hydroelectric Development of the Xingu River Basin." *Environmental Management* 38, no. 1 (2006): 16–27. https://doi.org/10.1007/s00267-005-00113-6.

———. "Dams with Big Reservoirs: Brazil's Hydroelectric Plans Threaten Its Paris Climate Commitments." *The Globalist*, January 29, 2017. http://www.theglobalist.com/dams-climate-change-global-warming-brazil-paris-agreement/.

———. "Decision-making on Amazon Dams: Politics Trumps Uncertainty in the Madeira River Sediments Controversy." *Water Alternatives* 6, no. 2 (2013): 313–25. http://www.water-alternatives.org/index.php/alldoc/articles/vol6/v6issue2/218-a6-2-15/file.

———. "Environmental and Social Impacts of Hydroelectric Dams in Brazilian Amazonia: Implications for the Aluminum Industry." *World Development* 77 (2016): 48–65. https://doi.org/10.1016/j.worlddev.2015.08.015.

———. "Environmental Impacts of Brazil's Tucuruí Dam: Unlearned Lessons for Hydroelectric Development in Amazonia." *Environmental Management* 27, no. 3 (2001): 377–96. https://doi.org/10.1007/s002670010156.

———. "Environmental Nightmare for the Amazon." *Alert*, August 22, 2017. http://alert-conservation.org/issues-research-highlights/2017/8/22/environmental-nightmare-for-the-amazon.

———. "Hydroelectric Dams in the Brazilian Amazon as Sources of 'Greenhouse' Gases." *Environmental Conservation* 22, no. 1 (1995): 7–19. https://doi.org/10.1017/s0376892900034020.

———. "Impacts of Brazil's Madeira River Dams: Unlearned Lessons for Hydroelectric Development in Amazonia." *Environmental Science & Policy* 38 (2014): 164–72. https://doi.org/10.1016/j.envsci.2013.11.004.

———. "Indigenous Lands at Risk, as Amazon Sellout by Brazil's Temer Continues." *Mongabay*, November 6, 2017. https://news.mongabay.com/2017/11/indigenous-lands-at-risk-as-amazon-sellout-by-brazils-temer-continues-commentary/.

———. "Indigenous Peoples as Providers of Environmental Services in Amazonia: Warning Signs from Mato Grosso." In *Global Impact, Local Action: New Environmental Policy in Latin America*, edited by A. Hall. London: University of London, School of Advanced Studies, Institute for the Study of the Americas, 2005.

———. "Planned Disinformation: The Example of the Belo Monte Dam as a Source of Greenhouse Gases." In *Brazil in the Anthropocene: Conflicts between Predatory Development and Environmental Policies*, edited by L.-R. Issberner and P. Lena. New York: Routledge, 2017.

———. "Possível mudança na política sobre barragens amazônicas." *Amazônia Real*, January 9, 2018. http://amazoniareal.com.br/possivel-mudanca-na-politica-sobre-barragens-amazonicas/.

———. "Social Impacts of Brazil's Tucuruí Dam." *Environmental Management* 24, no. 4 (1999): 483–95. https://doi.org/10.1007/s002679900248.

———. "Why Brazil's New President Poses an Unprecedented Threat to the Amazon." *Yale Environment 360*, November 8, 2018. https://e360.yale.edu/features/why-brazils-new-president-poses-an-unprecedented-threat-to-the-amazon.

———. "Will President Bolsonaro Withdraw Brazil from the Paris Agreement?" *Mongabay*, January 31, 2019. https://news.mongabay.com/2019/01/commentary-will-president-bolsonaro-withdraw-brazil-from-the-paris-agreement/.

Fearnside, P. M., and J. Ferraz. "A Conservation Gap Analysis of Brazil's Amazonian Vegetation." *Conservation Biology* 9, no. 5 (1995): 1134–47. https://doi.org/10.1046/j.1523-1739.1995.9051134.x.

Fearnside, P. M., and A. M. R. Figueiredo. "China's Influence on Deforestation in Brazilian Amazonia: A Growing Force in the State of Mato Grosso." In *China and Sustainable Development in Latin America: The Social and Environmental Dimension*, edited by R. Ray, K. Gallagher, A. López, and C. Sanborn. New York: Anthem Press, 2016.

Fearnside, P. M., and T. E. Lovejoy. "Political Attack on Amazonian Protected Areas." *Alert*, April 20, 2017. http://alert-conservation.org/issues-research-highlights/2017/4/20/political-attack-on-amazonian-protected-areas.

Fernandes, T. "Mourão diz que país não pode fugir da questão climática." *Folha de São Paulo*, 23 January 2019. https://www1.folha.uol.com.br/ambiente/2019/01/mourao-diz-que-brasil-nao-pode-fugir-da-questao-climatica.shtml.

Forest Comunicações. "Trailer do documentário 'O Complexo.'" YouTube, September 9, 2016. https://www.youtube.com/watch?v=1r53-axzv10.

Francis, T. "Delação da Odebrecht aponta irregularidades na construção das hidrelétricas do Madeira." *G1*, June 24, 2017. http://g1.globo.com/ro/rondonia/noticia/delacao-da-odebrecht-aponta-irregularidades-na-construcao-das-hidreletricas-do-madeira.ghtml.

Frias, M. C., L. Coelho, and L. Neves. "Leia a íntegra comentada do pronunciamento." *Folha de São Paulo*, January 23, 2019. https://www1.folha.uol.com.br/mundo/2019/01/leia-a-integra-do-discurso-de-jair-bolsonaro-no-forum-economico-mundial-em-davos.shtml.

Gaier, R. V. "Bolsonaro diz que pode retirar Brasil do Acordo de Paris se eleito." *Globo Extra*, September 3, 2018. https://extra.globo.com/noticias/brasil/bolsonaro-diz-que-pode-retirar-brasil-do-acordo-de-paris-se-eleito-23034957.html.

Gama, P. "Maiores doadores somam gasto de R$1 bi desde 2002: Construtores e bancos são principais financiadores de campanhas eleitorais." *Folha de São Paulo*, January 21, 2013. http://www1.folha.uol.com.br/fsp/poder/89730-maiores-doadoras-somam-gasto-de-r-1-bi-desde-2002.shtml.

Gonzales, J. "Brazil Wants to Legalize Agribusiness Leasing of Indigenous Lands." *Mongabay*, February 21, 2019. https://news.mongabay.com/2019/02/brazil-wants-to-legalize-agribusiness-leasing-of-indigenous-lands/.

Government of Brazil. "Intended Nationally Determined Contribution towards Achieving the Objective of the United Nations Framework Convention on Climate Change." 2016. https://www4.unfccc.int/sites/ndcstaging/PublishedDocuments/Brazil%20First/BRAZIL%20iNDC%20english%20FINAL.pdf.

Government of Brazil, Câmara dos Deputados. "PL-1.610/1996 Projeto de Lei." 1996. http://www.camara.gov.br/proposicoesWeb/fichadetramitacao?idProposicao=16969.

———. "Projeto de Lei Nº. . . , de 2004." 2004. http://www.camara.gov.br/proposicoesWeb/prop_mostrarintegra;jsessionid=E0835B55D081DCED29036318FAA6A7F1.proposicoesWebExterno2?codteor=225810&filename=PL+3729/2004.

Government of Brazil, Câmara dos Deputados, and Senado Federal. "Emenda Constitucional No–95." *Diário Oficial da União* 241 (December 16, 2016): 2–3. http://pesquisa.in.gov.br/imprensa/jsp/visualiza/index.jsp?jornal=1&pagina=2&data=16/12/2016.

Government of Brazil, Centrais Elétricas Brasileiras S/A. *Plano 2010: Relatório Geral. Plano Nacional de Energia Elétrica 1987/2010 (Dezembro de 1987).* Brasília: ELETROBRÁS, 1987.

Government of Brazil, Centrais Elétricas do Norte do Brasil. *Aproveitamento Hidrelétrico de Cachoeira Porteira. Relatório de Impacto Ambiental—RIMA.* Brasília: ELETRONORTE, 2 vols. 1988.

Government of Brazil, Comitê Interministerial sobre Mudança do Clima. *Plano Nacional sobre Mudança do Clima—PNMC—Brasil.* Brasília: Ministério do Meio Ambiente, 2008. http://www.mma.gov.br/estruturas/imprensa/_arquivos/96_01122008060233.pdf.

Government of Brazil, Empresa de Pesquisa Energética. *Plano Decenal de Expansão de Energia 2026.* 2 vols. Brasília: Ministério de Minas e Energia, 2017. http://www.epe.gov.br/pt/publicacoes-dados-abertos/publicacoes/Plano-Decenal-de-Expansao-de-Energia-2026.

Government of Brazil, Instituto Brasileiro do Meio Ambiente e dos Recursos Naturais Renováveis. "Parecer Técnico No. 114/2009–COHID/CGENE/DILIC/IBAMA, de 23/11/2009, Assunto: AHE Belo Monte. Ref: Análise técnica do Estudo de Impacto Ambiental do Aproveitamento Hidrelétrico Belo Monte, processo nº 02001.001848/2006-75." 2009. http://www.ibama.gov.br/licenciamento/index.php.

———. "Parecer Técnico No. 06/2010-COHID/CGENE/DILIC/IBAMA de 26 de janeiro de 2010. Assunto: Análise técnica das complementações solicitadas no Parecer nº 114/2009, referente ao Aproveitamento Hidrelétrico Belo Monte, processo nº 02001.001848/2006-75." 2010. http://www.ibama.gov.br/licenciamento/index.php.

———. "Parecer Técnico nº 93/2017-COHID/CGTEF/DILIC; Número do Processo: 02001.004420/2007-65; Interessado: EMPRESA DE ENERGIA SÃO MANOEL S.A." 2017. http://philip.inpa.gov.br/publ_livres/Dossie/S_Manoel/Docs_of/Parecer%2093_IBAMA_LO%20São%20Manoel_25ago2017.pdf.

Government of Brazil, Ministério do Planejamento. "Hidrelétricos-Marabá-MA PA TO." Programa de Aceleração do Crescimento, June 30, 2017. http://www.pac.gov.br/obra/8417.

Government of Brazil, Ministério dos Transportes. "Diretrizes da política nacional de transporte hidroviário." Secretaria de Política Nacional de Transportes. 2010. http://www2.transportes.gov.br/Modal/Hidroviario/PNHidroviario.pdf.

Government of Brazil, Presidência da Republica. *Constituição da República Federativa do Brasil de 1988.* Brasília: PR, 1988. http://www.planalto.gov.br/ccivil_03/constituicao/constituicaocompilado.htm.

———. "Decreto No 5.051, de 19 de abril de 2004." 2004. http://www.planalto.gov.br/ccivil_03/_ato2004–2006/2004/decreto/d5051.htm.

———. "Lei nº 4.348, de 26 de junho de 1964. Estabelece normas processuais relativas a mandado de segurança." 1964. https://arisp.files.wordpress.com/2008/12/boletim-ago-fev-1964-65_lei-n-4-348.pdf.

———. "Lei nº 8.437, de 30 de junho de 1992. Dispõe sobre a concessão de medidas cautelares contra atos do poder público e dá outras providências." 1992. http://legislacao.planalto.gov.br/legisla/legislacao.nsf/Viw_Identificacao/lei%208.437-1992?OpenDocument.

———. "Lei nº12.016, de 07 de agosto de 2009. Disciplina o mandado de segurança individual e coletivo e dá outras providências." 2009. http://www.planalto.gov.br/ccivil_03/_ato2007-2010/2009/lei/l12016.htm.

———. "Lei Complementar Nº 140, de 8 de dezembro de 2011. Fixa normas, . . . e altera a Lei No 6.938, de 31 de agosto de 1981." 2011. http://www.planalto.gov.br/CCIVIL_03/LEIS/LCP/Lcp140.htm.

———. "Lei Nº 13.334, de 13 de Setembro de 2016. Cria o Programa de Parcerias de Investimentos-PPI; altera a Lei nº 10.683, de 28 de maio de 2003, e dá outras providências." 2016. http://www.planalto.gov.br/ccivil_03/_ato2015-2018/2016/lei/L13334.htm.

Government of Brazil, Senado Federal. "Proposta de Emenda à Constituição Nº. . . , de 2012." 2016. http://www.senado.leg.br/atividade/rotinas/materia/getTexto.asp?t=120446&c=RTF&tp=1.

———. "Texto Final Projeto de Lei do Senado Nº 654, de 2015." 2015. http://legis.senado.leg.br/mateweb/arquivos/mate-pdf/183980.pdf.

Guerra, R., and A. Ribeiro. "Indicado para Meio Ambiente foi denunciado pelo MP por improbidade administrativa." *O Globo*, December 9, 2018. https://oglobo.globo.com/brasil/indicado-para-meio-ambiente-foi-denunciado-pelo-mp-por-improbidade-administrativa-23292920.

Hirata, T. "Custo socioambiental sobe e trava novas hidrelétricas." *Folha de São Paulo*, January 9, 2018. http://www1.folha.uol.com.br/mercado/2018/01/1948976-custo-ambiental-cresce-e-potencial-para-novas-hidreletricas-chega-ao-fim.shtml.

Hirata, T. "Sete em cada dez brasileiros são contra as privatizações." *Folha de São Paulo*, December 26, 2017. http://www1.folha.uol.com.br/mercado/2017/12/1945999-sete-em-cada-dez-brasileiros-sao-contra-as-privatizacoes.shtml.

"Índios em reservas são como animais em zoológicos, diz Bolsonaro." *G1*, November 30, 2018. https://g1.globo.com/sp/vale-do-paraiba-regiao/noticia/2018/11/30/indios-em-reservas-sao-como-animais-em-zoologicos-diz-bolsonaro.ghtml.

"Dilma veta energias renováveis não hidráulicas no Plano Plurianual." *Notícias Socioambientais*, January 21, 2016. http://www.socioambiental.org/pt-br/noticias-socioambientais/dilma-veta-energias-renovaveis-nao-hidraulicas-no-plano-plurianual.

———. "Impactos da PEC 215/200 sobre os povos indígenas, populações tradicionais e o meio ambiente." 2015. https://www.socioambiental.org/sites/blog.socioambiental.org/files/nsa/arquivos/isa_relatoriopec215-set2015.pdf.

International Labor Organization. "C169-Indigenous and Tribal Peoples Convention, 1989 (No. 169)." Geneva, Switzerland: ILO, 1989. http://www.ilo.org/dyn/normlex/en/f?p=normlexpub:12100:0::no::p12100_ilo_code:c169.

International Rivers. "Belo Monte Justice Now! Legal Campaign." November 6, 2012. https://www.internationalrivers.org/resources/belo-monte-justice-now-legal-campaign-7716.

———. "Era of Mega Hydropower in Brazilian Amazon Appears to Be Ending." January 4, 2018. https://www.internationalrivers.org/resources/press-release-era-of-mega-hydropower-in-brazilian-amazon-appears-to-be-over-16588.

———. "Supreme Court Judge Overturns Suspension of Belo Monte Dam." August 28, 2012. http://www.internationalrivers.org/resources/supreme-court-judge-overturns-suspension-of-belo-monte-dam-7656.

"Lava Jato: Empreiteiras pararam de doar a partidos." *Band Notícias*, July 19, 2015. http://noticias.band.uol.com.br/noticias/100000762100/lava-jato-empreiteiras-pararam-de-doar-a-partidos.html.

Leitão, M. "Belo Monte's Avatar." *International Rivers*, June 24, 2010. https://www.internationalrivers.org/resources/belo-monte%E2%80%99s-avatar-2762.

Lima, F., and L. Vettorazzo. "BNDES avisa não ter dinheiro para atender todos os pedidos da União." *Folha de São Paulo*, January 12, 2018. http://www1.folha.uol.com.br/mercado/2018/01/1949906-bndes-avisa-nao-ter-dinheiro-para-atender-todos-os-pedidos-da-uniao.shtml.

Lopes, R. J. "Datafolha indica que 80% rejeitam corte de proteção a matas." *Folha de São Paulo*, June 13, 2011. http://www1.folha.uol.com.br/ambiente/929142-datafolha-indica-que-80-rejeitam-corte-de-protecao-a-matas.shtml.

"Macron isola Bolsonaro: Sem acordo ambiental, não tem acordo comercial." *Brasil247*, November 29, 2018. https://www.brasil247.com/pt/247/mundo/376398/Macron-isola-Bolsonaro-sem-acordo-ambiental-não-tem-acordo-comercial.htm.

Magalhães, S. B., E. Castro, and R. Britto, eds. *Energia na Amazônia-Avaliação e Perspectivas Sócio-Ambientais*. 2 vols. Belém: Museu Paraense Emílio Goeldi-MPEG, Universidade Federal do Pará-UFPA and Associação de Universidades Amazônicas-UNAMAZ, 1996.

Magalhães, S. B., and M. C. da Cunha, eds. *A expulsão de Ribeirinhos em Belo Monte: Relatório da SBPC*. São Paulo: Sociedade Brasileira para o Progresso da Ciência-SBPC, 2017. http://portal.sbpcnet.org.br/livro/belomonte.pdf.

Margulis, S., and N. Untersell. "Shaping Up Brazil's Long-Term Development Considering Climate Change Impacts." In *Brazil in the Anthropocene: Conflicts between Predatory Development and Environmental Policies*, edited by L.-R. Issberner and P. Lena. New York: Routledge, 2017.

Marques, J., and T. Fernandes. "Bolsonaro diz ter pedido para não haver Conferência do Clima da ONU no Brasil." *Folha de São Paulo*, November 29, 2018. https://www1.folha.uol.com.br/ambiente/2018/11/bolsonaro-diz-que-pediu-para-nao-haver-conferencia-do-clima-no-brasil-em-2019.shtml.

Masionnave, F. "Bolsonaro Has Made Grim Threats to the Amazon and Its People." *Climate Home News*, October 8, 2018. http://www.climatechangenews.com/2018/10/08/bolsonaro-made-grim-threats-amazon-people/.

———. "Com cartazes pró-Bolsonaro, protesto de madeireiros obriga Ibama a fugir de cidade do AM." *Folha de São Paulo*, December 18, 2018. https://www1.folha.uol.com

.br/ambiente/2018/12/com-cartazes-pro-bolsonaro-protesto-de-madeireiros-obriga-ibama-a-fugir-de-cidade-do-am.shtml.

———. "Desmatamento na Amazônia explode durante período eleitoral" *Folha de São Paulo*, November 11, 2018. https://www1.folha.uol.com.br/ambiente/2018/11/desmatamento-na-amazonia-explode-durante-periodo-eleitoral.shtml.

Milanez, F. "Zé Claudio e Maria: Justiça histórica." *Carta Capital*, December 12, 2016. https://www.cartacapital.com.br/sociedade/ze-claudio-e-maria-justica-historica.

Miotto, K. "Norte Energia inicia obras de Belo Monte." OECO, March 9, 2011. http://www.oeco.com.br/salada-verde/24867-norte-energia-inicia-obras-de-belo-monte.

"'Nem um centímetro a mais para terras indígenas,' diz Bolsonaro." *De Olho nos Ruralistas*, February 8, 2018. https://deolhonosruralistas.com.br/2018/02/08/nem-um-centimetro-mais-para-terras-indigenas-diz-bolsonaro/.

Nepstad, D. C., S. Schwartzman, B. Bamberger, M. Santilli, D. Ray, P. Schlesinger, R. Lefebvre, A. Alencar, E. Prinz, G. Fiske, and A Rolla. "Inhibition of Amazon Deforestation and Fire by Parks and Indigenous Lands." *Conservation Biology* 20 (2006): 65–73. https://doi.org/10.1111/j.1523-1739.2006.00351.x.

Nitta, R., and L. N. Naka, eds. *Barragens do rio Tapajós: Uma avaliação crítica do Estudo e Relatório de Impacto Ambiental (EIA/RIMA) do Aproveitamento Hidrelétrico São Luiz do Tapajós*. São Paulo: Greenpeace Brasil, 2015. http://greenpeace.org.br/tapajos/docs/analise-eia-rima.pdf.

Nogueira, E. M., A. M. Yanai, S. S. Vasconcelos, P. M. L. A. Graça, and P. M. Fearnside. "Brazil's Amazonian Protected Areas as a Bulwark against Regional Climate Change." *Regional Environmental Change* 18, no. 2 (2018): 573–79. https://doi.org/10.1007/s10113-017-1209-2.

Nogueira E. M., A. M. Yanai, S. S. Vasconcelos. P. M. L. A. Graça, and P. M. Fearnside. "Carbon Stocks and Losses to Deforestation in Protected Areas in Brazilian Amazonia." *Regional Environmental Change* 18, no. 1 (2018): 261–70. https://doi.org/10.1007/s10113-017-1198-1.

Nunes, F., and V. Neder. "Eletrobrás quer retomar projeto de megahidrelétrica no Tapajós." *O Estado de São Paulo*, December 1, 2016. http://economia.estadao.com.br/noticias/geral,eletrobras-quer-retomar-projeto-de-megahidreletrica-no-tapajos,10000092046.

Observatório do Clima. "Desmatamento cresce 36% no período eleitoral." October 16, 2018. http://www.observatoriodoclima.eco.br/desmatamento-cresce-36-no-periodo-eleitoral/.

Palmquist, H. "Usina Teles Pires: Justiça ordena parar e governo federal libera operação, com base em suspensão de segurança." *Ponte*, November 27, 2014. http://ponte.org/usina-teles-pires-justica-ordena-parar-e-governo-federal-libera-operacao-com-base-em-suspensao-de-seguranca/.

Pamplona, N. "Produtores de energia querem indígenas como 'sócias' de hidrelétricas: Empresas vão entregar projeto para construção de usinas para equipe de Bolsonaro." *Folha de São Paulo*, November 12, 2018. https://www1.folha.uol.com.br/mercado/2018/11/produtores-de-energia-querem-indigenas-como-socias-de-hidreletricas.shtml.

Peres, C. "Ayres Britto acata pedido da AGU e obras de Belo Monte são retomadas." Instituto Socioambiental, August 28, 2012. https://acervo.socioambiental.org/acervo/noticias/ayres-britto-acata-pedido-da-agu-e-obras-de-belo-monte-sao-retomadas.

Phillips, D. "Jair Bolsonaro Launches Assault on Amazon Rainforest Protections." *The Guardian*, January 2, 2019. https://www.theguardian.com/world/2019/jan/02/brazil-jair-bolsonaro-amazon-rainforest-protections.

Pinto, L. F. "Corrigida, começa a terceira versão da usina de Belo Monte." *Jornal Pessoal*, November 28, 2003. http://philip.inpa.gov.br/publ_livres/Dossie/BM/Outros/Lúcio_Flávio_Pinto-Belo_Monte-terceira_versao.pdf.

Prado, A. P., S. Athayde, J. Mossa, S. Bohlman, F. Leite, and A. Oliver-Smith. "How Much Is Enough? An Integrated Examination of Energy Security, Economic Growth and Climate Change Related to Hydropower Expansion in Brazil." *Renewable and Sustainable Energy Reviews* 53 (2016): 1132–36. https://doi.org/10.1016/j.rser.2015.09.050.

Reese, G., and L. Jacob. "Principles of Environmental Justice and Pro-Environmental Action: A Two-Step Process Model of Moral Anger and Responsibility to Act." *Environmental Science & Policy* 51 (2015): 88–94. https://doi.org/10.1016/j.envsci.2015.03.011.

Reuters. "Liminar põe em risco prazo de privatização da Eletrobrás; governo vai recorrer." January 11, 2018. https://br.reuters.com/article/topNews/idBRKBN1F02FQ-OBRTP.

Rey, O. "Um olhar para as grandes perdas de energia nos sistema de transmissão elétrico brasileiro." In *Setor Elétrico Brasileiro e a Sustentabilidade no Século 21: Oportunidades e Desafios*, edited by P. F. Moreira. 2nd ed. Brasília: Rios Internacionais, 2012. http://www.internationalrivers.org/node/7525.

Rittle, C. "Discurso de Bolsonaro em Davos não combina com realidade de seu governo." Observatório do Clima, January 22, 2019. http://www.observatoriodoclima.eco.br/bolsonaros-davos-speech-hard-reconcile-reality-government/.

Rocha, J. "Bolsonaro Government Reveals Plan to Develop the 'Unproductive Amazon.'" *Mongabay*, January 28, 2019. https://news.mongabay.com/2019/01/bolsonaro-government-reveals-plan-to-develop-the-unproductive-amazon/.

Rodrigues, F. S., and R. Ribeiro Junior. "Construção do AHE Marabá: Uma abordagem sobre opções de desenvolvimento e o seu planejamento." III Encontro Latinoamericano de Ciências Sociais e Barragens. Belém: Universidade Federal do Pará-UFPA, 2010. http://www.ecsbarragens.ufpa.br/site/cd/ARQUIVOS/GT6-42-109-20101111185313.pdf.

Rodrigues, R. A., and P. M. Fearnside. "Índios Waimiri-Atroari impactados por tutela privada na Amazônia Central." *Novos Cadernos NAEA* 17, no. 1 (2014): 47–73. https://doi.org/10.5801/ncn.v17i1.1427.

Rojas, B., and B. Millikan. "El BNDES y el complejo hidroeléctrico Belo Monte." In *Casos Paradigmáticos: De inversión del Banco Nacional de Desarrollo Económico y Social de Brasil (BNDES) en Sur América. Necesidad y oportunidad para mejorar políticas*, edited by I. C. Carillo. Lima: Derecho, Ambiente y Recursos Naturales-DAR, 2014. https://www.internationalrivers.org/files/attached-files/137_casos_paradigmaticos.pdf.

Sanson, C. "Movimentos sociais repudiam Medida Provisória que diminui áreas protegidas na Amazônia." *Instituto Humanitas Unisinos Notícias*, May 31, 2012. http://www.ihu.unisinos.br/noticias/510033-movimentos-sociais-e-organizacoesda-sociedade-civil-lancam-carta-de-repudio-a-medida-provisoriaque-diminui-areas-protegidas-na-amazonia.

Sassine, V. "Ruralista vai cuidar de demarcação de terras indígenas e licença ambiental no governo Bolsonaro." *O Globo*, December 18, 2018. https://oglobo.globo.com/brasil/ruralista-vai-cuidar-de-demarcacao-de-terras-indigenas-licenca-ambiental-no-governo-bolsonaro-23313524?fbclid=IwAR14lldw4puNg5SeX1Z0f6Sgw6fZyQbcBinKvTq9PPEJJVhv7JrfuQkIFgE.

Schlosberg, D. *Defining Environmental Justice: Theories, Movements, and Nature*. New York: Oxford University Press, 2007.

———. "Theorising Environmental Justice: The Expanding Sphere of a Discourse." *Environmental Politics* 22, no. 1 (2013): 37–55. https://doi.org/10.1080/09644016.2013.755387.

Schlosberg, D., and D. Carruthers. "Indigenous Struggles, Environmental Justice, and Community Capabilities." *Global Environmental Politics* 10, no. 4 (2010): 12–35. https://doi.org/10.1162/GLEP_a_00029.

"Senador, Major Olímpio apoiará venda de reservas indígenas e fim de torcidas organizadas." Amazonia, October 9, 2018. http://amazonia.org.br/2018/10/senador-major-olimpio-apoiara-venda-de-reservas-indigenas-e-fim-de-torcidas-organizadas/.

Seto, G. "Bolsonaro diz que pretende acabar com 'ativismo ambiental xiita' se for presidente." *Folha de São Paulo*, October 9, 2018. https://www1.folha.uol.com.br/poder/2018/10/bolsonaro-diz-que-pretende-acabar-com-ativismo-ambiental-xiita-se-for-presidente.shtml.

Sevá Filho, A. O. "Profanação hidrelétrica de Btyre/Xingu: Fios condutores e armadilhas (até setembro de 2012)." In *Belo Monte e a Questão Indígena*, edited by J. P. de Oliveira and C. Cohn. Brasília: Associação Brasileira de Antropologia-ABA, 2014. http://jpoantropologia.com.br/pdfs/Belo_Monte_Questao_Indigena_-_Joao_Pacheco_de_Oliveira_&_Clarice_Cohn.pdf.

———. "Works on the Great Bend of the Xingu—A Historic Trauma?" In *Hydroelectric Dams on Brazil's Xingu River and Indigenous Peoples*, edited by L. A. O. Santos and L. M. M. de Andrade. Cultural Survival Report 30. Cambridge MA: Cultural Survival, 1990.

Silva, E. F., and A. M. Santos. "O caso Belo Monte: Desenvolvimento humano de povos indígenas e tecnopolítica de geração de energia." *Espaço Jurídico Journal of Law* 18, no. 1 (2017): 243–76. https://doi.org/10.18593/ejjl.v18i1.13105.

Silva, H. S. "Vídeo mostra momento do confronto entre PF e índios." YouTube, November 9, 2012. https://www.youtube.com/watch?v=3KF-aG30khg.

Soares, J., and R. Grandelle. "Bolsonaro afirma que vai sugerir mudanças no Acordo de Paris." *O Globo*, December 13, 2018. https://oglobo.globo.com/sociedade/bolsonaro-afirma-que-vai-sugerir-mudancas-no-acordo-de-paris-23301645?fbclid=IwAR3wNTtj2ZGDL_mBb5-usoAB05z1_Q1QSTLdVQjBKCgdalUGtE-7KDlFueo.

Sorribas, M. V., R. C. D. Paiva, J. M. Melack, J. M. Bravo, C. Jones, L. Carvalho, E. Beighley, B. Forsberg, and M. H. Costa. "Projections of Climate Change Effects on Discharge and Inundation in the Amazon Basin." *Climatic Change* 136, no. 3 (2016): 555–70. https://doi.org/10.1007/s10584-016-1640-2.

Soterroni, A. C., A. Mosnier, A. X. Y. Carvalho, G. Câmara, M. Obersteiner, P. R. Andrade, R. C. Souza , R. Brock, J. Pirker, F. Kraxner, P. Havlík, V. Kapos, E. K. H. J. zu Ermgassen, H. Valin, and F. M. Ramos. "Future Environmental and Agricultural Impacts of Brazil's Forest Code." *Environmental Research Letters* 13 (2018): art. 074021. http://iopscience.iop.org/article/10.1088/1748-9326/aaccbb/pdf.

Soterroni, A. C., F. M. Ramos, M. Obersteiner, and S. Polasky. "Fate of the Amazon Is on the Ballot in Brazil's Presidential Election." *Mongabay*, October 17, 2018. https://news.mongabay.com/2018/10/fate-of-the-amazon-is-on-the-ballot-in-brazils-presidential-election-commentary.

Sposati, R. "Por que a Polícia Federal matou Adenilson Munduruku?" *Brasil de Fato*, November 28, 2012. https://www.brasildefato.com.br/node/11236/.

Stickler, C. M., M. T. Coe, M. H. Costa, D. C. Nepstad, D. G. McGrath, L. C. Dias, H. O. Rodrigues, and B. S. Soares-Filho. "Dependence of Hydropower Energy Generation on Forests in the Amazon Basin at Local and Regional Scales." *Proceedings of the National Academy of Science USA* 110 (2013): 9601–6. https://doi.org/10.1073/pnas.1215331110.

Switkes, G. "Leader of Movement to Stop Amazon Dam Murdered." *World Rivers Review* 16, no. 5 (2001): 13. https://www.internationalrivers.org/files/attached-files/wrr.v16.n5.pdf.

Teixeira, M. G. C. *Energy Policy in Latin America: Social and Environmental Dimensions of Hydropower in Amazonia*. Aldershot, U.K.: Ashgate, 1996.

Toledo, M. "Após 5 meses, corpo de ativista é achado em lago de usina em RO." *Folha de São Paulo*, June 25, 2016. http://www1.folha.uol.com.br/poder/2016/06/1784814-apos-5-meses-corpo-de-ativista-e-achado-em-lago-da-usina-jirau-ro.shtml.

Tuffani, M. "Governo elimina área de Ambiente, Energia e Ciência e Tecnologia do Itamaraty." *Direto da Ciencia*, January 10, 2019. http://www.diretodaciencia.com/2019/01/10/governo-elimina-area-de-ambiente-energia-e-ciencia-e-tecnologia-do-itamaraty/.

"Veja como os deputados votaram o Código Florestal." *Congresso em Foco*, May 24, 2011. http://congressoemfoco.uol.com.br/noticias/veja-como-os-deputados-votaram-o-codigo-florestal/.

Ventura, M. 2018. "Fase de grandes hidrelétricas chega ao fim." *O Globo*, January 2, 2018. https://oglobo.globo.com/economia/fase-de-grandes-hidreletricas-chega-ao-fim-22245669.

Villas-Bôas, A., B. R. Garzón, C. Reis, L. Amorim, and L. Leite. *Dossiê Belo Monte: Não Há Condições para a Licença de Operação*. Brasília: Instituto Socioambiental-ISA, 2015. http://t.co/zjnVPhPecW.

Vitel, C. S. M. N., G. C. Carrero, M. C. Cenamo, M. Leroy, P. M. L. A. Graça, and P. M. Fearnside. "Land-Use Change Modeling in a Brazilian Indigenous Reserve: Construction a Reference Scenario for the Suruí REDD Project." *Human Ecology* 41, no. 6 (2013): 807–26. https://doi.org/10.1007/s10745-013-9613-9.

Vitel, C. S. M. N., P. M. Fearnside, and P. M. L. A. Graça. "Análise da inibição do desmatamento pelas áreas protegidas na parte Sudoeste do Arco de desmatamento." In *Anais XIV Simpósio Brasileiro de Sensoriamento Remoto, Natal, Brasil 2009*, edited by J. C.

N. Epiphanio and L. S. Galvão. São José dos Campos: Instituto Nacional de Pesquisas Espaciais-INPE, 2009. http://marte.sid.inpe.br/col/dpi.inpe.br/sbsr%4080/2008/11.13.14.42/doc/6377–6384.pdf.

Watanabe, P. "Bolsonaro recua de fusão de Ambiente e Agricultura e diz não querer xiita ambiental." *Folha de São Paulo*, November 1, 2018. https://www1.folha.uol.com.br/ambiente/2018/11/bolsonaro-recua-em-fusao-de-meio-ambiente-e-agricultura-e-diz-nao-querer-xiita-ambiental.shtml.

Zucco, C., and B. Lauderdale. "Distinguishing between Influences on Brazilian Legislative Behavior." *Legislative Studies Quarterly* 36, no. 3 (2011): 363–96. https://doi.org/10.1111/j.1939-9162.2011.00019.x.

4 When Plurinational States Undermine Indigenous Territories
TIPNIS in Bolivia

CARWIL BJORK-JAMES

While remolding itself as a plurinational state, the Bolivian government distinguished itself as one of the most enthusiastic proponents of indigenous and environmental rights in international diplomacy. This reputation was thrown radically into question in 2011, when the government's effort to build the Villa Tunari–San Ignacio de Moxos highway through the Isiboro Sécure National Park and Indigenous Territory (Territorio Indígena y Parque Nacional Isiboro Secure, TIPNIS) became a lightning rod for debate and protest. The controversy was the most visible sign of emerging cleavages between the Evo Morales' party, the Movement Towards Socialism–Political Instrument for the Sovereignty of the Peoples (Movimiento al Socialismo–Instrumento Político por la Soberanía de los Pueblos, MAS-IPSP), and important segments of the grassroots left, particularly around questions of the environment. After this moment, the government openly subordinated environmental protection and local autonomy to its economic agenda.

Two national indigenous alliances, the lowland Confederation of Indigenous Peoples of Bolivia (Confederación de Pueblos Indígenas de Bolivia, CIDOB) and the highland National Council of Ayllus and Marcas of Qollasuyu (Consejo Nacional de Ayllus y Markas del Qullasuyu, CONAMAQ), recognized TIPNIS as a bellwether case for local control over indigenous territories and organized a cross-country march to defend the territory. Conversely, *cocaleros*, agrarian colonizers of highland indigenous origins, and the local indigenous organization Indigenous Council of the South (Consejo Indígena del Sur, CONISUR) mobilized to back the highway. Pro-highway organizations in San Ignacio de Moxos and Yucumo both set up highway blockades against the march. National police reinforced the blockade and mass-arrested pro-TIPNIS marchers in September 2011, sparking a wave of solidarity actions that prevented the police from sending the marchers home. The regrouped march was backed by an unexpected outpouring, and the Morales government agreed to pass Law 180, protecting the park from further development. Still, by the end of the mobilization, there was an irreversible divide between the plurinational government and the ethno-ecological agenda advanced by CIDOB, CONAMAQ, environmentalists, and other activists. Events in 2017 and 2018, when the government moved to reactivate construction within the territory, confirmed the continuing symbolic resonance of the TIPNIS campaign for both the government and its grassroots critics.

This chapter argues not only that TIPNIS is an important flashpoint in twenty-first-century indigenous politics in Bolivia but that the territory itself, and indigenous communities' relation to it, helps us to understand the underlying conflict. Closer examination reveals an environmental conflict around the questions of road building and deforestation, but also about competing visions for the use of territory, the best future for its longtime inhabitants, and preferred systems of infrastructure. Locally some sixty communities affiliated with the Subcentral TIPNIS and the Subcentral Sécure (regional affiliates of the central indigenous federations) have frontally opposed the highway project, which is likely to lead to further deforestation and agricultural colonization within their territory. On the other hand, the Quechua and Aymara coca growers to their south are assuming a role long favored by the state: that of the agrarian settler. Regardless of the ethnic origins or culturally valued crop involved, agrarian settlement to produce a cash crop still drives deforestation and displacement of preexisting indigenous communities. This chapter explores how these communities' different subsistence strategies, relationship with the market, and approaches to land tenure underlie this fractious local conflict. Expanding outward from the local case, I examine how different indigenous conceptions of land and territory as well as distinct encounters with the environment can contribute to divergences among indigenous movements.

A Revealing Conflict

In the Andean region, global and hemispheric trends toward greater recognition of cultural diversity and indigenous rights, legal protection for sensitive natural areas, and accelerated extraction of natural resources are on a collision course. In Ecuador, Peru, Colombia, and Bolivia, socio-environmental conflicts around natural resource and infrastructure projects are increasingly visible parts of public life. These can usually be described as conflicts between the territorial visions of extractive industrialization (mining, oil, and gas extraction, large dams, and other infrastructure projects) and indigenous subsistence use. The spiraling chasms of open-pit mines may prove incompatible with locally useful watersheds and literally tear apart sacred landscapes, or the rectilinear petroleum concession blocks drawn in national capitals may cut across the boundaries of proposed native land claims. In short, worldviews, understandings of land and territory, and projects for using the earth's surface may simultaneously conflict, revealing a single intercultural divide, typically with indigenous perspectives arrayed on one side.

The TIPNIS highway conflict is somewhat different. First, the largest environmental danger associated with the highway is the locally organized conversion of forests into cropland. While this territorial use is oriented toward production for external use, it is organized locally by indigenous-identified cocaleros and is a predictable side effect rather than the ostensible purpose of the infrastructure project.[1] Second, therefore, the multiple indigenous positions within TIPNIS—Quechua or Aymara agrarian colonist; local Mojeño, Tsimané, or Yuracaré indigenous resident of TIPNIS; and indigenous resident within the colonized area—have different relations to the land and to the proposed highway project. (For convenience, I will refer to the Mojeño-Trinitario, Tsimané, and Yuracaré communities as the "TIPNIS indigenous" and the Quechua and Aymara coca growers as "cocaleros" throughout, without intending to withhold the designation of "indigenous" from the latter.) Whether indigenous people see themselves as flourishing through agrarian production for sale on the market or through endogenously directed development of the local territory influences their perspective.

High-profile environmental controversies such as the TIPNIS conflict help to draw out the divergences in perspective between the actors involved, as well as to clarify the priorities of the state that chooses sides among them. My approach here follows the extended case method long used in historical anthropology, "compiling situational knowledge into an account of social process" and observing how "a social order reveals itself in the way it responds to pressure."[2] By choosing high-profile conflicts, we may encounter cases in which political actors are more public and forthcoming about their views, with ample opportunities to "follow the conflict" into the multiple arenas where notions such as sacredness, contamination, and autonomy are envisioned and given social or legal force.[3] Prolonged protests, consultation processes, legal cases, human rights complaints, and press attention generate opportunities to engage with participants on all sides and with the paper trails that illustrate how different actors view the territory in dispute.

The material I present was gathered largely by documentary means, drawing on the public statements of actors, press coverage, interviews, urban presentations, records of consultations, and secondary use of the ethnographic literature. My access to these statements stems in part from participant observation in Cochabamba's urban campaign in defense of TIPNIS and from attending the 2013 Inter-American Commission on Human Rights with TIPNIS representatives. While there is always a risk of being misled by the self-presentation of actors involved in political conflict, much of the documentation of this case predates both the conflict and the Morales government.[4] The analysis I am presenting here rests on points corroborated by multiple sides of the conflict and between sources prior to and during the highway controversy.

The 2010 March and Limits on Autonomy

For the past three decades, CIDOB has pushed forward an agenda of transferring state functions to local community control on a territory-by-territory basis. This drive began with demands made during the 1990 March for Territory and Dignity, a thirty-four-day trek from Trinidad,

Beni, to La Paz in which the indigenous peoples demanded "an autonomous territory of [their] own."[5] The Isiboro-Sécure indigenous were the first of several Beni-based groups whose demands for territory came together in the march. They provided 190 of the marchers, and the march's spokesman was Marcial Fabricano, a Mojeño Trinitario.[6] He led the march as it grew from three hundred to eight hundred marchers in the long ascent to La Cumbre, the chilly high point of the road uphill from northern Bolivia to the capital, and then merged with a crowd of thousands of Quechua and Aymara supporters. After several days in La Paz, the marchers secured government recognition of four indigenous territories and ratification of International Labor Organization Convention 169 protecting indigenous and tribal peoples.

Such lowland–highland collaboration had lasting impacts, both advancing an indigenous agenda through the Bolivian political system in the 1990s and reorienting Bolivian highland peasant politics around the language of decolonization, autonomy, and indigenous rights. CIDOB would be joined by highland peasant organizations—the highland peasant Unified Union Confederation of Rural Bolivian Workers (Confederación Sindical Única de Trabajadores Campesinos de Bolivia, CSUTCB) and the Union Confederation of Agrarian Colonists of Bolivia (Confederación Sindical de Colonizadores de Bolivia, CSCB)—in further mobilizations. Later CIDOB marches introduced the demand for a national Constituent Assembly to redefine the nation and write a new Constitution. When such an Assembly gathered in 2006, CIDOB joined five other indigenous and campesino federations in a Pact of Unity, making common demands for a plurinational state and strong territorial autonomy for indigenous peoples.[7] In the 2006–7 Constituent Assembly they made already existing indigenous territories and indigenous municipalities that opted for autonomy into receptacles for administrative, executive, and judicial powers and ensured that minority indigenous peoples or nations received limited legislative representation.

CIDOB's Seventh March, held in July and August 2010, sought to solidify these positions within implementing legislation (primarily the Framework Law on Autonomy and Decentralization) to end externally imposed settlement and resource concessions within indigenous territories. The power of indigenous self-government is inherently limited when their territories are crosscut by oil, mining, and timber concessions. But the marchers' demand "that forest, mining, and other concessions . . . be annulled" was not heeded. Nor did the government address the problem of agrarian settlers, large-scale farmers, and ranchers' legal control over large chunks of indigenous lands, shrinking territories like Isiboro Sécure and turning many others in the Chiquitanía and Gran Chaco into a ragged patchwork. In the end, CIDOB (assisted by CONAMAQ) achieved changes in the autonomy legislation but couldn't reach a final deal with the executive branch. The chaotic final day of negotiation was filled with symbols of distrust, leaving fundamental issues unresolved and foreshadowing the conflict to come over TIPNIS.[8]

TIPNIS and the Conflict over the Highway

Isiboro-Sécure was established as a National Park in 1965. It spans 1,372,180 hectares and straddles the border between the Bolivian departments of Cochabamba and Beni. Its indigenous inhabitants are members of the Tsimané, Yuracaré, and Mojeño-Trinitario peoples. In 1990 members of these communities successfully demanded that the park be transformed into an indigenous territory.[9] Their organization, the Subcentral TIPNIS, was founded during a July 1988 meeting that also created a Women's Subcentral and began a tradition of Gatherings of Corregidors (indigenous traditional authorities).[10] Since the 1970s the southern portion of the park has been colonized by agricultural settlers, primarily coca farmers. Residents in this area now form a majority within the original boundaries of the park. The titling process for the Isiboro Sécure Native Community Land (Tierra Comunitaria de Origen), begun in 1997 and completed in 2009, excluded 124,000 hectares of the colonized zone from the title.[11]

There are three primary organizations of indigenous peoples in the territory: the Subcentral TIPNIS, Subcentral Sécure, and the CONISUR. In 2009 the Bolivian government handed over final title for the Isiboro Sécure Native Community Land to the Subcentral TIPNIS, recognizing its legitimacy.

Organizations in TIPNIS

The oldest and broadest organization in the territory is the Subcentral TIPNIS. (Indigenous organizations over large regions of the country are called Centrals; Subcentrals cover a smaller portion of a region.) It pertains to the Central de Pueblos Étnicos Mojeños del Beni. The Subcentral Sécure, created in 2003, includes nearly all communities on the Securé River itself and belongs to the Consejo de Pueblos Indígenas del Beni.[12] Census coverage has been limited, but a 1993 Indigenous Census conducted by the government found 4,563 inhabitants in forty-six communities, although, despite incomplete data, many factors suggest a substantially larger current population.[13]

CONISUR, founded in 1996, includes most but not all communities in the southernmost part of the territory, principally located in Polygon 7, the region around Isinuta, which has been steadily colonized since 1970 by agricultural settlers. Since 1992 the government has repeatedly drawn a Linea Roja (Red Line), intended to prevent the advance of further settlement into the park, but in practice it has been repeatedly redrawn to allow just such settlement. The database of the Indigenous and Community Conserved Areas estimates CONISUR's population at 915 people and lists fourteen CONISUR communities.[14] In some communities, such as Santísima Trinidad, indigenous residents with different political stances pledge loyalty to either the Subcentral or CONISUR. CONISUR affiliates with the Coordinadora de los Pueblos del Trópico de Cochabamba (CPITCO).

The organizational divide reflects different relationships to the departments of Beni and Cochabamba, the cocalero-controlled Villa Tunari Municipality, and more recently the highway. As the 2002 TIPNIS management plan acknowledges, "CONISUR is an organization basically created and supported by the Cochabamba Prefecture, which serves it as a mechanism for channeling aid to the communities of the south and through this to defend its sovereignty over the area."[15] The simple logistic accessibility of Villa Tunari to CONISUR and its distance from Subcentral meeting locations in northeast TIPNIS also contribute to the divide.[16]

Perhaps more than twenty thousand Aymara and Quechua colonists live in the 100,000-hectare Polygon 7, swamping the local indigenous population whose territory they have largely deforested.[17] As of 2005 there were 280 local unions organized into twenty-eight Centrales inside of this area.[18] These unions are members of the Federation of the Tropic of Cochabamba (Federación del Trópico de Cochabamba), itself one of the Six Federations (Seis Federaciones del Trópico de Cochabamba), the Chapare coca growers' union organization. The colonized zone is administered as District 8 of the Villa Tunari Municipality, which has been controlled by the Movement Towards Socialism since 1997.

The Isiboro-Sécure Park is located over forty kilometers from Villa Tunari and the Chapare roadway that is the heart of coca-growing country. Prior to 1986 forest loss in the southern portion of TIPNIS was insignificant. Over the next two decades, however, nearly 32,000 hectares of forest within the park were degraded by human activity, in a pattern densely clustered around the small roadways built into the park by agricultural colonists.[19] The resulting fields enveloped at least fourteen preexisting communities. From 1992 to 2009 cocaleros and the TIPNIS agreed on a series of boundaries, each known as the Red Line. Despite its serious name, the line has been repeatedly moved northward to accommodate growing agrarian colonization. In 2008 the National Agrarian Reform Institute formalized the titles of cocaleros in the region within the park to the south of the Red Line. This area, called Polygon 7, encompasses around 124,000 hectares, including both individually titled lands and TIPNIS indigenous communities.

The Villa Tunari–San Ignacio de Moxos Highway is a project connecting the towns of Villa Tunari (in the Chapare region of Cochabamba Department) and San Ignacio de Moxos (in Beni Department). The 306-kilometer road would pass through Isiboro Sécure along the 177-kilometer Segment II from Isinuta to Monte Grande, effectively dividing the park in half. In the process the Red Line limiting colonization would become indefensible. A study by the Bolivian Institute for Strategic Research projected that existing deforestation trends would lead to a 43 percent loss of forests in TIPNIS by 2030. But the road would accelerate deforestation by increasing access to the territory for illegal loggers as well as agricultural colonizers. It projected deforestation of 64 percent of the park by 2030 if the road was built.[20]

To understand the stance of TIPNIS communities on the highway, we have to look at their broader conceptions of the territory they inhabit and how it should be used.

TIPNIS Indigenous Conceptions of Territory

Gabriela Canedo Vásquez's 2011 study of the Moxos region investigates the "collective representations held by the indigenous peoples of territory as unlimited, the *Loma Santa*, the *casa grande*, collective property," in contrast to mestizo society's ideas of "delimited territory, with property line posts and boundaries, as individual property."[21] Loma Santa and Casa Grande, in particular, are emotionally potent and politically invoked terms in the speeches and resolutions of the TIPNIS indigenous, underscoring "our territory" as "our great house."

A quest for the Loma Santa, the sacred hill, oriented and motivated indigenous communities in Moxos beginning in the nineteenth century. Partly a religious quest and partly a form of cultural revival, the quest propelled numerous migrations fleeing harsh labor conditions in the rubber boom, the quinine industry, mestizo-owned cattle ranches, and governmental public works and as army recruits for the 1930s Chaco War, all of which relied on unpaid, forced, and sometimes dehumanizing work by indigenous people.[22] The Loma Santa was a way out, "a persistent dynamic of 'distancing' from contact with the *karayanas* [whites and mestizos], who progressively and gradually occupied the[ir] territory and submitted it to exploitation."[23] It was also a grounded utopian move toward "a sacred place where they can hunt and fish what they like and where they would be free of the presence of the whites."[24] For Adolfo Moye Rosendy, former president of the Subcentral TIPNIS, this quest arose because his ancestors took Christianity seriously:

> Someone said that we had been evangelized, that the Jesuit priests entered by force and managed to evangelize us. The [indigenous] brothers . . . decided to seek out a Holy Land, this idea was born of the people who began to say, "We believe in God, and God is going to give us a Holy Land." After being turned around on all sides, in the end they realized that the only holy place where there was really peace, tranquility, food

to live well, pure air, and to enjoy nature was in TIPNIS, and thus it is for us the Sacred Hill.[25]

Elsewhere in TIPNIS, including among many Yuracaré, the continuity of tradition and unity between the home and nature is expressed through the Casa Grande (Great House). Analogized to a multifamily collective house, the land serves as an abundant provider.

Most TIPNIS communities, whether Yuracaré, T'simane, or Mojeño-Trinitario, envision their economic and social future around the continuation of an economy largely separate from the dominant commodity-centered commerce of the Bolivian nation. Their engagement in trade has centered on a few high-value commodities (e.g., cacao for fair-trade chocolate, crocodile leather), ecotourism, and what they say is sustainable timber harvesting. These communities depend on the harvesting and growing of goods within the forest for their livelihood; even these exports would be impossible (or unsustainable) in a deforested territory.

Local ideas of development center on "possessing *collective* goods like a school, education, health posts . . . electricity, transportation . . . media, radio, that is all that belongs to the ambit of basic services."[26] (This contrasts with indigenous perceptions of the colonists as individualistic "people who are always 'counting their money.'"[27]) At the core of their strategy, however, is continuing to live on the land, with fishing and hunting as major sources of food and villages constructed from locally available materials. "The majority of the indigenous," writes Gabriela Canedo Vásquez, "have a conception of territory as the mother who gives them food; the territory includes the mountain that provides meat, wood for building houses, the field for cultivating, the spaces for raising domestic animals and fruit trees."[28]

This represents just one of the paths available to very peripheral communities in the tropics, one which depends on maintaining the relative ecological integrity of their land base. That is, it works only so long as they remain beyond the line of colonization. When this integrity is lost, the community must shift to commodity-based agriculture, landless labor, or export of their resources. Where large-scale cattle ranching predominates to the north of the three Mojeño indigenous territories

(TIPNIS, the Multiethnic Indigenous Territory, and the Mojeño-Ignaciano Indigenous Territory), there is little place for indigenous communities to maintain control of their homeland.[29] In such cases, indigenous people often end up instead on the lowest rung of the postfrontier labor market or as migrants to the cities.

Chapare Cocaleros' Conceptions of Territory

The Chapare cocaleros arrived in the region as agrarian colonists in the 1970s and 1980s; this influx began with Quechua-speaking peasants from elsewhere in Cochabamba Department, then Quechua- and Aymara-speaking rural highlanders, and finally miners made jobless by neoliberal shock therapy. By 1989 the Chapare population had reached 350,000, a fourteen-fold increase over 1967.[30] These migrants built new lives in the Chapare by clearing forested plots and planting crops for sale on the national market. Agricultural colonization in the Chapare was government policy, and the government rewarded colonists with title to the lands they cleared. Agrarian unions were the organizing mechanism for this colonization, orchestrating the opening of roads; numbering, delimiting, and awarding the plots to each family; and transmitting land claims to government titling agencies.[31]

Bolivian institutions sought to find economically viable products other than coca to produce there, including tea, fruit, and rice.[32] However, these crops could not compete with coca, traditionally chewed as a mild stimulant and remedy for altitude sickness and the internationally demanded raw material for cocaine. The Chapare became ground zero for a U.S.-backed drug war that targeted coca growing and coca growers, who responded by organizing a militant umbrella union, the Six Federations of the Tropics of Cochabamba, and defending the coca leaf as an emblem of traditional indigenous lifeways. Their political identity is militantly indigenous but constructed around the leaf, the various practices of using it—chewing, ritual offerings, garlands—and its place as a marker of indigeneity in Bolivian society rather than the land.[33] It is also anti-imperialist, shaped by the trauma of the U.S.-backed drug war that caused as many as ninety-five deaths among the Chapare cocaleros between 1980 and 2004.[34]

Improbably the union became the alliance-building vanguard of a national political party, the Movement Towards Socialism—Political Instrument for the Sovereignty of the Peoples.[35] President Morales, officially still the head of the Six Federations cocalero organization, recently (in 2017) remembered the unions as torchbearers of Andean community traditions from the highlands, specifically reciprocity, collective ownership, and living in harmony with nature.[36] While their form of collective ownership is unrecognizable to the TIPNIS indigenous, the agrarian union plays a role in meeting collective needs (like trails), reallocating land, dividing rights to sell coca, and progressively taking over the roles of trader and wholesaler of coca leaf.

The Overlap: CONISUR

Four decades of cocalero settlement have created a variety of relations between the cocaleros and the indigenous inhabitants of Polygon 7. Some live by selling fish and meat to the farmers who have deforested their lands or doing menial labor for the coca growers. Others have started their own coca plots, although only cocalero union members can sell their leaves in large, official markets, which are controlled by the union federations.[37] As prospects for hunting have declined, indigenous residents in Polygon 7 also increasingly rely on agriculture, selling their products to the coca-growing settlers to buy "salt, oil, sugar, clothes, school supplies, fuel," and other goods.[38] According to multiple reports, relations are not equitable.[39] Instead indigenous people are often dependent, landless laborers in their own land, earning around 20 bolivianos (a bit less than US$3) to harvest a coca plot or selling their fish or wild meat to colonists for around 300 bolivianos (~US$40) a month. Others earn income by authorizing the cutting of timber and the elimination of the forest on which their lives once depended.

Unlike those living in the intact sections of the park, the indigenous in the colonized areas have already moved from a way of living that is interdependent with the ecosystems of the park to one that is integrated with the national economy. This goes a long way to explain why they see a shared economic interest with the coca growers in the road.

Spatial Perspectives within the TIPNIS Conflict

The Subcentral TIPNIS, holders of collective land title to the territory, have stated and restated their opposition to the building of the Villa Tunari–San Ignacio de Moxos highway through their land dozens of times since 2003.[40] The Subcentral TIPNIS, Subcentral Sécure, and CONISUR all joined in condemning the highway in a May 2010 resolution, asserting, "Opening this highway would present a threat to our life as peoples who inhabit TIPNIS due to the loss of the natural resources and all the biodiversity upon [which] the Moxeños, Yuracarés, and Chimanes sustain their culture and life: a life and culture we have lived in our territory since before the creation of Bolivia and will continue to live in the future."[41] Citing this and other reasons, the meeting voted to "overwhelmingly and non-negotiably reject the construction of the Villa Tunari–San Ignacio de Moxos highway and of any highway segment that would affect our territory."

Conversely the cocaleros have long embraced the highway as a marker of progress and a pathway to new markets. "It is development for the municipality . . . because it is passing through its jurisdiction, and so it will benefit every community, every town," argued Feliciano Mamani, a union leader and the mayor of Villa Tunari.[42] The cocalero leader Leonilda Zurita spoke of "the historic dream of integration of the peoples of the Valleys and the Amazon."[43]

CONISUR initially opposed the highway in 2008, under the leadership of José Luis Moy Moye, and later of Marco Emilio Fabricano.[44] Under the leadership of Gumercindo Pradel, however, CONISUR aligned itself with the roadway and became one of its strongest advocates. In 2012 Pradel explained his position this way: "Those who live in the territory suffer. I want to go forward so that my poor brothers can progress so that in the future the next generations can be professionals and have good education and health services. We want good infrastructure. We want to have a good house like those leaders who have betrayed us."[45] Pradel has emphasized the ethnic continuity of CONISUR communities, while affiliates with the Subcentral TIPNIS see them as having fundamentally given up indigenous lifeways by accepting individual titles, affiliating with cocalero unions, and in some cases dissolving as separate communities.

The differing spatial perspectives of local groups have shaped the TIPNIS conflict in other ways. Briefly, these include the following:

Highway versus waterways. The government's plan to build a singular highway through the park is constantly touted as providing material development. As I mentioned earlier, the government has implicitly conditioned state services on approval of the highway, showering communities with boats, fuel, schools, and other development projects in advance of the consultation.[46] In most TIPNIS communities, however, the highway is irrelevant to the work of providing such services. Indeed it won't connect most TIPNIS indigenous communities to states, markets, or each other. "We the Yuracarés and the Trinitarios are people who live on the rivers, we make our communities by their banks," Silverio Muiba, a Moxeño Trinitario resident of Santíssima Trinidad observed to Sarela Paz a generation ago. "On the other hand, the Quechuas always live where there is a road: where the road runs out, so do the *collas* [highland indigenous people]."[47] Without the magic of overarching "development" discourse, the positive benefits of the road for them are difficult to perceive.

Presence of the state and cash economy. In public statements, the government and cocalero leaders described the "absence of the state" as causing a situation of "extreme poverty" in TIPNIS communities. By making cash income the sole measure of wealth, the government has devalued the Great House of naturally available goods that TIPNIS indigenous regard as their inheritance. From their point of view, the greater poverty is to be seen in the dependency, limited hunting and fishing, and wage labor of the CONISUR communities. For the state, however, economic development is measured by cash transactions and government revenue.

Collective versus individualized land titles. The TIPNIS indigenous hold property to their territory in common, through a collective title. This system accords with their use of the land, which relies on fishing, hunting, and gathering as key sources of food and goods, as well as cultivating crops. By contrast, each cocalero family has its own plot for growing coca and staples for their own use. When CONISUR members switched to a system of individual property, that was enough for the other TIPNIS indigenous to regard them as outside the collective. (This difference echoes a broader distinction between CIDOB's quest for "territory" and the peasant and intercultural communities' demand for "land," which I return to later.)

From the perspective of cocalero historical memory, these three points make sense. The Chapare is a place where they went from lives spurned by the absence of the state to eventual social movement control over the state. Their material progress was achieved through production for the market on plots allotted by family. And their development has been built around the Santa Cruz–Cochabamba highway. It's no surprise when Morales asks, "How can they oppose it? I cannot understand the indigenous brothers."

The different communities also have differing values for the core zone of the park, those upland portions under the highest level of environmental protection in the pre-2011 Management Plan. While the Subcentral TIPNIS communities are clustered in the east of the territory, this western area acts as a refuge for people and wildlife during the nearly annual flooding. TIPNIS indigenous perceive this area as vitally important because of their boundaryless vision of the landscape and its ecological relatedness.

Last but not least, the idea that indigenous people can be ranked by their degree of "civilization" and situated on a ladder that once led upward to a white, European ideal remains relevant in Bolivia today. Both Tathagatan Ravindran and Andrew Canessa have drawn attention to the continuum of indigenous statuses in Bolivia, which are hierarchically arranged beneath the mestizo urbanite.[48] For Canessa, this hierarchy descends through towns and rural villages to the still denigrated and still excluded *ayllu*, or highland rural community. For Ravindran, the hierarchy continues down the Andean slopes to the presumably backward residents of the Amazon lowland. At the bottom of this hierarchy lurk the foolish savages, labeled CIDOB, mocked in an editorial cartoon in the state-run daily newspaper for being remotely controlled by USAID and the Center for Juridical Studies and Social Investigation, a Bolivian human rights NGO.[49]

The notion of savagery arose repeatedly in the TIPNIS conflict. At a cocalero youth gathering on July 31, 2011, President Morales told the audience, "Just because they [the TIPNIS indigenous] say no, doesn't mean it won't be done." Later in the talk, he urged one-on-one efforts to win their support for the road and, failing that, to seduce the women among them: "You, comrades, have to explain, to orient the indigenous comrades. . . . If I had the time, I would go seduce the Yuracaré women. So then, youth, you have instructions from the president to conquer the Trinitario and Yuracaré female comrades so that they do not oppose the construction of the road." His off-color "proposal" was met by applause from the crowd but public criticism from feminists.[50] Just over a month later, the CSUTCB leader Ricardo Coraite sparked further controversy when he defended the road by saying, "We don't want the indigenous to live like savages anymore."[51] The hierarchy from savage to civilized is alive and well.

What makes the plurinational era different, according to Ravindran, is that upwardly mobile indigenous merchants and urban migrants have ceased to whiten their own identities as they ascend the ladder; instead they are recasting urban spaces, "from the café to the ministry," as indigenous spaces. Ravindran goes further, identifying a current in indigenous politics that also accepts extractive development while demanding a redistribution of its cash proceeds to the indigenous majority. This he terms the "expansionist spatial imaginary," a view embraced by the CSCIB, CSUTCB, Bartolina Sisa Federation, and many urban indigenous people. Within this view, decolonization and extractive development are linked: there is no greater sign of self-possession of the country than the capacity to sell its resources and reap the corresponding rewards.

Mobilization and Countermobilization

The TIPNIS indigenous were not consulted prior to a series of government actions to make the Villa Tunari–San Ignacio de Moxos highway a reality, including its design (2008), a loan from Brazil's National Bank for Economic and Social Development, a bilateral Brazil-Bolivia agreement on the loan (early 2011), and the start of construction (June 2011). In response, three Subcentrals—TIPNIS, Sécure, and the Women of

TIPNIS—and three Beni indigenous organizations declared themselves in a state of emergency and asked CIDOB to mobilize nationally on their behalf.[52] Meanwhile Morales and other government officials explicitly denied local indigenous communities the right to choose, saying the road would be built "whether the indigenous want it or not." With these broader issues at stake, CIDOB and CONAMAQ embarked on Bolivia's first national ethno-ecological campaign of the twenty-first century. By mid-August 2011 they had gathered some five hundred of their members for a cross-country march in Trinidad, Beni.

Blessed by a Catholic opening ceremony and a highland Andean ritual, the marchers began their long walk to Bolivia's capital on August 15. Ernesto Noe, a leader in the 1990 indigenous march, accused the Morales government of doubletalk on the environment: "I want to say to the President of the Republic, who is decorated as a defender of the lands of the world. . . . Why in Bolivia, then, does this government of ours want to destroy our territory, our habitat, want to destroy our environment and our communities? This cannot be!"[53]

While the 2011 Grand National Indigenous March brought together lowland and highland organizations and summoned the memory of the 1990 march, it also raised contradictions within the grassroots left coalition to the surface and sparked several counterprotests along the march route. These countermobilizations against the indigenous march included the tactics of road blockades as a means of pressure and attacking "certain leaders" as "traitors" to the broader cause. Prior to 2010, however, these tactics were deployed against powerful external forces or as a part of intra-union leadership battles. From 2006 to 2009 the MAS-IPSP and its social movement allies had navigated the lengthy stalemate between their government and the right-wing opposition by judiciously deploying massive crowds rather than police. Elsewhere I argue that this "nonviolent statecraft" allowed the MAS government to avoid more direct confrontations between state forces and right-wing opponents, undercut the opposition narrative of MAS authoritarianism, and put the right wing's violence and racism in the spotlight.[54]

However, while in the field in 2010 and 2011, the same forces began to countermobilize against other left grassroots movements who the government alleged were "endangering the process of change" with their demands. The government threatened blockades against the June–July 2010 CIDOB March, backed a peasant road blockade of the Potosí regional strike's negotiating delegation in August 2010, and convened a mass march in Cochabamba to repudiate the April 2011 general strike by the national labor federation. The Pact of Unity and the National Coordination for Change (CONALCAM) took the lead in these pro-government actions.[55]

During the first phase of the Eighth March there were two episodes of mobilization against the march. Pro-highway organizations in San Ignacio de Moxos and the agricultural colonist federation in Yucumo both set up blockades. The blockaders in San Ignacio de Moxos, who included members of the regional workers' federation (COR San Ignacio de Moxos) and other residents, blocked passage out of their town for three days beginning on August 19. An attempted blockade in San Borja, Beni, was outflanked by more massive efforts by march supporters. Finally, however, the opposition coalesced into a road blockade on the Beni–La Paz border in Yucumo. The national government sent police troops to guard the space between marchers and blockaders "to prevent violence," thereby reinforcing the cordon and blocking the march from September 15 on.

During a ten-day delay, food shortages and limited access to water bedeviled the march, while a new wave of activists mobilized in solidarity. A cross-country delegation from Cochabamba, La Paz, and Santa Cruz arrived to join and supply the march.[56] Simultaneously indigenous organizations began a vigil in the San Francisco Plaza of central La Paz. Finally, on September 24 a women's delegation physically attached themselves to the government negotiator Foreign Minister David Choquehuanca and insisted he march through the police blockade with them. Even prior to this controversial event, the government had made preparations to break up the indigenous encampment. That night Marcos Farfán, vice minister of police and the interior regime, addressed police commanders with a frank assessment of the government's motives:

> One must have a global, integrated vision of the matter, although it can be reduced to just the operation that we will carry forward tomorrow. . . . Because if we don't produce, if we don't create industries, if we

don't build highways, if we don't diversify our production, if we don't transform our gas into other products, if we don't generate added value from our raw materials, this government will fail. Everything will be ruined. . . . I believe it's important to clarify those elements that have to do with all Bolivians, police officers or not, with the whole country. But it is within this framework that we, lamentably, are entering into carrying out this operation. . . . Tomorrow, we will be working.[57]

The police repression of the march on Sunday, September 25, brought worldwide visibility to the conflict. In the afternoon, police mounted an attack on the indigenous marchers, beating and tear-gassing them, seeking out and arresting prominent leaders, and leaving behind a chaotic scene of injury, flight, and fear. Marchers, including prominent leaders, were grabbed, tackled, and handcuffed in front of network television cameras. Between 70 and 280 were injured in the assault, including Celso Padilla, hospitalized with multiple hematomas, and Wilson Melgar, who suffered an embolism. At least 152 protesters were treated for injuries or high blood pressure in San Borja or Rurrenabaque; a Prosecutor's Office report in September referred to 74 wounded; a legal complaint presented by the marchers asserts 280 were injured.[58]

Security forces evacuated hundreds of marchers in buses and vans. The passage of this police caravan set off clashes with protesters in San Borja and at the Rurrenabaque airport, where the government tried to fly the indigenous marchers back to the point of origin of the march. Tacana, Tsimané, and Mosetén indigenous people and Rurrenabaque townspeople organized overnight, rallied at the airport, and finally stormed the runway, disrupting the prisoner transport. Released prisoners regrouped at Quiquibey, thirty-six kilometers closer to La Paz than Yucumo—the site of the standoff and raid—where they restarted the march on September 30. Despite the predictions of possible confrontations with members of the *colonos* union in the towns of Caranavi and Palos Blancos, the marchers were publicly welcomed. Within the first week after the raid, street protests occurred in eight of nine departmental capitals, a twenty-four-hour general strike of the Bolivian Workers' Central labor union confederation was held in solidarity, and a department-wide strike was held in Beni.

A second march, led by affiliates of CONAMAQ, traveled from Oruro to La Paz, growing from 75 to 220 participants along the way. A three-day civic strike in Beni and a one-day general strike by the Bolivian Worker's Central unfolded within a week, including road blockades in Cochabamba and numerous Beni towns.

Prominent members of the Morales government resigned in protest, including Defense Minister Cecilia Chacón and Marcos Farfán, the vice minister quoted earlier. Pablo Solón, a former UN ambassador and an organizer of the Cochabamba Climate Summit, wrote an open letter to Morales, stating, "One cannot speak of defending Mother Earth and at the same time promote the construction of a road that will harm Mother Earth, doesn't respect indigenous rights and violates human rights in an 'unforgiveable' way."[59] The president made a televised statement late on September 26, announcing the suspension of Segment 2 of the highway and asking the families of those attacked for forgiveness.[60] He disavowed ordering any violence and started an investigation of the incident. However, Morales soon reiterated that projects like the Cochabamba–Beni highway are "questions of state" not subject to a local veto. Interior Minister Sacha Llorenti resigned while also disavowing responsibility for the raid.

As the marchers climbed toward La Paz, the seven minority indigenous members of the Bolivian Chamber of Deputies drafted legislation to permanently protect TIPNIS from highway construction. Two of them, Pedro Nuni and Bienvenido Zacu, had been among those arrested at Chaparina.[61] When the 1,200 marchers reached La Paz on October 19, tens of thousands lined the route to applaud or briefly join in their journey. That evening several hundred indigenous protesters set up camp in the Plaza Murillo opposite the Presidential Palace. The arrival of the marchers in La Paz and their mass backing by Paceños propelled the government to accept the CIDOB-backed legislation, which passed as Law 180 on October 25, 2011. This law declared TIPNIS an "intangible zone" and prohibited any highway from crossing the park. Fifteen separate agreements on a range of issues were signed between the government and indigenous representatives within a week.[62]

Consultation

Despite the agreement, the government and its allies pressed for renewed construction of the highway. From December 20, 2011, to January 30, 2012, CONISUR led a pro-highway march from Insinuta through the Chapare and Cochabamba to La Paz. Marchers came from communities in Polygon 7, several communities just north of it, and San Ignacio de Moxos.[63] Campesinos and transport workers from Cochabamba, La Paz campesinos, and El Alto neighbors were among the comparatively smaller crowds that came out in support of the march.[64] In both Cochabamba and La Paz, government officials were quick to promote the march and begin a dialogue on legislation. On February 10, 2012, the government enacted Law 222, authorizing a government-organized consultation on the future of the territory, despite the opposition of the Subcentral TIPNIS.

The government took full advantage of its power to frame the consultation. It organized the process around four questions:

1. Should TIPNIS be made an "intangible zone," as stated in Law 180, or should the law be abrogated?
2. What safeguards against environmental degradation should be taken?
3. What development priorities do communities have?
4. What conditions should be placed upon the design and construction of the highway?[65]

Intangibility had not been defined in Law 180, but government ministries and the Subcentral had agreed on implementing language in December 2011. The government shelved this agreement and unilaterally decided that "intangibility" means that nearly all economic activities—including ecotourism, sustainable nut and cacao harvesting, and other projects previously approved—must be suspended until Law 180 is revoked.[66] The consultation was accompanied by the public bestowing of gifts and development assistance that were implicitly or explicitly conditioned on acceptance of the highway.[67] Questions 2 and 3 implicitly linked future government development assistance and restrictions on new land invasions to the construction of the highway. Unless communities refused to state any conditions under which they would accept it, their answers to question 4 were counted as support for the highway.

The consultation was conducted from July 29 to November 25, 2012. The government produced a final report on the consultation claiming that fifty-seven of fifty-eight consulted communities rejected Law 180, and fifty-five accepted the highway, but these headline numbers are disputed and conceal radical disagreements.[68] The same report recognized eleven communities that refused the consultation altogether.[69] Six solidarity activists, led by a lawyer named Ivan Bascopé, accompanied a CIDOB–Subcentral TIPNIS delegation on a tour of the territory from August 29 to September 26, where the delegation held a series of workshops on the consultation process in seventeen communities. They provided a list of thirty communities that rejected the consultation process altogether.[70] A joint delegation of the Catholic Church, the Permanent Assembly of Human Rights–Bolivia, and the International Federation of Human Rights visited thirty-six communities between November 29 and December 13. Contradicting the government report, they found that sixteen of these communities had not received the government consultation team and that just three of the nineteen consulted communities accepted the highway; community leaders were bypassed, the consultation team would arrive suddenly and unannounced, and consultation meetings often involved only those in agreement with the process.[71] Despite these limitations, the official report itself provides evidence of local opposition: a half-dozen communities that demanded that indigenous authorities decide on the final route, two said "yes, but not through the middle of TIPNIS," and others accepted an "ecological highway" that wouldn't damage the park; all these were counted as "in favor of the highway." Fifty-three of the communities called for deploying more personnel, whether from National Protected Areas Service (Servicio Nacional de Áreas Protegidas, SERNAP), the military, or the communities themselves, to defend the territory from new settlements. At least thirty-nine called for increased penalties on illegal settlers, often demanding thirty-year prison sentences.[72]

While the government organized and defended its consultation process, funding for the project fell apart. The Brazilian contractor OAS suspended work in December 2011 over questions of payment, and Morales canceled

OAS's contract in April 2012.[73] Brazil's National Bank for Economic and Social Development immediately suspended its loan for the project, leaving the project in limbo.[74] MAS officials took the officially positive but controversial results of the consultation as reason enough for a new law to build Segment 2 of the road, but hesitated before enacting it. Instead they pledged to focus on "eradicating extreme poverty" in the territory, at least through the next round of national elections.[75] Construction of Segment 1, which is largely inside the colonized Polygon 7, was contracted to a public corporation and a local builder in Villa Tunari and completed in September 2016. Segment 3 has proved more challenging, with the task changing hands twice and an earthen platform only partially complete as of May 2017.

Divisions

The TIPNIS conflict has deepened existing tensions among grassroots indigenous movements into profound organizational divides. In November 2011 a truncated Unity Pact met in Sucre without CIDOB and with CONAMAQ largely absent and endorsed the construction of the highway.[76] Differences in perspective between the three peasant confederations (highland peasants in the CSUTCB, intercultural colonist communities in the Syndicalist Confederation of Intercultural Communities of Bolivia [Confederación Sindical de Comunidades Interculturales de Bolivia, CSCIB, the renamed agrarian colonist federation], and the Bartolina Sisa women's confederation), known as "the triplets," and CIDOB and CONAMAQ ("the twins") had simmered below the surface for years. The twins' political vision emphasized collective landownership and reconstitution of ancestral territory, and they remained more aloof from a permanent alliance with any political party, even the MAS. In June 2007, amid the Constituent Assembly, they had rebelled against Morales's insistence that "indigenous peoples could not have dominion over or property in the natural resources on their territories." Yet when they did so, they found the triplets—who had previously signed on to their calls for autonomy—aligned with the government, "speaking from the perspective of the national interest, rather than the indigenous point of view," as Salvador Schavelzon described it.[77]

In mid-2012 CIDOB and the Subcentral organized a new sixty-two-day march to reject Law 222 and the consultation it proposed. While they waited fruitlessly in La Paz for the government to meet their demands, pro-government CIDOB members met in Santa Cruz and elected Melva Hurtado as the organization's new president, setting off a schism in the organization that has yet to heal.

A similar split emerged in CONAMAQ in 2013, when Hilarión Mamani Navarro and Gregorio Choque Quispe led a campaign to realign the organization with the government. After six attempts to occupy the organization's headquarters in 2013, the pro-MAS faction organized their own Jach'a T'antachawi, a large gathering that is the organization's highest authority, and elected Mamani as leader on December 10.[78] That night his supporters burst into the organization's La Paz headquarters and beat up three CONAMAQ leaders. A standoff followed between an "organic CONAMAQ" led by Freddy Bernabé and Mamani's pro-government CONAMAQ. Organic CONAMAQ activists maintained a month-long vigil outside the police-controlled headquarters, until Mamani's supporters attacked them and took control of the building on January 14, with police acquiescence.

By early 2014 "the twins" were each themselves divided in half. The blows inflicted by the police at Chaparina and during the struggle for the CIDOB and CONAMAQ headquarters in Santa Cruz and La Paz would not soon be forgotten. Moreover the wrenching battles within the national organizations were followed by quieter forms of contestation. The government refused to meet with critical leaders and held up development funds previously promised.[79] Rosa Chao, leader of the Cochabamba affiliate of CIDOB, CPITCO, found herself unwelcome at government ministries, which worked instead through CONISUR's Gumercindo Pradel. She stepped aside and was replaced by a pro-Hurtado leader, as did Celín Quenevo, the leader of the Central Federation of Indigenous Peoples of La Paz (Central de Pueblos Indígenas de La Paz, CPILAP).[80] In June 2015 a CIDOB technical worker explained to me that any government meeting requested by an organization affiliated with organic CIDOB would be denied.

In December 2013, during the leadership crisis at CONAMAQ, the Bolivian government expelled Ibis, a largely publicly funded Danish development NGO that had just published a guide on indigenous communities' right of free, prior, and informed consent.[81] Announcing the expulsion, Minister Juan Ramón Quintana said, "We have grown tired of tolerating the political interference of Ibis in Bolivia, tired of tolerating that it promotes internal conflict in Bolivia's organizations, CIDOB, CONAMAQ, and other indigenous organizations."[82] The expulsion put aid organizations on notice to stay away from providing operational support for government critics, even if this support was part of long-standing backing for the indigenous movement. Meanwhile the Indigenous Fund, a purportedly independent mechanism for indigenous organizations to receive a share of national hydrocarbon taxes, fell under the shadow of a massive corruption and mismanagement scandal; a government audit found a loss of 71 million bolivianos on unexecuted projects, as well as numerous cases where project funds were transferred into the personal bank accounts of indigenous leaders.[83] The Fund had been a major demand of indigenous organizations and was overseen by the leaders of the Unity Pact. When it fell, criminal accusations sent Rural Development Minister Nemesia Achacolla and pro-government CIDOB leader Melva Hurtado to jail, while Adolfo Chávez sought political asylum in Ecuador rather than face charges related to the Fund.[84] Beyond its effect on many existing leaders, the scandal at the Fund left indigenous organizations without a dedicated funding source just as the international funders were being intimidated against supporting them.

By 2014 the government had moved to full-throated support for extractivism as an economic strategy. Morales has floated proposals for over 48,000 megawatts of hydroelectric generating capacity in the Bolivian lowlands, including the El Bala/El Chepete dam project in Madidi National Park, for an oil refinery in the Amazon region of northern La Paz, and for quintupling the land under cultivation in the next decade, mostly by expanding mechanized monoculture. Supreme Decree 2366 (issued May 20, 2015) opens protected areas—national parks and biosphere reserves, many of which are also indigenous territories—to oil and gas drilling. A 2015 Incentives Law transfers 12 percent of hydrocarbon tax revenue to a special fund to reward companies that make large investments in oil and gas extraction. Even rhetorically, "extractivism" has gone from being an external critique to a self-defined goal of the government and an explanation for its success thus far. "We will use extractivism . . . for another two decades," Vice President Álvaro García Linera told an audience convened by the UN Development Program in 2016.[85]

Renewed Conflict (and Setbacks, Delays, and Complications)

In 2017, after years of a tense stalemate, the government reactivated construction of the most controversial segment of the highway. Plurinational Legislative Assembly members Ramona Moye (MAS-IPSP, Beni) and Patricia Chávez (MAS-IPSP, Cochabamba) sponsored legislation to reverse Law 180 and authorize roadways within TIPNIS, arguing, "It's the request of the people, the community members, their *corregidores* [a traditional leadership post] and leaders, through the [2012] prior consultation carried out within the territory itself."[86] When the Subcentral TIPNIS president Domingo Nogales stated his support for the bill, community members convened an emergency meeting to de-recognize him and elevate Vice President Fabián Gil to leadership. Marqueza Teco, head of the Women's Subcentral within TIPNIS, had already become a prominent voice against both the road and the new legislation.

As the legislative assembly debated the bill, contractors slipped inside the territory and began construction on bridges. The push for the new law revived coalitions of indigenous peoples, environmentalists, and government critics from both left and right. Despite nationwide protests, Law 266 was promulgated on August 13, 2017. While the text of the law calls for the elaboration of an "integral protection plan" and a transportation plan for TIPNIS, construction of the roadway continued without any further consultation. By August 2018 work on at least two of three major bridges within the park had been completed, but the route between them remains unpaved.[87]

The TIPNIS leaders Teco and Gil brought the situation to the International Rights of Nature Tribunal, a nongovernmental body endorsed by the 2010 World People's Summit on Climate Change and the Rights of Mother Earth, which the Bolivian government hosted in Tiquipaya, Cochabamba. Three environmentalists formed the Tribunal's August 2018 delegation to investigate the TIPNIS situation. In January 2019 they issued a stinging rebuke of the government's actions in TIPNIS, which they characterized as marked by "growing disrespect for the country's legal and institutional framework," and demanded that the highway be suspended, continuing colonization be stopped within TIPNIS, and the August 2017 law be repealed.[88]

Meanwhile new complications emerged for the project. In August 2018 Jacinto Noza, cacique mayor of the previously pro-road CONISUR, demanded a new consultation "to decide how the road will go, how [the communities] want the road to be, because we have to protect the territory and they cannot destroy the forest."[89] (Noza sought a new route that would not go through the unoccupied center of TIPNIS, but along the rivers where most communities lie.) In February 2019 the head of Bolivia's highway administration declared, "The government has decided to put the brakes on the project. We are not starting construction on the highway."[90]

Conclusions

The TIPNIS conflict served to illuminate and deepen other important conflicts within the Bolivian left grassroots, especially between the government and ecologically minded communities. Xavier Albó, the most prominent anthropologist among Bolivia's public intellectuals, characterized the government's strategy in 2012 in four words: "repression, division, exhaustion, deception."[91] However, it also raises three important complications about the alignment of communities within those conflicts.

First, as the split between Subcentral TIPNIS and CONISUR demonstrates, ethnic identity alone is insufficient to predict the political positioning of indigenous peoples. While the transnational indigenous movement has been a prolific generator of ethno-environmental claims, many indigenous peoples and communities are poorly positioned to speak primarily in defense of their land base. This is particularly true in cases of detachment from adequate subsistence (occasioned in this case by deforestation), dependency on outside actors, or intentional reliance on natural resource rents as a local strategy.[92] Moreover, despite the idealization of indigenous integration with the local environment, indigenous autonomy may in fact rely on an economic base of commercial and aid relationships with outsiders. The simultaneous suppression of economic activity (through the government's implementation of Law 180) and organizational funding (through intimidation of development donors such as Ibis) has greatly reduced indigenous organizations' capacity to maneuver independently. In TIPNIS, as in other sites, the indigenous critique of extractive development is grounded in an alternative vision for economic integration rather than a return to (a perhaps idealized) ancestral autarky.

In other words, there is no purely ethical critique of extractivism. Rather, ethno-environmental critiques must be paired with ethno-economic approaches that sustain the resistant community. Accordingly, strategies of collective withdrawal (the search for the Loma Santa, the recovery of land-based subsistence, the creation of traditionalist villages, etc.) and alternative engagement with outsiders (ecotourism, fair trade agriculture, indigenous ownership of environmental knowledge, and reparations) are fundamental parts of indigenous resistance to destructive patterns of development that may be too easily overlooked by theorists of social movements.[93]

Second, the Morales government's decision to prioritize national economic development and export revenues over local indigenous autonomy is a defining moment. It sets a limit on just how far plurinational governance will go, leaving major choices with dramatic local repercussions out of the hands of indigenous traditional authorities. As a Bolivian human rights worker (whom I anonymize here on request) explained to me in June 2015:

> The economy of our state, as in the majority of Latin America, survives based on mining and hydrocarbons. Consequently, from this point of view, it's very complicated to find this equilibrium, this meeting point between the rights of indigenous peoples and the economic development of the national state, or of the plurinational state—which has not changed in absolutely any way when it comes to the subject of development, [whether labeled as] "friendly," "in equilibrium," "with respect for nature and Mother Earth." . . . They say [*speaking in a cute voice*] "indigenous peoples," "the rights of indigenous peoples" [*switching to speaking boldly*] exactly to the point where this does not limit; disadvantage, or hinder this development that the states have.

In a place where there are diverse ways of being indigenous, with different relations of these ways of being and worldviews to the land, this means that some indigenous communities find themselves excluded from the promise of effective self-governance offered by the plurinational state.

Third (and this should temper my first point), economics does not always dictate politics. While the agrarian colonists of Yucumo showed solidarity with the Chapare cocaleros by blocking the march, the colonist communities of San Borja and Caranavi displayed their solidarity with the CIDOB-CONAMAQ march. The fracturing of Bolivia's agrarian colonizers reveals the fundamental ambiguity of their position. As I've suggested in my description of the cocaleros, traditional agricultural communities are both assemblages of individual proprietors committed to selling for the market and achieving greater economic rewards *and* collective endeavors seeking legalization, land titling, and the preservation of an integral (and perhaps sustainable) form of culturally significant use of the land. It is no accident that global organizations of small-scale peasants, led by La Vía Campesina, are seeking rights modeled after those recognized for indigenous peoples, or that the same movements emphasize the cultural and ecological distinctiveness of small-scale agriculture.[94] There are multiple bases for solidarity among grassroots groups, and these go beyond a shared class position to include similar forms of movement organization, reclaiming of cultural traditions, and shared critiques of the centralization of power.

NOTES

1. While exactly which Bolivian government priorities lie behind the project is a complicated and uncertain question, government officials have described it as a way of integrating Beni into the national economy, reducing transportation times and costs between Beni and Cochabamba, linking central Bolivian producers to external markets, limiting the power of right-leaning Santa Cruz over Beni, integrating people living along the road into state-provided services, and generally extending the presence of the state throughout the national territory. Overall, infrastructural and commercial integration is presumed to bring progress and economic development.
2. Gluckman, *Analysis*; Van Velsen, "Extended-Case Method"; Garbett, "Analysis"; Burawoy, "Extended Case Method," 18, 17.
3. Marcus, "Ethnography."

4. One study also bears the imprint of Vice President Álvaro García Linera as research coordinator but challenges positions he later took in government.
5. Ernesto Noe quoted in Canedo Vásquez, *La Loma Santa*, 98.
6. Fabricano would later be head of CIDOB from 1994 to 1996. Marchers estimate provided in García Linera, Chávez León, and Costas Monje, *Sociología*, citing Lehm Ardaya, *Milenarismo y movimientos*.
7. Seven points of the twelve-point declaration issued by the Pact and its allies at the II Encuentro de Organizaciones Campesinas, Indígenas y Originarias in Santa Cruz (on June 1–3, 2006) concerned indigenous autonomy. The Pact's answer to "What Is a Plurinacional State?" is spelled out in their proposed draft constitution issued on August 5, 2006. Pacto de Unidad, *El Pacto de Unidad*, 45–46, 145–46.
8. This reportedly included the refusal of some CIDOB marchers to sign any document with the signature of Autonomies Minister Carlos Romero and the physical ripping of the agreement by a government negotiator. Peña y Lillo, "Autonomía"; CIPCA, "Concluye."
9. By Supreme Decree 22610 on September 24, 1990.
10. Orozco Ramírez, García Linera, and Stefanoni, *No somos juguete de nadie*, 62–65; Canedo Vásquez, *La Loma Santa*, 135.
11. An additional 137,783 hectares, part of twenty-five ranching properties in the north of TIPNIS, remained to be titled to third-party owners as of 2010. Fundación Tierra, *Territorios*, 265–73.
12. The Central de Pueblos Étnicos Mojeños del Beni split from the Central de los Pueblos Indígenas de Beni following a conflict detailed in Canedo Vásquez, *La Loma Santa*, 137–38. Orozco Ramírez, García Linera, and Stefanoni report there "isn't much clarity in the territorial reach that it [the Subcentral Sécure] would articulate" (*No somos juguete de nadie*, 66). See also Albó, "Una mirada," 32–34. Both organizations affiliate with CIDOB and both were present in the 2011 and 2012 national indigenous marches.
13. Canedo Vásquez, *La Loma Santa*, 71. Census data have remained incomplete: there were some thirteen uncounted communities during that census; figures in 2001 were not disaggregated; and in 2012 the conflict with the government led many TIPNIS communities to refuse the presence of census officials. In addition the large family size in 1993, over six per household, predicts high population growth. On the limits of the 1993 count, see Movimiento Regional por la Tierra and Colectivo Autogestionario Territorios en Resistencia, "TIPNIS," 5. On community concerns with the 2012 census, see Subcentral TIPNIS, Comisión Recorrido, and Bascopé Sanjinés, *Informe*, 51–52.
14. Indigenous and Community Conserved Areas, "Global ICCA Database."
15. "CONISUR es una organización básicamente creado y apoyado por la Prefectura de Cochabamba a la cual le sirve como mecanismo para canalizar ayuda a las comunidades del sur y de este modo defender su soberanía sobre el area." Surkin, *Caracterización*, 28. This description was used on CPITCO's official website in 2011: http://www.cpitco.org/consejos/trinitario2.html (accessed December 29, 2011). In June 2012 CPITCO de-recognized CONISUR's affiliation with it. "CPITCO desconoce al Conisur."

16. Orozco Ramírez, García Linera, and Stefanoni, *No somos juguete de nadie*, 68.
17. Bolivia's Fundación Tierra estimated without explanation that some twenty thousand cocalero families live in the zone, a high percentage of the overall Chapare cocaleros population. In 1995 an expert estimate of the cocaleros population was 9,800. Orozco Ramírez, García Linera, and Stefanoni, *No somos juguete de nadie*, 87.
18. Orozco Ramírez, García Linera, and Stefanoni, *No somos juguete de nadie*, 74.
19. Throughout this period small patches of land clearing dominated the process of deforestation. Siles, "Spatial Modelling," 36–38.
20. Programa de Investigación Ambiental, *Viabilidad Economica*.
21. Canedo Vásquez, *La Loma Santa*, 58.
22. Canedo Vásquez, *La Loma Santa*, 81–83.
23. Canedo Vásquez, *La Loma Santa*, 55.
24. Canedo Vásquez, *La Loma Santa*, 143. On "grounded utopian movements," see Price, Nonini, and Tree, "Grounded Utopian Movements."
25. Printed in Subcentral TIPNIS and Campaña en Defensa del TIPNIS, *Memoria*, 33. For a parallel argument for other Moxeño communities, see Pedro Nuni as quoted in Canedo Vásquez, *La Loma Santa*, 150.
26. Wilder Molina, quoted in Orozco Ramírez, García Linera, and Stefanoni, *No somos juguete de nadie*, 97; emphasis mine.
27. Orozco Ramírez, García Linera, and Stefanoni, *No somos juguete de nadie*, 95.
28. Canedo Vásquez, *La Loma Santa*, 151.
29. Canedo Vásquez, *La Loma Santa*, 121–30.
30. Grisaffi, "We Are Originarios," 429.
31. An oral historical description appears in Komadina and Geffroy, *El Poder*, 81–82.
32. A representative sample of these efforts is detailed in the following newspaper articles: "Comisión de Reforma Agraria Realiza Una Inspección En Chapare"; "Iniciaron Estudios Estadísticos de Áreas Cultivables En Chapare"; "Se Inició Racionalización de Cultivos de Coca En Chapare"; "Cultivo de Té En Chapare Se Inciará Próximamente."
33. Grisaffi, "We Are Originarios."
34. Ortuño, *De la coca*, 214.
35. Komadina and Geffroy, *El Poder*.
36. Evo Morales, speech at UN high-level event commemorating the tenth anniversary of the UN Declaration on the Rights of Indigenous Peoples, April 25, 2017, https://www.youtube.com/watch?v=9c-9-heblls.
37. Orozco Ramírez, García Linera, and Stefanoni, *No somos juguete de nadie*, 93.
38. Orozco Ramírez, García Linera, and Stefanoni, *No somos juguete de nadie*, 93.
39. "Indígenas En El TIPNIS Viven de Vender Productos a Colonos"; Orozco Ramírez, García Linera, and Stefanoni, *No somos juguete de nadie*, 90–102.
40. According to Leonardo Tamburini of the Ceter for Juridical Studies and Social Investigation, "La Carretera."

41. Resolution 0001/2010 of the 29th Extraordinary Meeting of Corregidores of the Autonomous Isiboro-Securé Indigenous Territory and National Park from the Moxeño, Yuracaré, and Chimane Indigenous Peoples, May 18, 2010.
42. Quoted in Orozco Ramírez, García Linera, and Stefanoni, *No somos juguete de nadie*, 110.
43. "Se reactiva demanda para vía por el Tipnis."
44. Albó, "Una mirada," 35.
45. Kenner, "Interviews," translated by Kenner.
46. Thereby violating norms of free consultation.
47. Paz Patiño, "El Limite Yuracaré."
48. Ravindran, "Geographies"; Canessa, *Intimate Indigeneities*.
49. No one was more flummoxed to be labeled antigovernmental conspirators than the staff of CEJIS, an NGO based in Santa Cruz. CEJIS had advised the MAS-IPSP since the 1990s, and no fewer than five former CEJIS staffers served as ministers during Morales's first term.
50. "La consulta por el TIPNIS no será vinculante."
51. "Defensor del Pueblo."
52. Subcentral TIPNIS, "Resolución No. 1/2011."
53. Quoted in Fundación Tierra, *Marcha*, 55.
54. Bjork-James, "Claiming Space," 266–68.
55. In September 2010 documents (Declaración de Principios and Reglamento), CONALCAM redefined itself as "the strategic alliance of the Plurinational State," including the executive and legislative branches of government, the MAS-IPSP, and social movement organizations; that it "recognizes and backs the leadership of brother Evo Morales Ayma," and that it takes as one of its purposes to "impede, denounce, and sanction the infiltration of counter-revolutionary enemies within CONALCAM." Under Article 14 of the regulations, organizations within CONALCAM are obliged to "defend and make respected the leadership of the President." (These documents are in my possession.)
56. Participants also included Gustavo Guzmán (former Bolivian ambassador to the United States), Lino Villca (former MAS senator), Omar Fernández (former irrigators' union leader in Cochabamba), Moíses Torres and Vladimir Machicado (Movimiento Sin Tierra, or Landless Movement), feminists, university students, and El Alto community leaders. "Cerca de 50 Personas Parten de La Paz a Reforzar La Marcha."
57. The vice minister's comments were video-recorded and leaked to the press in 2017; they are archived online at https://www.youtube.com/watch?v=ec7td4fyeai. My transcription and translation are available at https://woborders.blog/2017/08/06/marcos-farfan-2011/.
58. "Tras Intervención, Heridos Se Recuperan Por Cuenta Propia"; "Indígenas Presentan Demanda Penal Contra Evo, Álvaro y Otros Por Genocidio."
59. Solón, "Carta."
60. Other than preliminary work on a bridge, it is not clear that work on Segment 2 had started.

61. In January 2012 Nuni, Zacu, and Julio Cortez distanced themselves from the MAS-IPSP and formed an independent indigenous bloc in the Chamber of Deputies. Before the 2014 national election, the MAS-IPSP reorganized the election of special indigenous representatives, and its own party rules, to avoid future defections.
62. For analysis of the content and prospects of these agreements see Fundación Tierra, *Marcha*, 11–42.
63. Albó, "Una mirada," 19–20, 46–48.
64. Mita, "La marcha."
65. Ministerio de Obras Públicas, Servicios y Vivienda, *Informe*, 68.
66. The Subcentral TIPNIS and the government reached an agreement on December 4, 2011. The government's unilateral Supreme Decree 1146, which forbids all commercial use of natural resources, was issued on February 24, 2012.
67. "La consulta del Gobierno fue precedida y acompañada por regalos, prebendas y promesas de desarrollo y servicios" (The government's consultation was preceded and accompanied by gifts, bribes, and promises of development and services). "Documentan Las Denuncias Contra Consulta En Tipnis."
68. Guarachi, "En el Tipnis"; Ministerio de Obras Públicas, Servicios y Vivienda, *Informe*.
69. It labels them "comunidad[es] que no desea[n] ejercer su derecho." Ministerio de Obras Públicas, Servicios y Vivienda, *Informe*, 85–97.
70. Subcentral TIPNIS, Comisión Recorrido, and Bascopé Sanjinés, *Informe*, 67–74.
71. Asamblea Permanente de Derechos Humanos de Bolivia and Federación Internacional de Derechos Humanos, *Informe*, 13, 20.
72. Ministerio de Obras Públicas, Servicios y Vivienda, *Informe*, 300–307, 280–95.
73. "Evo anuncia decisión de anular contrato con OAS para carretera."
74. "Carretera por el TIPNIS ya no tiene financiamiento."
75. Mealla, "La Asamblea."
76. The only signature on the Unity Pact's resolution from within CIDOB was that of CONISUR leader Gumercindo Pradel. (Document archived at https://woborders.files.wordpress.com/2019/08/pactodeunidad-nov2011.pdf.)
77. Schavelzon, *El nacimiento*, 182, 183.
78. A MAS deputy denounced that the government funded this gathering and activists circulated an approved budget for its expenses. "Persiste conflicto en Conamaq y se convoca a nuevo congreso."
79. Erbol cited the Central de Pueblos Indígenas de La Paz organizers' explanation that "the government has a list of indigenous leaders that challenge [it] by putting forward their rights and their demands, to whom the ministries shut their doors and do not respond to the demands that their communities have" Apaza, "El movimiento."
80. Apaza, "El movimiento."
81. Lundsgaarde, *Domestic Politics*.
82. Corz, "Gobierno."

83. Layme, "Dirigentes." Formally, the Development Fund for Indigenous and Native Peoples and Campesino Communities (Fondo de Desarrollo para los Pueblos Indígenas Originarios y Comunidades Campesinas). Given the small scale of many recipient organizations, this kind of bank transfer did not always indicate that the project was carried out in a corrupt manner.
84. "Ratifican detención de Melva Hurtado por caso Fondioc"; Luizaga and Cusicanqui, "Justicia"; "COICA no consigue asilo para el tacana Adolfo Chávez."
85. Corz, "García."
86. "La Ley de Desarrollo Integral Del TIPNIS Nació de La Consulta a Sus Propios Habitantes."
87. "Empresas ya concluyeron los tres puentes en el Polígono 7 del Tipnis."
88. International Tribunal for the Rights of Nature, "Report," 21–23.
89. "Conisur exige otra consulta para ejecutar carretera por el Tipnis."
90. "La ABC afirma que el Gobierno frenó la carretera por el Tipnis."
91. "Albó: El Gobierno."
92. For discussion of how different Yuracaré communities approach market relations, see Paz Patiño, "El Limite Yuracaré." For a look at how Guaraní indigenous communities have sought to incorporate hydrocarbon extraction into their quest for territorial autonomy, see Anthias, "From 'Territories.'"
93. For broader arguments in this direction, see Simpson, *Dancing*; LaDuke, *All Our Relations*; Price, Nonini, and Tree, "Grounded Utopian Movements."
94. See Edelman and James, "Peasants' Rights."

BIBLIOGRAPHY

"Albó: El Gobierno Usó La Estrategia Del Desgaste, La Confrontación y El Engaño Para Validar Su Consulta." *Erbol Digital*, July 5, 2012. Archived at https://cedib.org/post_type_titulares/albo-el-gobierno-uso-la-estrategia-del-desgaste-la-confrontacion-y-el-engano-para-validar-su-consulta-erbol-6-7-2012/.

Albó, Xavier. "Una mirada profunda del TIPNIS." In *Coraje, Memorias de la Octava Marcha Indígena por la Defensa del TIPNIS*, edited by Alex Contreras Baspineiro. Cochabamba: Industria Gráfica "J.V." Editora, 2012. https://www.alainet.org/images/CORAJE%20tipnis.pdf.

Anthias, Penelope. "From 'Territories of Difference' to 'Ethnic Governable Spaces'? Shifting Visions of Territory and Autonomy in the Bolivian Chaco." Paper presented at LASA 2015, San Juan, Puerto Rico, 2015. https://www.academia.edu/12755655/From_Territories_of_Difference_to_Ethnic_Governable_Spaces_Shifting_Visions_of_Territory_and_Autonomy_in_the_Bolivian_Chaco_paper_presented_at_LASA_2015_.

Apaza, Ana. "El movimiento indígena está en crisis, pero no destruido." Erbol Digital, January 27, 2014. http://www.erbol.com.bo/noticia/indigenas/27012014/el_movimiento_indigena_esta_en_crisis_pero_no_destruido.

Asamblea Permanente de Derechos Humanos de Bolivia and Federación Internacional de Derechos Humanos. "Informe de Verificación de La Consulta Realizada En El Territorio Indígena Parque Nacional Isiboro Sécure." 2013. https://www.fidh.org/img/pdf/bolivia609esp2013.pdf.

Bjork-James, Carwil. "Claiming Space, Redefining Politics: Urban Protest and Grassroots Power in Bolivia." City University of New York, 2013. https://academicworks.cuny.edu/gc_etds/3329/.

Burawoy, Michael. "The Extended Case Method." *Sociological Theory* 16, no. 1 (March 1998): 4–33. https://doi.org/10.2307/202212.

Canedo Vásquez, Gabriela. *La Loma Santa una utopía cercada: Territorio, cultura y estado en la Amazonía boliviana*. La Paz: Plural Editores, 2011.

Canessa, Andrew. *Intimate Indigeneities: Race, Sex, and History in the Small Spaces of Andean Life*. Narrating Native Histories. Durham NC: Duke University Press, 2012.

"Carretera por el TIPNIS ya no tiene financiamiento." *Opinión*, June 15, 2012. http://www.opinion.com.bo/opinion/articulos/2012/0615/noticias.php?id=60112.

"Cerca de 50 Personas Parten de La Paz a Reforzar La Marcha." *Página Siete*, September 21, 2011.

Centro de Investigación y Promoción del Campesinado. "Concluye La VII Marcha de Los Pueblos Indígenas." July 23, 2010. http://www.cipca.org.bo/index.php/quienes-somos/category.php?id=85. Site no longer active.

"COICA no consigue asilo para el tacana Adolfo Chávez." *Erbol Digital*, June 24, 2016. http://www.erbol.com.bo/noticia/indigenas/24062016/coica_no_consigue_asilo_para_el_tacana_adolfo_chavez.

"Comisión de Reforma Agraria Realiza Una Inspección En Chapare." *Los Tiempos*, April 27, 1974.

"Conisur exige otra consulta para ejecutar carretera por el Tipnis." *Los Tiempos*, August 24, 2018. http://www.lostiempos.com/actualidad/economia/20180824/conisur-exige-otra-consulta-ejecutar-carretera-tipnis.

Corz, Carlos. "García perfila 2 décadas de extractivismo y afianzar la producción frente al informe del PNUD." *La Razón*, August 18, 2016. http://la-razon.com/economia/Garcia-extractivismo-produccion-PNUD-Bolivia-pobreza_0_2547945273.html.

———. "Gobierno expulsa de Bolivia a la ONG IBIS por injerencia política." *La Razón*, December 20, 2013. http://www.la-razon.com/nacional/Gobierno-Bolivia-ONG-IBIS-injerencia_0_1964803571.html.

"CPITCO desconoce al Conisur; Pradel dice que abandonan esa regional." *Opinión*, June 18, 2012. http://www.opinion.com.bo/opinion/articulos/2012/0618/noticias.php?id=60524.

"Cultivo de Té En Chapare Se Inciará Próximamente." *Los Tiempos*, April 14, 1976.

"Defensor del Pueblo: 'Coraite debe pedir perdón a los indígenas.'" *El Día*, September 7, 2011. https://www.eldia.com.bo/index.php?c=&cat=1&pla=3&id_articulo=73150.

"Documentan Las Denuncias Contra Consulta En Tipnis." *Los Tiempos*, March 27, 2013. Archived at https://cedib.org/post_type_titulares/documentan-las-denuncias-contra-consulta-en-tipnis-los-tiempos-27-3-13/.

Edelman, Marc, and Carwil James. "Peasants' Rights and the UN System: Quixotic Struggle? Or Emancipatory Idea Whose Time Has Come?" *Journal of Peasant Studies* 38, no. 1 (January 2011): 81–108.

"Empresas ya concluyeron los tres puentes en el Polígono 7 del Tipnis." *Los Tiempos*, August 11, 2018. http://www.lostiempos.com/actualidad/economia/20180812/empresas-ya-concluyeron-tres-puentes-poligono-7-del-tipnis.

"Evo anuncia decisión de anular contrato con OAS para carretera." *La Razón*, April 11, 2012. http://www.la-razon.com/nacional/Evo-decision-contrato-oas-carretera_0_1594040630.html.

Fundación Tierra. *Marcha Indígena Por El TIPNIS: La Lucha En Defensa de Los Territorios*. La Paz: Fundación Tierra, 2012.

———. *Territorios Indígena Originario Campesinos En Bolivia: Entre La Loma Santa y La Pachamama*. La Paz: Fundación Tierra, 2011.

Garbett, G. Kingsley. "The Analysis of Social Situations." *Man*, new series, 5, no. 2 (June 1970): 214–27.

García Linera, Álvaro, Marxa Chávez León, and Patricia Costas Monje. *Sociología de Los Movimientos Sociales En Bolivia: Estructuras de Movilización, Repertorios Culturales y Acción Política*. La Paz: DIAKONIA, Accion Ecuménica Sueca, 2004.

Gluckman, Max. *Analysis of a Social Situation in Modern Zululand*. Manchester, U.K.: Published on behalf of the Rhodes-Livingstone Institute by the Manchester University Press, 1958.

Grisaffi, Thomas. "We Are Originarios . . . 'We Just Aren't from Here': Coca Leaf and Identity Politics in the Chapare, Bolivia." *Bulletin of Latin American Research* 29, no. 4 (2010): 425–39. https://doi.org/10.1111/j.1470–9856.2010.00385.x.

Guarachi, Angel. "En el Tipnis 55 comunidades apoyan vía por el parque; El Gobierno anuncia construir una carretera ecológica." *La Razón*, December 7, 2012. http://www.la-razon.com/index.php?_url=/nacional/Tipnis-comunidades-Gobierno-construir-carretera_0_1738026274.html.

"Indígenas En El TIPNIS Viven de Vender Productos a Colonos." *La Razón*, October 14, 2011. http://www.la-razon.com/index.php?_url=/nacional/Indigenas-TIPNIS-vender-productos-colonos_0_1484251581.html.

"Indígenas Presentan Demanda Penal Contra Evo, Álvaro y Otros Por Genocidio." *Erbol Comunicaciones*, October 15, 2011.

Indigenous and Community Conserved Areas. "Global ICCA Database: CONISUR—TIPNIS." Accessed June 25. 2017. https://www.iccaconsortium.org/wp-content/uploads/images/stories/Database/1-%20isiboro_secure_bolivia_icca_database.pdf. Site no longer active.

"Iniciaron Estudios Estadísticos de Áreas Cultivables En Chapare." *Los Tiempos*, February 20, 1976.

International Tribunal for the Rights of Nature. "Report of the Commission of the International Rights of Nature Tribunal on the Case of the Isiboro Sécure Indigenous Territory and National Park (TIPNIS-Bolivia)." January 16, 2019. https://therightsofnature.org/wp-content/uploads/2019/01/Report-TIPNIS-Commission-final-pdf.pdf.

Kenner, Dario. "Interviews: Bolivian Indigenous Leaders from CONISUR and CIDOB on TIPNIS Conflict." *Bolivia Diary* (blog), February 6, 2012. https://boliviadiary.wordpress.com/2012/02/06/interviews-bolivian-indigenous-leaders-from-conisur-and-cidob-on-tipnis-conflict/.

Komadina, Jorge, and Céline Geffroy. *El Poder Del Movimiento Político: Estrategia, Tramas Organizativas e Identidad Del MAS En Cochabamba (1999–2005)*. La Paz: Fundación PIEB, 2007.

"La ABC afirma que el Gobierno frenó la carretera por el Tipnis." *Los Tiempos*, February 16, 2019. http://www.lostiempos.com/actualidad/pais/20190216/abc-afirma-que-gobierno-freno-carretera-tipnis.

"La consulta por el TIPNIS no será vinculante." *La Razón*, August 1, 2011. http://www.la-razon.com/index.php?_url=/economia/consulta-TIPNIS-vinculante_0_1441655852.html.

"La Ley de Desarrollo Integral Del TIPNIS Nació de La Consulta a Sus Propios Habitantes." *Tiempo y Materia (Cámera de Diputados)* 22 (2017). http://www.diputados.bo/sites/default/files/tiempo22.pdf.pdf.

LaDuke, Winona. *All Our Relations: Native Struggles for Land and Life*. Cambridge MA: South End Press, 1999.

Layme, Beatriz. "Dirigentes fueron parte del desfalco al Fondo Indígena." *Página Siete*, August 15, 2016. http://www.paginasiete.bo/economia/2016/8/15/dirigentes-fueron-parte-desfalco-fondo-indigena-106228.html.

Lehm Ardaya, Zulema. *Milenarismo y movimientos sociales en la Amazonia boliviana: La búsqueda de la loma santa y la marcha indígena por el territorio y la dignidad.* Santa Cruz, Bolivia: Centro de Investigación y Documentación para el Desarrollo del Beni, 1999.

Luizaga, Dennis, and Juan José Cusicanqui. "Justicia rechaza el cese de La detención preventiva de exministra Achacollo." *La Razón*, May 23, 2017. http://www.la-razon.com/nacional/Justicia-detencion-preventiva-exministra-Achacollo-Bolivia_0_2714728591.html.

Lundsgaarde, Erik. *The Domestic Politics of Foreign Aid*. London: Routledge, 2013.

Marcus, George E. "Ethnography in/of the World System: The Emergence of Multi-Sited Ethnography." *Annual Review of Anthropology* 24, no. 1 (1995): 95–117. https://doi.org/10.1146/annurev.an.24.100195.000523.

Mealla, Luis. "La Asamblea no tratará aún leyes para el TIPNIS." *La Razón*, April 11, 2013. http://www.la-razon.com/nacional/Asamblea-tratara-leyes-tipnis_0_1813018726.html.

Ministerio de Obras Públicas, Servicios y Vivienda. *Informe Final Del Proceso de Consulta Previa, Libre e Informada a Los Pueblos Moxeño-Trinitario, Yuracaré y Chimane Del Territorio Indígena y Parque Nacional Isiboro Sécure (TIPNIS) 2012*. La Paz, 2013.

Mita, Gladys. "La marcha del Conisur llega a La Paz sin lograr apoyo multitudinario." *Opinión*, January 31, 2012. http://www.opinion.com.bo/opinion/articulos/2012/0131/noticias.php?id=41816.

Movimiento Regional por la Tierra and Colectivo Autogestionario Territorios en Resistencia. "TIPNIS: Defendiendo Nuestra Casa Grande." Estudios de Caso. Movimiento Regional por la Tierra, 2016. https://porlatierra.org/casos/90.

Orozco Ramírez, Shirley, Álvaro García Linera, and Pablo Stefanoni. *No somos juguete de nadie: Análisis de la relación de movimientos sociales, recursos naturales, Estado y descentralización*. Cochabamba: Plural Editores, 2006.

Ortuño, Fernando Salazar. *De la coca al poder: políticas públicas de sustitución de la economía de la coca y pobreza en Bolivia, 1975–2004*. Buenos Aires: Consejo Latinoamericano de Ciencias Sociales, 2008.

Pacto de Unidad. *El Pacto de Unidad y El Proceso de Construcción de Una Propuesta de Constitución Política Del Estado: Sistematización de La Experiencia*. Edited by Fernando Garcés. La Paz: Unitas, 2010. http://redunitas.org/wp-content/uploads/2019/04/PACTO_UNIDAD.pdf.

Paz Patiño, Sarela. "El Limite Yuracaré." *Facetas*, July 24, 1994. http://www.bibvirtual.ucb.edu.bo/etnias/digital/106000425.pdf.

Peña y Lillo, Christian. "Autonomía. Gobierno vuelve a frenar la marcha de indígenas con nuevo acuerdo." *El Deber*, July 23, 2010. http://eju.tv/2010/07/autonoma-gobierno-vuelve-a-frenar-la-marcha-de-indgenas-con-nuevo-acuerdo/.

"Persiste conflicto en Conamaq y se convoca a nuevo congreso." *La Patria*, December 12, 2013. http://www.lapatriaenlinea.com/?nota=166614.

Price, Charles, Donald Nonini, and Erich Fox Tree. "Grounded Utopian Movements: Subjects of Neglect." *Anthropological Quarterly* 81, no. 1 (2008): 127–59. https://doi.org/10.1353/anq.2008.0005.

Programa de Investigación Ambiental. *Viabilidad Economica Para El Desarrollo de Iniciativa Que Reduzcan La Deforestacion En El TIPNIS*. La Paz: PIEB, 2011. https://www.scribd.com/document/68630238/Investigacion-TIPNIS-PIEB.

"Ratifican detención de Melva Hurtado por caso Fondioc." *Erbol Digital*, January 29, 2016. http://www.erbol.com.bo/noticia/seguridad/29012016/ratifican_detencion_de_melva_hurtado_por_caso_fondioc.

Ravindran, Tathagatan. "Geographies of Indigenous Identity: Spatial Imaginaries and Racialised Power Struggles in Bolivia." *Antipode* 51, no. 3 (2019): 949–67. https://doi.org/10.1111/anti.12517.

Schavelzon, Salvador. *El nacimiento del estado plurinacional de Bolivia: Etnografía de una Asamblea Constituyente*. La Paz: Plural, 2012.

"Se Inició Racionalización de Cultivos de Coca En Chapare." *La Prensa*, June 20, 1976.

"Se reactiva demanda para vía por el Tipnis." *El Día*, November 20, 2011. https://www.eldia.com.bo/index.php?c=&cat=1&pla=3&id_articulo=79094.

Siles, Nelson Jery Sanabria. "Spatial Modelling and Prediction of Tropical Forest Conversion in the Isiboro Sécure National Park and Indigenous Territory (TIPNIS), Bolivia." Master's thesis, International Institute for Geo-Information Science and Earth Observation, 2009. http://www.itc.nl/library/papers_2009/msc/nrm/siles.pdf.

Simpson, Leanne. *Dancing on Our Turtle's Back: Stories of Nishnaabeg Re-Creation, Resurgence, and a New Emergence*. Winnipeg: Arbeiter Ring, 2014.

Solón, Pablo. "Carta abierta sobre TIPNIS a Presidente Evo Morales." Upside Down World, September 28, 2011. http://upsidedownworld.org/noticias-en-espa/noticias-en-espa-noticias-en-espa/carta-abierta-sobre-tipnis-a-presidente-evo-morales-pablo-solon/.

Subcentral TIPNIS. "Resolución No. 1/2011." June 9, 2011. https://www.scribd.com/document/352691719/Subcentral-tipnis-Resolucio-n-No-01–2011.

Subcentral TIPNIS and Campaña en Defensa del TIPNIS. *Memoria: Foro Departamental: Territorio Indígena y Parque Nacional Isiboro Sécure (Cochabamba, 23 de Septiembre de 2010)*. Cochabamba: CENDA, 2010.

Subcentral TIPNIS, Comisión Recorrido, and Iván Bascopé Sanjinés. *Informe Del Recorrido Realizado Por Las Comunidades Del TIPNIS: El Proceso de Consulta Propio y La Visión de Desarrollo Para El Territorio Indígena*. La Paz: Red Jurídica Amazónica—Fundación Construir, 2012.

Surkin, Jordi Beneria. *Caracterización Técnica de Aspectos Históricos, Culturales, Socioeconómicos, Organizativos y de Gestión Del Territorio Indígena Parque Nacional Isiboro Secure (TIPNIS)*. La Paz: GTZ, 2002.

Tamburini, Leonardo. "La Carretera Del 'Progreso.'" *La Razón*, July 31, 2011.

"Tras Intervención, Heridos Se Recuperan Por Cuenta Propia." *Página Siete*, October 6, 2011.

Van Velsen, Jaap. "Extended-Case Method and Situational Analysis." In *The Craft of Social Anthropology*, edited by A. L. Epstein. London: Tavistock, 1967.

5 Environmental Justice in the REDD+ Frontier
Experiences from the Amazon and Beyond

JUAN PABLO SARMIENTO BARLETTI AND ANNE M. LARSON

Prologue: From Rights to Justice

The deforestation and degradation of forests in the Global South have become central to current global efforts to address climate change. The United Nations Framework Convention on Climate Change's (UNFCCC) initiative for reducing emissions from deforestation and forest degradation (REDD+) is perhaps the most notable in this regard. The initiative intends to reward participating nations economically in five areas: (1) reducing emissions from deforestation; (2) reducing emissions from forest degradation; (3) conservation of forest carbon stocks; (4) sustainable management of forests; and (5) enhancement of forest carbon stocks.[1]

As with most global solutions, there are marked debates for and against REDD+ in the specialized literature.[2] Some analysts see it as a transformative solution with the potential to benefit the environment, forest-dependent communities, and governments of the Global South, while contributing significantly to climate mitigation. Others argue that it is only a "greenwashing" governance mechanism that reifies government control over forests, stripping local communities of their rights and livelihoods and excluding them from forest governance while forcing them to solve a problem created mainly in the industrialized North. Much of the critique of REDD+ centers on sorting out winners and losers, commonly seen as private and/or government interests (winners) versus those of local and indigenous communities (losers). On the one hand, this position has been strengthened by a series of reports on transgressions of the rights of indigenous peoples in the wider context of REDD+. On the other, indigenous peoples and local communities have taken advantage of REDD+ as a platform for putting their rights on the global agenda, and in some cases they have been able to exert significant influence over national policy.[3]

This chapter builds on a preliminary review of published allegations of abuses of the rights of indigenous peoples in the context of REDD+ readiness and implementation, which was undertaken by the authors.[4] In this chapter we approach the issue through the lens of environmental justice. Currently victories in terms of justice for indigenous and local populations in relation to REDD+ cannot be attributed to its design but rather occurred because social movements have been able to strategically and creatively make gains within the context of REDD+ negotiations, preparation or readiness, and implementation.

Although REDD+ may be an innovative global solution for a global problem, it must still be implemented locally.[5] This may sound straightforward, but it is worth emphasizing, as it reveals the way in which REDD+, like other global solutions before it, has been conceived and is being implemented. In light of the evidence in the scholarly literature, we argued that this awareness of local contexts is especially important considering the safeguards for protecting indigenous and local communities that were adopted in Cancún, at the UNFCCC's Conference of the Parties 16, as a sort of "do no harm" guideline.[6]

The lack of attention to the different political, economic, environmental, and social issues at play in the national and local contexts in which REDD+ is being rolled out (to which we refer hereafter as the REDD+ frontier) reveals a need for a more reflective engagement with this specificity.[7] As the criticisms of mainstream approaches in international development show, this is not a new problem but one commonly faced when global solutions are applied to local peoples.[8]

The literature also notes a challenge in how parties to the UNFCCC are being asked to engage with REDD+ safeguards. The UNFCCC currently requires that, to be considered for results-based payments, countries promote and support safeguards in their REDD+ activities.[9] This is currently addressed by the Warsaw Framework's requirement for parties to maintain Safeguard Information Systems, but it remains to be seen how this will work in practice.[10] Some observers note that safeguards will be based on national legal frameworks and country-specific political, economic, and social priorities, as well as existing legal interpretations of relevant rights.[11] All of these, in turn, are based on different histories of interactions between states and between states and their citizens. For indigenous citizens in nations of the Global South, these histories have often been characterized by forced removal from their territories, discriminatory decision-making practices, and structural racism.[12] These historical experiences guide most legal systems in REDD+ countries and, concomitantly, determine how rights are practiced for indigenous peoples.[13]

Our review revealed a number of situations related to REDD+ in which the rights of indigenous peoples have been transgressed. This is notable, as these transgressed rights, and those mentioned in the Cancún Safeguards, are among those granted to indigenous peoples by the United Nations Declaration on the Rights of Indigenous Peoples (UNDRIP) and other relevant agreements, such as International Labor Organization Convention 169 (ILO 169). While the reference to UNDRIP in the Cancún Safeguards is positive, "the Declaration itself does not provide specific legal mechanisms to achieve recognition of local community rights and interests that are directly applicable in a REDD+ context."[14] In other words, whether or not a rights-based REDD+ is applied depends not on how it has been conceived at the level of the UNFCCC but on the regulatory framework of the state in the varied contexts in which it is being readied and implemented.

To some observers, referring to actions that contravene these international agreements as abuses or violations of rights may seem excessive. However, the rights of indigenous peoples recognized by UNDRIP include self-determination, protection from cultural destruction, and participation in decisions that affect them.[15] Although the most common allegations of rights abuses in the context of REDD+ national readiness and/or subnational implementation were related to self-determination, there were also cases of open violence against indigenous groups by government and private forces. Thus, though all the allegations of rights abuses in the literature may not have been caused by REDD+, there is still a need for greater awareness that the REDD+ frontier overlaps contexts of abuses, displacement, and dispossession, as well as the repression and assassination of environmental activists. REDD+ has often proven ill-prepared to navigate these tensions.

It is precisely because of its focus on forests of the Global South that REDD+ can become a tool for environmental justice. Its frontier overlaps spaces of large-scale extractive activity and its concomitant land degradation and pollution as well as a wide scope of rights abuse.[16] These are spaces of "ecologically unequal exchange" in which disempowered nations have little option to export goods that may have a lower socioecological impact on local forests and communities.[17] For scholars and activists working under the banner of environmental justice, this unequal economic and ecological exchange becomes an ecological debt that is one of the key manifestations of Global North-South environmental injustice.[18] In discussions about REDD+, this unequal exchange takes the form of the potential commodification of carbon, also known as carbonization or carbonification in the specialized literature, by which forests become primarily valued for their carbon content.[19] This concern is related to the criticism leveled at REDD+ since its early days, of the initiative's potential to become an offset mechanism for developed countries to compensate for their own role in climate change.[20] Indigenous peoples are already dependent on ecosystem services and run the risk of suffering further injustice from a mechanism designed to address a global problem that is not of their making.[21]

This is a prime example of the kind of potential environmental injustice occurring in the REDD+ frontier, which REDD+ can address through an emphasis on the rights of indigenous and local populations. Perhaps even more important, we argue that REDD+ can check environmental injustice by promoting justice at the local level, by ensuring access to the material and nonmaterial benefits that come with these rights.[22] In this chapter we explore how REDD+ can move from a "do no harm" approach to one that enables the struggle for justice by communities in the Amazon and beyond. We start by examining environmental justice as an analytical tool and concept based on experience, in light of wider theoretical discussions of the redistributive, recognitional, and representational characteristics of injustice in postcapitalist societies.[23] We then explore the unequal power relations at play in REDD+ as a frontier of interactions among actors, discourses, and knowledge systems marked by power inequalities. We illustrate these different characteristics of injustice by examining three key themes of injustice in the context of REDD+: land tenure, gender, and participation.

The exploration that follows is set in the Amazon for four main reasons: (1) it is the traditional territory of a large number of indigenous peoples; (2) it is within the administrative and legal power of nine different nations and thus demonstrates a variety of development agendas that may be at odds with REDD+ and addressing climate change; (3) it has the largest remaining stock of tropical forests; and (4) it faces an ongoing and violent process of deforestation. The combination of these characteristics has made it a discursively and symbolically powerful space in the global struggle to address climate change.

The Nuances of Environmental Justice

"Environmental justice," as an analytical term, emerged in the early 1980s from the experience of African American communities of waste dumping in the United States. Since then, analysts, activists, and practitioners have consistently shown how low-income and nonwhite communities suffer "environmental racism," as they bear the brunt of the environmental impacts from, for example, incinerators, refineries, and waste dumps.[24] Analysts note the unequal burden of pollution placed on these communities, which are also commonly denied access to resources and decision-making processes related to the environment where they live. Currently, as Schroeder et al. note, the term has been used in the Global North in relation to the unequal distribution of environmental impacts and in the Global South with regard to the unequal distribution of benefits.[25] This awareness of inequality is related to the now widely acknowledged warning that the negative impacts of climate change are more likely to affect local populations, and especially women within those populations, who are not historically large contributors to greenhouse emissions yet who already live in precarious environments and suffer from political disempowerment and social exclusion.[26]

The shortcomings of previous initiatives for global forest governance foreshadow the potential impact that REDD+ may have on the lives of forest-dependent peoples today.[27] Prior research demonstrates that forest-dependent communities have suffered distributive, procedural, and recognitional injustices (including displacement, an unequal burden of conservation, and the dismissal of local experiences of the forest) as a result of conservation initiatives.[28] For most of the specialized literature, the failures of forest initiatives to improve the lives of forest-dependent communities lie in a lack of both capacity and political will.[29] From this perspective, the proposed solution to these challenges is for REDD+ to learn from the past and to "do no harm" to the communities in the forests within the REDD+ frontier by adhering to the safeguards that have been set out by the UNFCCC.

Analysts, however, worry that REDD+ safeguards will serve instead as guidelines from which each party to the UNFCCC will develop and report their own interpretation rather than as a set to be followed equally by all.[30] Unlike safeguards in previous global initiatives, those established for REDD+ allow the parties to the UNFCCC to identify their own specific approaches to addressing injustices that may arise from REDD+ readiness and implementation, thus allowing them to decide what is just and what is not.[31] As noted earlier, this has proven problematic in political contexts where legal and political systems have historically limited the rights and participation of indigenous peoples in the decision-making processes that affect their lives and futures.[32] This may lead to national REDD+ safeguards that reproduce these structures of discrimination and address only the injustices that are recognized by current legal systems, in spite of international conventions.

Our current exploration is set within scholarly debates calling for more attention to the aspects of justice in environmental governance that go beyond the current emphasis on distributive and procedural justice, commonly expressed in terms of equity, which is characteristic of much of the literature on REDD+.[33] This distinction is inspired by Schlosberg's work on the distributive, procedural, and recognitional dimensions of environmental justice, which in turn builds on Fraser's seminal work on injustice within capitalist systems.[34]

Fraser's approach typifies three kinds of claims of injustice: redistribution, recognition, and representation.[35] The first describes the unjust distribution of resources (as applied to environmental justice, to the sharing of the costs and benefits of ecosystem services) between powerful and less-powerful actors (e.g., the Global North and South, the economically rich and poor, etc.). The second has to do with demands for recognition of (and respect for) the different identities of groups and people within groups that is carried in identity politics (e.g., in terms of ethnicity, race, gender, sexual orientation, etc.). These demands seek to eliminate forms of sociocultural domination of some actors over others. The third describes the participation that people are allowed in the processes for making decisions that affect their lives (e.g., an intervention in their territory). Fraser notes that to address redistribution, neither recognition nor representation alone is an effective remedy: these "dimensions of justice . . . are mutually irreducible although practically intertwined," and singling out these dimensions is therefore helpful to understand how to resolve them.[36]

Olson's summary of the characteristics of injustice in Fraser's position, and of the potential remedies for each, is helpful.[37] According to his summary, the maldistribution of resources can be remedied by their redistribution (an economic dimension), the misrecognition of misrepresented groups through their recognition (a sociocultural dimension), and the marginalization of groups from political processes through their inclusion (a political dimension). In what follows, we understand these characteristics of injustice as ideal types by which to examine the possibility of REDD+ as a justice-enabling tool.[38] Within this context, scholars have identified four different aspects of how local communities like those that find themselves inside the REDD+ frontier experience environmental injustice. Kuehn typifies these as (1) distributive injustice (through unequal exposure to environmental hazards and access to natural resources); (2) procedural unfairness (exclusion from decision making); (3) corrective injustice (poor enforcement of environmental legislation); and (4) social injustice (the link between degradation and structural issues like poverty and ethnic discrimination).[39] Distinctions like those made by Fraser, Schlosberg, and Kuehn underline the multiple nuances of environmental injustice at play in the REDD+ frontier, either because they were already at play before REDD+ was rolled out or because they were caused by its implementation.

REDD+, Environmental Governance, and Environmental Justice

Much of the critique of REDD+ reflects wider concerns associated with environmental governance at large. In recent decades political ecologists have observed that forest-dependent communities may end up with limited access to their traditional territories and livelihood practices because of natural resource conservation initiatives.[40] These early discussions over access transformed into a discussion about justice, partly because of demands made by grassroots organizations. As we have noted, grassroots organizations have provided us with most of the evidence that REDD+ can enable justice because they have been able to use REDD+ and discussions around it strategically to make gains and win attention to their cause, not because REDD+ was intentionally designed to allow for this.

Four aspects of REDD+ raise justice concerns.[41] The first is the conception of property rights inherent to REDD+, which has led to an emphasis on redistributive justice by benefit sharing and a failure to address the recognition-related and procedural aspects of injustice.[42] The second criticism notes an emphasis on control by national governments, to the detriment of the views of indigenous peoples and local communities.[43] This emphasis means that safeguards must be filtered through national legal structures and priorities, which are themselves the product of the ways in which national governments have historically engaged with their indigenous citizens.[44] As Turnhout et al. note, this also raises awareness of questions of injustice related to REDD+'s state-centered design and the possibility that funds derived from REDD+ will remain at the national level.[45] Yet it is notable that there are ways to counter this centralization, as set out by (for example) the World Bank's Dedicated Grant Mechanism for Indigenous Peoples and Local Communities and the several grants available for indigenous peoples to develop their own approaches to REDD+.[46] Third, some scholars warn that REDD+ risks prioritizing carbon over all other values of forests; they point out that although parties to the UNFCCC are encouraged to pursue cobenefits in their readying and implementation of REDD+, these cobenefits will not be included in the monitoring, reporting, and verification of their programs. Thus cobenefits may be a condition for payment but are not to be used to calculate the payment itself.[47] The fourth position criticizes the ways in which REDD+ engages with local communities. It finds a discursive emphasis on their rights in decision making, but in practice less on community members' access to spaces for decision making or the way these processes are established and conducted.[48] Most lack the full force of free, prior, and informed consent (FPIC) as set out in international agreements over the rights of indigenous peoples, such as ILO 169.[49]

In what follows, we engage with three cross-cutting issues—land tenure, gender, and participation—all of which present potential roadblocks for an environmentally just REDD+. Going back to the types of injustice set out by Fraser, these represent issues of redistribution, recognition, and representation. These three broad themes are all deeply political and informed by historical structures of injustice that are both global and local in nature. Together they help us problematize the economic, sociocultural, and governance structures behind injustice and provide more evidence that a lack of explicit attention to justice may place REDD+ on the same path as the kind of mainstream development initiatives that were described by Ferguson as an "anti-politics machine." At the same time, determining how these multiple layers of injustice play out in the REDD+ frontier points to the issues that must be addressed to ensure that forest carbon does not become yet another economic commodity boom that negatively impacts the lives of forest-dependent communities.

Our analysis highlights three key themes. The first is that any exploration of justice related to initiatives like REDD+ requires an understanding of the multiple dimensions of justice. We do not conceive of justice only as a matter of redistribution but also in terms of the settling of historical debts in terms of tenure security for indigenous communities, equity in their participation in decision-making processes, and the intersectionality of local experiences through attention to gendered experiences. The second theme is that, as has been discussed in the scholarly literature, the focus of REDD+ is slanted toward the symptoms of injustice rather than what McDermott et al. call "contextual equity" and the structural causes of inequality.[50] This is related to work by Li on the way development initiatives make the solutions to the problems they identify, in the contexts on which they act, technical and thus nonpolitical (e.g., ignoring marginalization and focusing on capacities, and thus not challenging the status quo).[51] The third theme is related to what Ribot and Peluso call "access."[52] This takes us beyond rights to participation or the use of resources in a forested area, for example, or whether a person or group is truly able to benefit from material or symbolic resources despite those rights.

Land Tenure and Redistributive Justice

Over half of all tropical forests in the world fall within the traditional territories of indigenous peoples, many of whom are struggling to defend their rights to those territories or to have them recognized as theirs.[53] The land tenure situation for indigenous peoples in Ecuador is illustrative of this situation. Although indigenous peoples occupy 65 percent of that country's forests, 37.5 percent of these groups still hold no legal ownership over their land.[54] Similarly, although indigenous peoples have a mix of formal titles and disputed claims to 40 percent of the Peruvian Amazon, more than half of these territories are overlapped by extraction concessions, as are one-fifth of protected natural areas.[55]

Despite this, a recent systematic review of the literature on REDD+ projects throughout the world found that although REDD+ discourse places great emphasis on rights recognition and on tenure clarity and security, this is not reflected in practice.[56] Similarly Sunderlin et al., based on research that included twenty-one subnational initiatives and 141 villages in five countries (including Brazil and Peru), revealed that tenure has been addressed only marginally in REDD+ initiatives.[57] Importantly, they asserted that this is the fault not of the proponents but of the national frameworks through which land tenure is recognized. This mismatch between discourse and practice corresponds with earlier reports stating that REDD+ is being readied and implemented in contexts where land tenure is neither clearly defined nor enforced.[58] There is also evidence that the establishment of REDD+ projects has led to the expulsion of forest-dependent peoples.[59] Concern over the relationship between indigenous land tenure and conservation initiatives runs more widely, as evidenced in the 2016 report by the United Nations Special Rapporteur on the Rights of Indigenous Peoples, which affirmed that the violation of indigenous rights in the name of conservation is a global constant.[60]

Following pressure from grassroots organizations and their allies throughout the world, the importance of clear land tenure for indigenous peoples is now widely recognized as a key enabling factor for REDD+ readiness and the achievement of its targets.[61] In the Amazon this has been reflected in an ongoing initiative to formalize five million hectares for indigenous peoples under the REDD+-related Joint Declaration of Intent signed by Peru, Norway, and Germany.[62] Although current titling processes are laudable, however, they have not signaled a pro-justice transformation of the relationship between the Peruvian state and its indigenous citizens. In fact there are large swaths of territory claimed by indigenous communities that have been marked for concessions to extractive companies; others that have been taken over by migrants from the country's Andean region, commonly with support from the central or local government; and yet others that have been made into protected areas where indigenous people cannot live or access resources.[63]

In Peru the titling and formalization process has proven to be aimed more at satisfying the requirements of donors than at transforming policy.[64] This "means to an end" attitude toward titling is best reflected in the way indigenous communities in areas with open, violent conflict with private interests have been left out of the titling process. This has most worryingly been reported for communities in the Ucayali region of the Peruvian Amazon, where the Federación de Comunidades Nativas de Ucayali, a grassroots organization that represents indigenous communities in the Ucayali Valley, asserts that the demands for land of those communities, pressured (through dispossession and open violence) by formal and informal logging and agro-industrial expansion, remain unaddressed.[65] Other grassroots organizations and national media outlets have reported invasions of titled indigenous territories by groups that were granted overlapping individual property titles by subnational governments.[66] This is a context in which, over the past two decades, indigenous groups in the Peruvian Amazon have been able to have their land rights respected only by organizing large-scale strikes that threatened extraction sites or by taking their demands directly to donor countries.[67] Even in the cases where communities have obtained a title, this has not granted them automatic use rights over the resources in the forests and subsoil of their territory, as the state retains ownership over these.[68]

Crucially for REDD+, current discussions in Peru seem to foreshadow that, as in Ecuador, the state will also retain forest carbon rights. At the moment the only point of comparison we have about how this might work in Peru in relation to forest carbon is that for indigenous communities to commercialize forest products, they must follow an overregulated and expensive set of steps to obtain permits from the government. Communities are seldom able to pay for these permits and end up in unequal partnerships with timber companies.[69] In a similar vein, Loayza et al. write that the implementation of Ecuador's REDD+ approach will be complicated, as "political determination is needed to clarify processes regarding land rights."[70] This is related to a gap between de jure and de facto land rights, which further complicates matters, as even if "existing legal frameworks provide constitutional rights . . . , holding a land title does not mean secure administration and control over the corresponding territories."[71] Thus in both Peru and Ecuador we find the injustice of access in spite of land rights.

The process of redistribution to address historical injustice in these contexts is incomplete. Despite donor demands and some positive responses from a few national government offices, the REDD+ effort in Peru has not yet been able to address such concerns and therefore risks exacerbating the situation of unrecognized claims to land by indigenous communities. This is especially worrisome in light of assertions, like those of Jacob and Brockington, that the monetary incentives of REDD+ could enable powerful private and state actors to claim untitled indigenous lands or marginalize indigenous communities' claims.[72] This was seen early on in the REDD+ process in the Amazon, with cases of so-called carbon cowboys in both the Peruvian and the Ecuadorean Amazon.[73] At the same time, it is worth noting that lessons from other countries, such as Tanzania, show that even in places where REDD+ may have initially caused open conflict between and within local communities over access to forested areas and due to potential REDD+ payments for the formalization of previously customary boundaries, some project REDD+ initiatives still, directly and sometimes indirectly, led to clearer community tenure rights.[74]

Generally the centrality of land rights in both UNDRIP and ILO 169 is not clearly reflected in the implementation of REDD+ in national legal systems. This is true even when the issue of clear land tenure for indigenous peoples and REDD+ goes beyond rights and justice and is related to the effectiveness of REDD+ itself, as recent academic studies conclude that areas occupied by indigenous peoples are more likely to be conserved than those that are not.[75] As we have shown, however, REDD+ has the potential to help close the gap between rights and access in terms of land.

Gender and Recognition Justice

Thinking of gender in terms of recognition justice highlights a key issue in how initiatives related to mainstream development, and now climate-related activities, often fail to recognize and understand the intersectionality of indigenous experiences, especially when it comes to gender.[76] While attention to the experiences of indigenous and local communities is key to any intervention in forests of the Global South, failure to consider difference internal to these communities may reproduce the structures of inequality within them. Perhaps the most important distinction within forest-dependent communities is related to power asymmetries along gender lines. This differentiated experience is recognized in UN Sustainable Development Goal 5, which addresses the need for policies to strengthen women's landownership and rights.

Current research on REDD+ notes that even in cases in which benefits have accrued from conservation, these are not distributed equitably along gender lines because of sociocultural practices in those communities.[77] These inequalities can exacerbate the already precarious positions held by women within their communities. A research project that included communities in Peruvian Amazonia (as well as Indonesia and Tanzania) has shown that women have fewer opportunities to participate in REDD+ and other initiatives and that they seek different kinds of benefits (even when it comes to noncash benefits) than those sought by men.[78] To illustrate, whereas women preferred objects and tools for their homes, animals to raise, and seedlings of nontimber species, men preferred technical and legal assistance, construction materials, and seedlings of timber species. The study also found an even greater difference along gender lines in terms of differentiated access to information and decision making about the distribution of benefits. In the communities where the study took place, men were more likely to have greater access to both, despite equal rights to those spaces and discursive emphasis on gender equality. This asymmetry is explained by two linked issues: men dominate both forest resources and decision-making spaces, both part of the public sphere, whereas women are commonly associated with the private sphere of home and family life. This constitutes a double bias: the gendered distinctions made by community members themselves, as well as the gendered expectations of those implementing projects.[79]

In a study by Larson et al. that included communities of the Brazilian and Peruvian Amazon, this state of affairs has led to women in villages with REDD+ projects experiencing a decline in subjective well-being relative to a control group.[80] Emphasizing the differentiated experiences along gender lines, when mixed-gender groups were asked to talk about their understanding of well-being, they agreed that it was based on good health, education, and diet, but women-only groups tended to add that having their own source of income was a key aspect of their well-being. This is not solely a problem of channeling money to households but also of recognizing the different access that women have to benefits accrued from projects like REDD+. A large body of work in the field of international development has shown that failure to pay serious attention to the access limitations suffered by women, going beyond considering gender as "a means to improve programme efficiency" or seeing the "gender problem" as one that can be solved by solving women's economic poverty, creates the risk of reinforcing or exacerbating power inequalities and the situation of women in their communities.[81]

As noted by the UNFCCC, women are already more vulnerable to climate change, hence they require a development and implementation of REDD+ that goes beyond "adding and stirring women into the REDD+ pot" or merely serving to legitimize an initiative.[82] Recognition of women's differentiated experience and status, in a way that goes beyond thinking of the "community" or the "household" as an undifferentiated unit or of women as intrinsically being the fabric of "social capital," is necessary for addressing their experience of injustice.[83]

Participation and Procedural Justice

It is widely accepted that the participation of forest-dependent communities throughout the REDD+ process is necessary for the program to reach its objectives.[84] A systematic review by Saeed et al., however, notes that most REDD+ projects have failed to apply FPIC, have made decisions before consulting local communities, and have strategically withheld information from those communities in order to manage their expectations.[85] This resonates with the analysis by Di Gregorio et al. of media coverage of equity under REDD+ in various countries were REDD+ is being readied and implemented.[86] Their study, which includes Brazil and Peru, the two countries with the largest share of the Amazon rainforest, reveals an emphasis on national equity concerns, to the detriment of participation by local and indigenous peoples and recognition of their rights.

Although it is laudable that the Cancún Safeguards call for "the full and effective participation of . . . indigenous peoples and local communities," indigenous people's right to FPIC has been legislated and operationalized in different countries, whether because of UNDRIP, ILO 169, or donor responsibilities, more as a tool for communication of something that has already been decided.[87] It therefore is not a process of inclusive decision making that recognizes and seeks to transform the unjust degree of participation that indigenous peoples have historically been allowed in decisions about their territories and futures throughout the Global South, or the fact that some of these groups are not even recognized as being indigenous. This is not only an issue of the way countries have interpreted FPIC; for example, the World Bank's operational policies for the Forest Carbon Partnership Facility currently do not require FPIC, applying a lower standard of free, prior, and informed *consultation*. More generally, where participation has been recognized, it has commonly been applied only to implementation of REDD+ activities and not to the previous steps of the process.[88]

We have found that without strict guidelines, FPIC application in REDD+ readying and implementation processes will present great, even contradictory variation from country to country.[89] In Peru this concern stems from the manner in which the state has included in legislation the responsibilities that it assumed by signing ILO 169. Despite its name, the Law of Prior Consultation does not follow the spirit of FPIC and seeks only to consult (and, in practice, communicate to) communities before the implementation of a law or development initiative that the government deems will affect their lives.[90] Not only does passage of these laws and initiatives not require their consent, but the consultation is not carried out throughout the process of planning and implementing the initiative. We have seen that the greater (though still limited) access that indigenous representatives have to decision making within the REDD+ process in Peru results more from donor pressure than from the government's adherence to its own laws. This reflects a predominance, in environment-related initiatives in the Amazon, of a paradigm that favors outsider technical knowledge over local knowledge, as demonstrated by Pokorny et al.[91] It is worth noting that the right to FPIC has not been passed on to nonindigenous communities that would also be affected by an initiative or project, despite their demands.

In a similar vein, a study of Socio Bosque, a conservation incentive initiative in Ecuador's national REDD+ program, revealed that even where attention is paid to broadening local communities' participation in decisions about the implementation of a program, local people were still uninformed about what the actual program involved, its design, and its benefits, and did not participate throughout the entire implementation of the program.[92] More generally, and illustrating the wider feminist critique of participatory processes, studies elsewhere show that when participatory decision making is applied, it tends to reproduce sociocultural norms that grant women unequal access to decision making.[93] In other settings, such processes may omit key community actors from REDD+ processes in countries where indigenous peoples are not recognized as such by national law or where local communities are not recognized as having the right to FPIC. The issue of remedying procedural injustices therefore is not limited to providing the right to participation but requires ensuring that access to decision making is inherent in the design of the initiative or the process.

Conclusion: Moving from "Doing No Harm" to Enabling Justice

In this chapter we have outlined some of the key theoretical perspectives on environmental justice from which we have analyzed injustice within the REDD+ frontier. As with rights, we note that safeguards are the key vehicle for an environmentally just REDD+. We have explored this by examining land tenure, gender, and participation, three broad, interrelated areas of injustice faced by Amazonian communities. These three themes, in turn, represent the distributive, recognition-related, and procedural characteristics of injustice noted by analysts like Fraser and Schlosberg.

The various chapters in this volume illustrate the spaces of deep environmental injustice into which the REDD+ frontier is expanding. These include various kinds of conflict, from open violence over the control of forests to epistemological conflicts over their meaning. This historical development, which is seen throughout the world, is also reflected in scholarly reports of social conflicts over forest resources within the context of REDD+ in various parts of the Global South.[94] For REDD+ to fulfill its promise as a win-win situation for the environment and people, it must be attentive to environmental injustice and the violations of indigenous rights, whether or not it causes them; otherwise it runs the risk of catalyzing further injustice.

In the context of REDD+ in the Amazon, these tenure, gender, and participation injustices could worsen if REDD+ safeguards maintain their current "do no harm" approach instead of being redirected to "do better." The historical maldistribution of land since the colonial era can be remedied by its redistribution; the failure to recognize the experience of women in indigenous communities can be addressed by policies that actively take into account their differentiated position within their societies and the structures that reify that position; and the marginalization of groups from decision making can be addressed with policies that actively include them and develop their capacities to fully participate. Yet this goes beyond recognition of rights, as indigenous peoples may have recognized differentiated *rights* to land and participation, and women may have the *right* to enjoy the same advantages as men. The opportunity that REDD+ offers to address injustice in the Amazon and beyond lies in ensuring that participating countries and projects follow safeguards that ensure the ability to *exercise* these rights. This is an intrinsically political and sociocultural issue that cannot be resolved by an overtly technical approach. As it stands, REDD+ is a prime example of the development complex, of the meeting point between measures for monitoring the change brought about by an initiative and the deeply historical and sociopolitical nature of the dynamics of exclusion and inclusion in the contexts where these initiatives are implemented.

Ultimately, as with other global environmental initiatives (and development initiatives in general), the possibility of an environmentally just REDD+ lies in whether or not it can acquire a deep awareness of how previous similar initiatives are transformed by place. In the final analysis, the different experiences of REDD+, in its top-down form, are not simply erroneous national or local interpretations of a carefully thought-out framework for action by the parties to the UNFCCC and other related actors. They are examples of the different degrees of power inequalities and histories of interactions between forest-dependent communities and governments in the Global South and the inability of global initiatives to deal with the structures that create such inequalities. At the same time, the justice potential of REDD+ is alive at the grassroots level, at least in certain locations.[95] These efforts may not have transformed REDD+ more globally, but there have been advances in winning special attention through multilateral vehicles, such as the World Bank's dedicated mechanism and the Green Climate Fund and REDD+ proposals presented by indigenous organizations themselves.

We have presented three broad categories of injustice that cut across different experiences in the REDD+ frontier. However, these do not encompass wider issues of injustice in the same frontier that REDD+ may be able to address if it transitions from reliance on safeguards that require practitioners to "do no harm" and becomes a tool for accompanying and enabling local struggles against injustice in the forests of the Global South. In the Amazon this would constitute a major step forward in terms of more serious engagement with the proposals for REDD+ Indigena Amazonica (Indigenous Amazonian REDD+), which have been conceived and implemented by organizations that represent indigenous Amazonian peoples.[96] REDD+ Indigena Amazonica places the rights recognized by

UNDRIP at the core of REDD+ and encourages noncarbon benefits and the holistic management of territories in accordance with indigenous approaches to environmental management. Representatives of indigenous Amazonians have proposed their own alternative, and we emphasize the need for international bodies to fund pilot studies and continue the work to scale up similar transformative proposals from the grassroots, which are motivated by rights and justice.

NOTES

This work was supported by the Norwegian Agency for Development Cooperation; the European Commission; the International Climate Initiative of the German Federal Ministry for the Environment, Nature Conservation, Building, and Nuclear Safety; the United Kingdom's Department for International Development; and the CGIAR Research Programs on Forest, Trees, and Agroforestry (FTA) and Policies, Institutions, and Markets (PIM).

1. United Nations, "Framework Convention."
2. See, e.g., Phelps, Webb, and Agrawal, "Does REDD+ Threaten"; Agrawal, Nepstad, and Chhatre, "Reducing"; Birrell, Godden, and Tehan, "Climate Change"; Gupta, "Glocal Forest."
3. Jodoin, *Forest.*
4. Sarmiento Barletti and Larson, "Rights Abuse."
5. Sarmiento Barletti and Larson, "Rights Abuse."
6. The Cancún Safeguards, which "should be promoted and supported" by Parties to the UNFCCC, state, "Respect for the *knowledge and rights of indigenous peoples and members of local communities*, by taking into account relevant international obligations, national circumstances and laws, and noting that the United Nations General Assembly has adopted the United Nations Declaration on the Rights of Indigenous Peoples' and 'The full and effective participation of relevant stakeholders, *in particular, indigenous peoples and local communities* [the four REDD+ actions]". United Nations, "Framework Convention," 26.
7. See Sanders et al., "Guinea-Pig."
8. See, among many others, Ferguson, *Anti-Politics Machine*; Escobar *Encountering Development.*
9. United Nations Framework Convention on Climate Change Secretariat, "Key Decisions."
10. Other international safeguard initiatives include the Social and Environmental Principles and Criteria developed by the UN-REDD Programme and the World Bank's Operational Policies and Procedures, which have been established for the activities of the Forest Carbon Partnership Facility (FCPF 2010). UN-REDD and the World Bank have also developed related guidance materials, such as the "UN-REDD Programme Guidelines on Free, Prior and Informed Consent" and the "UN-REDD/ FCPF Joint

Guidelines on Stakeholder Engagement." There are also a series of safeguard initiatives for jurisdictional REDD+ (e.g., "REDD+ Social and Environmental Standards Initiative") and for project-based REDD+ activities (e.g., "Climate, Community and Biodiversity Standard").

11. See Jodoin, *Forest*; Pham et al., "Local Preferences."
12. The term "Global South" "refers broadly to the regions . . . outside Europe and North America, mostly (though not all) low-income and often politically or culturally marginalized. The use of the phrase Global South marks a shift from a central focus on development or cultural difference toward an emphasis on geopolitical relations of power" (Dados and Connell, "Global South," 12). Our usage of this term also reflects those nations that analysts of international development, such as Escobar, *Encountering Development*, have described as having been "underdeveloped" because of a history of colonial and postcolonial interactions.
13. Sarmiento Barletti and Larson, "Rights Abuse."
14. Godden and Tehan, "REDD+."
15. United Nations, "Declaration."
16. Gardner, "Peru."
17. Martinez-Alier, *Environmentalism.*
18. See, e.g., Mickelson, "Leading."
19. See, e.g., Gupta et al. "In Pursuit"; Kosoy and Corbera, "Payments"; Vijge, "Carbonizing"; Vijge and Gupta, "Framing REDD+"; Turnhout et al., "Envisioning." We do not expand on this further because of lack of space, but this reflects the "fortress carbon" (Leach and Scoones, Political Ecologies," 7) approach to forest conservation for carbon mitigation, which potentially leads to the "carbonized exclusion" (see Asiyanbi, "Political Ecology"; Lord, "Displacement") of indigenous and local populations (see Mahanty et al., "Unravelling"; Nel and Hill, "Constructing").
20. Gupta et al., "In Pursuit."
21. Espinoza and Feather, *Reality.*
22. See also Myers et al., "Messiness."
23. See Fraser, *Scales of Justice*; Fraser, *Justice Interruptus*; Fraser, *Unruly Practices*. See also Schlosberg, *Defining*; Schlosberg, "Reconceiving."
24. See, e.g., Heiman, "Race"; Smith and McDonough, "Beyond Public"; Shrader-Frechette. *Environmental Justice.*
25. Schroeder et al., "Third World."
26. United Nations Human Rights, Office of the High Commissioner, "Human Rights"; United Nations Development Programme, "Human Development Report."
27. See Chomba et al., "Roots"; Marion Suiseeya, "Transforming."
28. For a review, see Marion Suiseeya, "Contesting."
29. See Angelsen et al., *Analysing*; Brockhaus, Di Gregorio, and Carmenta, "REDD+"; Vongvisouk et al., "REDD+."
30. See Marion Suiseeya and Caplow, "In Pursuit."

31. See Marion Suiseeya, "Contesting."
32. See Chomba et al., "Of Mice"; Chomba et al., "Roots."
33. On the current scholarly debates, see, e.g., Marion Suiseeya. "Transforming"; Martin et al., "Justice"; Myers et al., "Messiness." Although we do not discuss this further in this chapter, it is worth noting that equity in REDD+ is commonly discussed in terms of distribution through benefit-sharing mechanisms (see, e.g., Luttrell et al., "Who Should"; Pham et al., *Approaches*). See Brown, Seymour, and Peskett, "How"; Marion Suiseeya and Caplow, "In Pursuit"; Dunlop and Corbera, "Incentivizing."
34. Schlosberg, *Defining*; Schlosberg, "Reconceiving"; Fraser, *Unruly Practices*; Fraser, *Justice Interruptus*; Fraser, *Scales of Justice*. See also McDermott and Schreckenberg, "Examining"; McDermott, Mahanty, and Schreckenberg, "Examining"; Sikor, *Justices*.
35. See also Schlosberg, *Defining*.
36. Fraser et al., "Recognition," 376.
37. Olson, "Participatory."
38. Following Weber's "The Essentials of Bureaucratic Organization," these are concepts derived from observable reality that describe the most essential aspects of a social phenomenon. These concepts serve as tools for exploring social reality, but they are also permeable and seldom exist in their pure form in practice.
39. Kuehn, "Taxonomy." See also Gonzalez, "Environmental Justice."
40. See, e.g., Peluso, "Coercing."
41. For a review, see Marion Suiseeya, "Transforming."
42. See Marion Suiseeya, "Transforming"; Sikor et al., "Toward an Empirical Analysis"; Myers et al., "Messiness."
43. See, e.g., Hiraldo and Tanner, "Forest Voices"; Bolin and Taku Tassa, "Exploring."
44. See Chomba et al., "Of Mice."
45. Turnhout et al., "Envisioning."
46. World Bank, "Dedicated."
47. See Dickson and Kapos, "Biodiversity"; Jagger et al., "Multilevel"; Vijge, "Carbonizing"; Vijge et al., "Framing National REDD+."
48. See Marion Suiseeya, "Transforming."
49. Although we do not engage with this explicitly, this reflects the recognition in the environmental justice literature of how different actors may have different notions of equity and justice (see, e.g., Walker, *Environmental*; Sikor, *Justices*). For example, Rantala and Di Gregorio, "Multistakeholder," note how a central element of the development of Tanzania's national REDD+ strategy involved designing a benefit-sharing mechanism that centered on finding equitable outcomes. Stakeholders, however, had different ideas about how this equity was to be achieved. On the one hand, the central government posited that the best way to ensure equitable outcomes was through a fund that it would control. On the other hand, civil society organizations and nongovernmental organizations argued that such a model would reify power inequalities in terms of who benefits from forests, as the fund would follow the practices of previous approaches to

natural resource management in which the government had distributed little to local communities. This point is relevant for all REDD+ contexts and worth exploring in those cases where national governments have already decreed that carbon, like most key natural resources (minerals, hydrocarbons, and timber), is to be managed by the government as a proxy of the nation. In Peru, where this is almost certain to be the case, indigenous and local communities in areas of large-scale extractive activity receive little benefit despite the decentralization of tax revenues from extraction, which the central government has implemented over the past decade. Funds allocated to the regions and provinces where large-scale extraction sites are located are commonly spent according to the development priorities of the nonindigenous elites who lead regional and local governments (see Sarmiento Barletti, "Kametsa Asaiki"). This takes us to another experience of distributive injustice that we do not explore further in this chapter.

50. McDermott, Mahanty, and Schreckenberg, "Examining"; Di Gregorio et al., "Equity"; Myers et al., "Messiness."
51. Li, "Rendering."
52. Ribot and Peluso, "Theory."
53. See the website of Landmark: http://www.landmarkmap.org/.
54. Loayza et al., "Analysis."
55. Huertas, "Agua"; White, "Perfect Storm?"
56. Saeed, McDermott, and Boyd, "Are REDD+." See also Godden and Tehan, "REDD+."
57. Sunderlin et al., "Creating."
58. See, e.g., Larson et al., "Land Tenure"; Sunderlin et al., *Challenge*.
59. See, e.g., Dwyer, Ingalls, and Baird, "Security"; Lord, "Displacement."
60. United Nations General Assembly, "Rights."
61. See, e.g., Awono et al., "Tenure"; Resosudarmo et al., "Does Tenure."
62. Bosques, "Declaración."
63. See Sarmiento Barletti, "Kametsa Asaiki."
64. SERVINDI, "Norwegian." On titling, see Monterroso et al., *Reinvindicacion*.
65. Federacion de comunidades nativas del Ucayali y Afluentes, "Memorial."
66. "Otro comunero."
67. See Hughes, "Indigenous"; Monterroso et al., *Reinvindicacion*.
68. See Monterroso et al., *Reinvindicacion*.
69. See Sarmiento Barletti, "Angry Earth."
70. Loayza et al., "Analysis," 27.
71. Loayza et al., "REDD+," 76.
72. Jacob and Brockington. "Learning."
73. Espinoza and Feather, *Reality*; De Jong, del Castillo Torres, and Salazar, "Carbon Cowboys"; Reed, "REDD+." Carbon cowboys are "actors who are willing to push the limits of established negotiation mechanisms to gain control over forest areas" (Aguilar-Støen, "Better Safe," 91).
74. Raftopoulos, "REDD+"; Scheba and Rakotonarivo, "Territorialising REDD+." See also Jodoin, *Forest*; Kijazi et al., "Multilevel Governance."

75. For example, Blackman et al., "Titling"; Schleicher et al., "Conservation."

76. Bee and Sijapati Basnett, "Engendering."

77. See Chan and Satterfield, "Justice." It is worth noting that the same can be said about these costs in regard to nations as most biological "hotspots" (e.g., the Andes-Amazon region) are in the Global South (see, e.g., Balmford and Whitten, "Who Should Pay"). But this also works in the inverse, as many of the natural resources imported by countries in the industrialized Global North come from the Global South.

78. Andersson et al. "Wealth."

79. See, e.g., Wilson, "Towards a Radical."

80. Larson et al., "Gender." Although representing a different set of relations, this is reminiscent of the results of work by Izquierdo ("Well-Being") of the subjective well-being of indigenous Matsigenka people living in close proximity to Peru's flagship, large-scale extractive gas project at Camisea. Izquierdo reports a decrease in their subjective well-being despite statistics presented by the government and PlusPetrol (the company leading the Camisea consortium) that Matsigenka people have never been healthier or more affluent. See Sarmiento Barletti, "Angry Earth," for a similar experience among Ashaninka people, also in the Peruvian Amazon.

81. Bee and Sijapati Basnett. "Engendering"; Larson et al., "Land Tenure."

82. Bee and Sijapati Basnett. "Engendering." See, e.g., Westholm and Arora-Jonsson, "Defining." See also United Nations, "Introduction."

83. Cornwall, "Whose Voices?"; Cornwall, "Making"; Molyneux, "Gender"; Wilson, "Towards a Radical."

84. See United Nations Framework Convention on Climate Change Secretariat, "Key Decisions."

85. Saeed, McDermott, and Boyd, "Are REDD+."

86. Di Gregorio et al., "Equity."

87. Airey and Krause, "Georgetown"; Bayrak and Marafa, "Ten Years"; United Nations, "Framework Convention," 13.

88. Saeed, McDermott, and Boyd, "Are REDD+."

89. Sarmiento Barletti and Larson, "Rights Abuse."

90. Sarmiento Barletti and Seedhouse, "Truth."

91. Pokorny, Scholz, and de Jong, "REDD+."

92. Krause, Collen, and Nicholas, "Evaluating."

93. See, e.g., Cornwall, "Whose Voices?"; Cornwall, "Making."

94. See, among many others, Sills et al., *REDD+ on the Ground*; Sunderlin et al., *Challenge*; Visseren-Hamakers et al. "Will REDD+ Work?"

95. See, e.g., Jodoin, *Forest*.

96. AIDESEP. "RIA."

BIBLIOGRAPHY

Aguilar-Støen, Mariel. "Better Safe Than Sorry? Indigenous Peoples, Carbon Cowboys and the Governance of REDD in the Amazon." *Forum for Development Studies* 44, no. 1 (2017): 91–108.

Agrawal, Arun, Daniel Nepstad, and Ashwini Chhatre. "Reducing Emissions from Deforestation and Forest Degradation." *Annual Review of Environmental Resources* 36 (2011): 373–96.

AIDESEP. "RIA: Mas que carbono es vida y cultura." Accessed 21 August 2019. http://www.aidesep.org.pe/noticias/ria-mas-que-carbono-es-vida-y-cultura.

Airey, Sam, and Torsten Krause. "'Georgetown Ain't Got a Tree. We Got the Trees': Amerindian Power and Participation in Guyana's Low Carbon Development Strategy." *Forests* 8, no. 3 (2017): 51–75.

Andersson, Krister P., Steven M. Smith, Lee J. Alston, Amy E. Duchelle, Esther Mwangi, Anne M. Larson, Claudio de Sassi, Erin O. Sills, William D. Sunderlin, and Grace Y. Wong. "Wealth and the Distrubtion of Benefits from Tropical Forests: Implications for REDD+." *Land Use Policy* 72 (2018): 510–22.

Angelsen, Arild, Maria Brockhaus, William D. Sunderlin, and Louis Verchot, eds. *Analysing Redd+: Challenges and Choices*. Bogor, Indonesia: CIFOR, 2012.

Asiyanbi, Adeniyi. "A Political Ecology of REDD+: Property Rights, Militarised Protectionism, and Carbonised Exclusion in Cross River." *Geoforum* 77 (2016): 146–56.

Ayari, Ines. "Commentary: The Dynamics between Indigenous Rights and Environmental Governance: A Preliminary Analysis and Focus on the Impact of Climate Change Governance through the Reducing Emissions from Deforestation and Forest Degradation (REDD) Programme." *Alternative* 10, no. 1 (2014): 81–86.

Awono, Abdon, Olofunso Somorin, Richard Atyi, and Patrice Levang. "Tenure and Participation in Local REDD+ Projects: Insights from Southern Cameroon. Environment." *Science and Policy* 35 (2014): 76–86.

Balmford, A. and T. Whitten. "Who Should Pay for Tropical Conservation, and How Could the Costs Be Met?" *Oryx* 37 (2003): 238–50.

Bayrak, Mucahid Mustafa, and Lawal Mohammed Marafa. "Ten Years of REDD+: A Critical Review of the Impact of REDD+ on Forest-Dependent Communities." *Sustainability* 8 (2016): 620–42.

Bee, Beth, and Bimbika Sijapati Basnett. "Engendering Social and Environmental Safeguards in REDD+: Lessons from Feminist and Development Research." *Third World Quarterly* 38, no. 4 (2017): 1–18.

Birrell, Kathleen, Lee Godden, and Maureen Tehan. "Climate Change and REDD+: Property as a Prism for Conceiving Indigenous Peoples' Engagement." *Journal of Human Rights and the Environment* 3, no. 2 (2012): 196–216.

Blackman Allen, Leonardo Corral, Eirivelthon Santos Lima, and Gregory Asner. "Titling Indigenous Communities Protects Forests in the Peruvian Amazon." *Proceedings of the National Academy of Sciences* 114, no. 16 (2017): 4123–2418.

Bolin, Anna, and Dominic Taku Tassa. "Exploring Climate Justice for Forest Communities Engaging in REDD+: Experiences from Tanzania." *Forum for Development Studies* 39, no. 1 (2012): 5–29.

Bosques. "Declaración conjunta de intención (DCI) entre el Gobierno de la República del Perú, el Gobierno del Reino de Noruega y el Gobierno de la República Federal de Alemania." Accessed July 12, 2019. http://www.bosques.gob.pe/declaracion-conjunta-de-intencion.

Brockhaus, Maria, Monica Di Gregorio, and Rachel Carmenta. "REDD+ Policy Networks: Exploring Actors and Power Structures in an Emerging Policy Domain." *Ecology and Society* 19, no. 4 (2014): 29–61.

Brown, David, Frances Seymour, and Leo Peskett. "How Do We Achieve REDD Co-benefits and Avoid Doing Harm?" In *Moving Ahead with REDD: Issues, Options and Implications*, edited by Arild Angelsen. Bogor, Indonesia: Center for International Forestry Research, 2008.

Chan, Kai, and Terre Satterfield. "Justice, Equity, and Biodiversity." In *The Encyclopedia of Biodiversity*, edited by S. A. Levin. Oxford: Elsevier, 2013.

Chomba, Susan, Juliet Kariuki, Jens Friis Lund, and Fergus Sinclair. "Of Mice and Men: Why the Unintended Consequences of Carbon Markets Matter." *Land Use Policy* 61 (2017): 99–102.

Chomba, Susan, Juliet Kariuki, Jens Friis Lund, and Fergus Sinclair. "Roots of Inequity: How the Implementation of REDD+ Reinforces Past Injustices." *Land Use Policy* 50 (2016): 202–13.

Cornwall, Andrea. "Making a Difference? Gender and Participatory Development." IDS Discussion Paper 378. Institute of Development Studies, University of Sussex, 2001.

———. "Whose Voices? Whose Choices? Reflections on Gender and Participatory Development." *World Development* 31, no. 8 (2003): 1325–42.

Dados, Nour, and Raewyn Connell. "The Global South." *Contexts* 11, no. 1 (2012): 12–13.

De Jong, Will, Dennis del Castillo Torres, and Angel Salazar. "Carbon Cowboys in Peru and the Prospects of Local REDD Governance." *Portes: Revista Mexicana de Estudios Sobre la Cuenca del Pacifico* 8 (2014): 61–83.

Dickson, Barney, and Valerie Kapos. "Biodiversity Monitoring for REDD+." *Current Opinions in Environmental Sustainability* 4 (2012): 717–25.

Di Gregorio, Monica, Maria Brockhaus, Tim Cronin, Efrian Muharrom, Levania Santoso, Sofi Mardiah, and Mirjam Büdenbender. "Equity and REDD+ in the Media: A Comparative Analysis of Policy Discourses." *Ecology and Society* 18, no. 2 (2013): 39–54.

Dunlop, Tessa, and Esteve Corbera. "Incentivizing REDD+: How Developing Countries Are Laying the Groundwork for Benefit-Sharing." *Environmental Science Policy* 63, no. 2 (2016): 44–54.

Dwyer, Michael, Micah Ingalls, and Ian Baird. "The Security Exception: Development and Militarization in Laos's Protected Areas." *Geophysical Journal of the Royal Astronomical Society* 69 (2016): 207–17.

Escobar, Arturo. *Encountering Development: The Making and Unmaking of the Third World*. Princeton NJ: Princeton University Press, 1995.

Espinoza, Roberto, and Conrad Feather. *The Reality of REDD+ in Peru: Between Theory and Practice*. Lima: AIDESEP, FENAMAD, CARE, and FPP, 2012.

Federacion de comunidades nativas del Ucayali y Afluentes. "Memorial Feconau 050617." June 5, 2017. Internet Archive. archive.org/stream/MemorialFeconau050617/MEMORIAL.FECONAU_05.06.2017_djvu.txt.

Ferguson, James. *The Anti-Politics Machine: "Development," Depoliticization, and Bureaucratic Power in Lesotho*. Cambridge, U.K.: Cambridge University Press, 1990.

Fraser, Nancy. *Justice Interruptus: Critical Reflections on the "Postsocialist" Condition*. London: Routledge, 1997.

Fraser Nancy. *Scales of Justice: Reimagining Political Space in a Globalizing World*. New York: Columbia University Press, 2009.

Fraser, Nancy. *Unruly Practices: Power, Discourse and Gender in Contemporary Social Theory*. Cambridge, U.K.: Polity Press, 1989.

Fraser, Nancy, Hanne Dahl, Pauline Stoltz, and Rasmus Willig. "Recognition, Redistribution and Representation in Capitalist Global Society: An Interview with Nancy Fraser." *Acta Sociologica* 47, no. 4 (2004): 374–82.

Gardner, Elie. "Peru Battles the Golden Curse of Madre de Dios." *Nature* 486 (2012): 306–7.

Godden, Lee, and Maureen Tehan. "REDD+: Climate Justice and Indigenous and Local Community Rights in an Era of Climate Disruption." *Journal of Energy & Natural Resources Law* 34, no. 1 (2016): 95–108.

Gonzalez, Carmen. "Environmental Justice, Human Rights, and the Global South." *Santa Clara Journal of International Law* 151 (2015): 151–95.

Gupta, Aari, Eva Lövbrand, Esther Turnhout, and Marjanneke Vijge. "In Pursuit of Carbon Accountability: The Politics of REDD+ Measuring, Reporting and Verification Systems." *Current Opinions in Environmental Sustainability* 4 (2012): 726–31.

Gupta, Joyeeta. "Glocal Forest and REDD+ Governance: Win-win or lose-lose?" *Current Opinions in Environmental Sustainability* 4 (2012): 620–27.

Heiman, Michael. "Race, Waste, and Class: New Perspectives on Environmental Justice." *Antipode* 28 (1996): 111–21.

Hiraldo, Rocio, and Thomas Tanner. "Forest Voices: Competing Narratives over REDD+." *IDS Bulletin* 42 (2011): 42–51.

Huertas, Beatriz. "Agua e identidad cultural: La defensa de la Reserva Comunal Amarakaeri frente a la actividad hidrocarburífera, Madre de Dios, Perú." In *Agua e industrias extractivas: Cambios y continuidades en los Andes*, edited by Patricia Urteaga. Lima: Instituto de Estudios Peruanos, Justicia Hídrica, Concertación, 2011.

Hughes, Neil. "Indigenous Protest in Peru: The 'Orchard Dog' Bites Back." *Social Movement Studies* 9, no. 1 (2010): 85–90.

Izquierdo, Carolina. "Well-Being among the Matsigenka of the Peruvian Amazon: Health, Missionsm Oil and 'Progress.'" In *Pursuits of Happiness: Well-Being in*

Anthropological Perspective, edited by Gordon Mathews and Carolina Izquierdo. Oxford: Berghahn.

Jacob, Thabit, and Dan Brockington. "Learning from the Other: Benefit Sharing Lessons for REDD+ Implementation Based on CBFM Experience in Northern Tanzania." *Land Use Policy*, October 23, 2017. http://dx.doi.org/10.1016/j.landusepol.2017.10.028.

Jagger, P., Maria Brockhaus, Amy Duchelle, Maria Fernanda, Gebara, K. Lawlor, Ida Resosudarmo, and William Sunderlin. "Multilevel Policy Dialogues, Processes, and Actions: Challenges and Opportunities for National REDD+ Safeguards Measurement, Reporting, and Verification (MRV)." *Forests* 5 (2014): 2136–62.

Jodoin, S. *Forest Preservation in a Changing Climate: REDD+ and Indigenous and Community Rights in Indonesia and Tanzania*. Cambridge, U.K.: Cambridge University Press, 2017.

Kijazi, Martin H., Joshue Ivan Joel, Anne M. Larson, and Natalia Cisneros. "Multilevel Governance, Carbon Management and Decisions in Tanzania." CIFOR Working Paper 226. Bogor, Indonesia.

Kosoy, Nicolas, and Esteve Corbera. "Payments for Ecosystem Services as Commodity Fetishism." *Ecological Economics* 69 (2010): 1228–36.

Krause, Torsten, Wain Collen, and Kimberly Nicholas. "Evaluating Safeguards in a Conservation Incentive Program: Participation, Consent, and Benefit Sharing in Indigenous Communities of the Ecuadorian Amazon." *Ecology and Society* 18, no. 4 (2013): 1–16.

Kuehn, Robert. "A Taxonomy of Environmental Justice." *Environmental Law Reporter* 30 (2000): 10681–703.

Larson, Anne, Maria Brockhaus, William Sunderlin, Amy Duchelle, Andrea Babond, Therese Dokken, Thuy Thuy Pham, Ida Resosudarmo, Galia Selaya, and Abdon Awono. "Land Tenure and REDD+: The Good, the Bad and the Ugly." *Global Environmental Change* 23 (2013): 678–89.

Larson, Anne, David Solis, Amy Duchelle, Stibniati Atmadja, Ida Resosudarmo, Therese Dokken, and Mella Komalasri. "Gender Lessoms for Climate Initiatives: A Comparative Study of REDD+ Impacts on Subjective Wellbeing." *World Development* 108 (2018): 86–102.

Leach, Melissa and Ian Scoones. "Political Ecologies of Carbon in Africa." In *Carbon Conflicts and Forest Landscapes in Africa*, edited by Melissa Leach and Ian Scoones. New York: Routledge, 2015.

Li, Tania Murray. "Rendering Society Technical: Government through Community and the Ethnographic Turn at the World Bank in Indonesia." In *Adventures in Aidland: The Anthropology of Professionals in International Development*, edited by David Mosse. Oxford: Berghahn, 2011.

Loayza, T., et al. "Analysis of Land Management and Legal Arrangements in the Ecuadorian Northeastern Amazon as Preconditions for REDD+ Implementation." *Forest Policy and Economics* 83 (2017): 19–28.

Loayza, T., et al. "REDD+ Implementation in the Ecuadorian Amazon: Why Land Configuration and Common-Pool Resources Management Matter." *Forest Policy and Economics* 70 (2016): 67–79.

Locatelli, Bruno, Vanessa Evans, Andrew Wardell, Angela Andrade, and Raffaele Vignola. "Forests and Climate Change in Latin America: Linking Adaptation and Mitigation." *Forests* 2 (2011): 431–50.

Lord, Emma. "Displacement, Power and REDD+: A Forest History of Carbonized Exclusion." In *Global Forest Governance and Climate Change*, edited by Emmanuel Nuesrini. London: Palgrave, 2018.

Luttrell, Constance, Lasse Loft, Maria Fernanda Gebara, Demetrius Kweka, Maria Brockhaus, Arild Angelsen, and William Sunderlin. "Who Should Benefit from REDD+? Rationales and Realities." *Ecology and Society* 18 (2013): 52.

Mahanty, Sango, Wolfram Dressler, Sarah Milne, and Colin Filer. "Unravelling Property Relations around Forest Carbon." *Singapore Journal of Tropical Geography* 34 (2013): 188–205.

Marion Suiseeya, Kimberley. "Contesting Justice in Global Forest Governance: The Promises and Pitfalls of REDD+." *Conservation and Society* 15, no. 2 (2017): 189–200.

Marion Suiseeya, Kimberley. "Transforming Justice in REDD+ through a Politics of Difference Approach." *Forests*, no. 7 (2016): 300–313.

Marion Suiseeya, Kimberley, and Susan Caplow. "In Pursuit of Procedural Justice: Lessons from an Analysis of 56 Forest Carbon Project Designs." *Global Environmental Change* 23 (2013): 968–79.

Martin, Adrian, Brendan Coolsaet, Esteve Corbera, Neil Dawson, James Fraser, Ina Lehman, and Iokiñe Rodriguez. "Justice and Conservation: The Need to Incorporate Recognition." *Biological Conservation* 197 (2016): 254–61.

Martinez-Alier, Joan. *The Environmentalism of the Poor: A Study of Ecological Conflicts and Valuation.* London: Elgar, 2002.

Martinez-Alier, Joan, Isabelle Anguelovski, Patrick Bond, Daniela Del Bene, F. Demaria, Julien-Francois Gerber, Lucie Greyl, Willi Haas, Hali Healy, Victoria Marín-Burgos, Godwin Ojo, Marcelo Porto, Leida Rijnhout, Beatriz Rodríguez-Labajos, Joachim Spangenberg, Leah Temper, Rikard Warlenius, and Ivonne Yánez. "Between Activism and Science: Grassroots Concepts for Sustainability Coined by Environmental Justice Organizations." *Journal of Political Ecology* 21 (2014): 19–60.

McDermott, Constance, and Kate Schreckenberg. "Examining Equity: A Multidimensional Framework for Assessing Equity in the Context of REDD+" In *Proceedings of the Oxford Conference 2012 Beyond Carbon: Ensuring Justice and Equity in REDD+ across Levels of Governance.* Oxford: Oxford Centre for Tropical Forests, 2012.

McDermott, Melanie, Sango Mahanty, and Kate Schreckenberg. "Examining Equity: A Multidimensional Framework for Assessing Equity in Payments for Ecosystem Services." *Environmental Science and Policy* 33 (2013): 416–27.

Mickelson, Karin. "Leading toward a Level Playing Field, Repaying Ecological Debt, or Making Environmental Space: Three Stories about International Environmental Cooperation." *Osgoode Hall Law Journal* 43, nos. 1–2 (2005): 137–70.

Molyneux, Maxine. "Gender and the Silences and Social Capital: Lessons from Latin America." *Development and Change* 33, no. 2 (2002): 167–88.

Monterroso, Iliana, Peter Cronkleton, Daniel Pinedo, and Anne Larson. *Reinvindicacion de derechos colectivos: Reformas de tenencia de tierras y bosques en el Peru (1960–2016)*. Bogor, Indonesia: CIFOR, 2017.

Murdiyarso, Daniel, Maria Brockhaus, William Sunderlin, and Lou Verchot. "Some Lessons Learned from the First Generation of REDD+ Activities." *Current Opinions in Environmental Sustainability* 4 (2012): 678–85.

Myers, Rodd, Anne Larson, Ashwin Ravikumar, Laura Kowler, Anastasia Yang, and Tim Trench. "Messiness of Forest Governance: How Technical Approaches Suppress Politics in REDD+ and Conservation Projects." *Global Environmental Change* 50 (2018): 314–24.

Naughton-Treves, Lisa, and Kelly Wendland. "Land Tenure and Tropical Forest Carbon Management." *World Development* 55 (2014): 1–6.

Nel, Adrian, and Douglas Hill. "Constructing Walls of Carbon—The Complexities of Community, Carbon Sequestration and Protected Areas in Uganda." *Journal of Contemporary African Studies* 31 (2013): 421–40.

Olson, Kevin. "Participatory Parity and Democratic Justice." In *Adding Insult to Injury: Nancy Fraser Debates Her Critics*, edited by Kevin Olson. London: Verso, 2008.

"Otro comunero es amenazado de muerte en Ucayali." *La Republica*. Accessed August 21, 2019. https://larepublica.pe/politica/1104216-otro-comunero-es-amenazado-de-muerte-en-ucayali/.

Peluso, Nancy "Coercing Conservation: The Politics of State Resource Control." *Global Environmental Change* 3 (1993): 199–217.

Pham Thuy Thuy, Maria Brockhaus, Grace Wong, Le Ngoc Dung, Januarti Sinara Tjajadi, Lasse Loft, Cecilia Luttrell, and S. Assembe Mvondo. *Approaches to Benefit Sharing: A Preliminary Comparative Analysis of 13 REDD+ Countries*. Bogor, Indonesia: CIFOR, 2013.

Pham Thuy Thuy, Moira Moeliono, Maria Brockhaus, Le Ngoc Dung, and Grace Wong. "Local Preferences and Strategies for Effective, Efficient, and Equitable Distribution of PES Revenues in Vietnam: Lessons for REDD+." *Human Ecology* 42, no. 6 (2014): 885–99.

Phelps, Jacob, Edward Webb, and Arun Agrawal. "Does REDD+ Threaten to Recentralize Forest Governance?" *Science* 328 (2010): 312–13.

Pokorny, Benno, Imme Scholz, and Wil de Jong. "REDD+ for the Poor or the Poor for REDD+? About the Limitations of Environmental Policies in the Amazon and the Potential of Achieving Environmental Goals through Pro-Poor Policies." *Ecology and Society* 18, no. 2 (2013): 3–18.

Raftopoulos, M. "REDD+ and Human Rights: Addressing the Urgent Need for a Full Community-Based Human Rights Impact Assessment." *International Journal of Human Rights* 20, no. 4 (2016): 509–30.

Rantala, Salla, and Monica Di Gregorio. "Multistakeholder Environmental Governance in Action: REDD+ Discourse Coalitions in Tanzania." *Ecology and Society* 19, no. 2 (2014): 66–77.

Reed, Pablo. "REDD+ and the Indigenous Question: A Case Study from Ecuador." *Forests* 2, no. 4 (2011): 525–49.

Resosudarmo Ida, Stibniati Atmadja, Andini Ekaputri, Dian Intarini, Yaya Indriamoko, and Pangestuti Astri. "Does Tenure Security Lead to REDD+ Project Effectiveness? Reflections from Five Emerging Sites in Indonesia." *World Development* 55 (2014): 68–83.

Ribot, Jesse, and Nancy Peluso. "A Theory of Access." *Rural Sociology* 68, no. 2 (2009): 153–81.

Saeed, Abdul-Razak, Constance McDermott, and Emily Boyd. "Are REDD+ Community Forest Projects Following the Principles for Collective Action, as Proposed by Ostrom?" *International Journal of the Commons* 11, no. 1 (2017): 572–96.

Sanders, A. J. P., H. da Silva Hyldmo, R. D. Prasti, R. M. Ford, A. M. Larson, and R. J. Keenan. "Guinea-Pig or Pioneer? Translating Global Environmental Objectives through Local Actions. The Case of Central Kalimantan as Indonesia's Official REDD+ Pilot Province." *Global Environmental Change* 42 (2017): 68–81.

Sarmiento Barletti, Juan Pablo. "The Angry Earth: Wellbeing, Place, and Extractive Development in the Amazon." *Anthropology in Action* 22, no. 4 (2016): 43–53.

———. "Kametsa Asaiki: The Pursuit of the 'Good Life' in an Ashaninka Village." PhD dissertation, University of St. Andrews, 2011.

Sarmiento Barletti, Juan Pablo, and Anne Larson. "Rights Abuse Allegations in the Context of REDD+ Readiness and Implementation: A Preliminary Review and Proposal for Moving Forward." CIFOR *Infobrief* 190 (2017): 1–8.

Sarmiento Barletti, Juan Pablo, and Lexy Seedhouse. "The Truth and Reconciliation Commission and Law of Prior Consultation: Of Obstacles and Unintended Opportunities for Democratization and Political Participation in Peru." *Latin American Perspectives* (2019): 1–17.

Scheba, Andreas, and O. Sarobidy Rakotonarivo. "Territorialising REDD+: Conflicts over Market-Based Forest Conservation in Lindi, Tanzania." *Land Use Policy* 57 (2016): 625–37.

Schleicher, Judith, Carlos Peres, Tatsuya Amano, William Llactayo, and Nigel Leader-Williams. "Conservation Performance of Different Conservation Governance Regimes in the Peruvian Amazon." *Scientific Reports* 7 (2017): 1–10.

Schlosberg, David. *Defining Environmental Justice: Theories, Movements, and Nature*. Oxford: Oxford University Press, 2009.

Schlosberg, David. "Reconceiving Environmental Justice: Global Movements and Political Theories." *Environmental Politics* 13 (2004): 517–40.

Schroeder, Heike, and Constance McDermott. "Beyond Carbon: Enabling Justice and Equity in REDD+ across Levels of Governance." *Ecology and Society* 19, no. 1 (2014): 31–33.

Schroeder, Richard, Kevin St. Martin, Bradley Wison, and Debarati Sen. "Third World Environmental Justice." *Society and Natural Resources* 21, no. 7 (2008): 547–55.

SERVINDI. "Norwegian Government to Peru on the JDI: 'We Are Waiting for Results.'" Accessed August 21, 2019. https://www.servindi.org/actualidad-articles-english/23/04/2019/norwegian-government-peru-jdi-we-are-waiting-results.

Shrader-Frechette, Kristin. *Environmental Justice: Creating Equality, Reclaiming Democracy*. Oxford: Oxford University Press, 2002.

Sikor, Thomas. *The Justices and Injustices of Ecosystem Services*. London: Routledge, 2013.

Sikor, Thomas, and Hoang Cam. 2016. "REDD+ on the Rocks? Conflict over Forest and Politics of Justice in Vietnam." *Human Ecology* 44 (2016): 217–27.

Sikor, Thomas, Adrian Martin, Jante Fisher, and Jun He. "Toward an Empirical Analysis of Justice in Ecosystem Governance." *Conservation Letters* 7 (2014): 524–32.

Sills, Erin, Stibniati Atmadja, C. de Sassi, Amy E. Duchelle, D. Kweka, Iap Resosudarmo, and Wiliam D. Sunderlin. *REDD+ on the Ground: A Case Book of Subnational Initiatives across the Globe*. Bogor, Indoensia: CIFOR, 2014.

Smith, P. D., and M. H. McDonough. 2001. "Beyond Public Participation: Fairness in Natural Resource Decision Making." *Society and Natural Resources* 14 (2001): 239–49.

Sunderlin, William, Claudeio de Sassi, Erin Sills, Amy Duchelle, Anne Larson, Ida Aju Pradnja Resosudarmo, Abdon Awono, Demetrius Leo Kweka, and Thu Ba Huynh. "Creating an Appropriate Tenure Foundation for REDD+: The Record to Date and Prospects for the Future." *World Development* 106 (2018): 376–92.

Sunderlin, William, Andini Ekaputri, Erin Sills, Amy Duchelle, Demetrius Kweka, and Rachael Diprose. *The Challenge of Establishing redd+ on the Ground: Insights from 23 Subnational Initiatives in Six Countries*. Occasional Paper No. 104. Bogor, Indonesia: CIFOR, 2014.

Turnhout, Esther, Aarti Gupta, Janice Weatherley-Singh, Marjanneke Vijge, Jessica De Koning, Ingrid Visseren-Hamakers, Martin Herold, and Markus Lederer. "Envisioning REDD+ in a Post-Paris Era: Between Evolving Expectations and Current Practice." *Wiley Interdisciplinary Reviews: Climate Change* 8, no. 1 (2017): e425.

United Nations. "Declaration on the Rights of Indigenous Peoples." March 2008. http://www.un.org/esa/socdev/unpfii/documents/DRIPS_en.pdf.

———. "Framework Convention on Climate Change." March 15, 2011. https://unfccc.int/resource/docs/2010/cop16/eng/07a01.pdf.

———. "Introduction to Gender and Climate Change." Accessed July 12, 2009. https://unfccc.int/topics/gender/the-big-picture/introduction-to-gender-and-climate-change.

United Nations Development Programme. "Human Development Report 2011: Sustainability and Equity. A Better Future for All." 2011. http://hdr.undp.org/sites/default/files/reports/271/hdr_2011_en_complete.pdf.

United Nations Framework Convention on Climate Change Secretariat. "Key Decisions Relevant for Reducing Emissions from Deforestation and Forest Degradation in Developing Countries (REDD+)." June 2014. https://unfccc.int/files/land_use_and_climate_change/redd/application/pdf/compilation_redd_decision_booklet_v1.1.pdf.

———. "Report of the Conference of the Parties on Its Sixteenth Session, Held in Cancun from 29 November to 10 December 2010." March 2011. https://unfccc.int/resource/docs/2010/cop16/eng/07a01.pdf.

United Nations General Assembly. "Rights of Indigenous Peoples." July 29, 2016. http://www.un.org/en/ga/search/view_doc.asp?symbol=A/71/229.

United Nations Human Rights, Office of the High Commissioner. "Human Rights and Climate Change." Accessed July 12, 2019. http://www.ohchr.org/EN/Issues/HRAndClimateChange/Pages/HRClimateChangeIndex.aspx.

Vijge, Marjanneke. "Carbonizing Forest Governance: Analyzing the Consequences of REDD+ for Multilevel Forest Governance." PhD dissertation, Wageningen University, 2016.

Vijge, Marjanneke, Maria Brockhaus, Monica Di Gregorio, and Esther Muharrom. "Framing National REDD+ Benefits, Monitoring, Governance and Finance: A Comparative Analysis of Seven Countries." *Global Environmental Change* 39 (2016): 57–68.

Vijge, Marjanneke, and Aarti Gupta. "Framing REDD+ in India: Carbonizing and Centralizing Indian Forest governance?" *Environmental Science and Policy* 38 (2014): 17–27.

Visseren-Hamakers Ingrid, Aarti Gupta, Martin Herold, and Marielos Peña-Claros. "Will REDD+ Work? The Need for Interdisciplinary Research to Address Key Challenges." *Current Opinion in Environmental Sustainability* 4 (2012): 590–96.

Vongvisouk, Thoumthone, Guillaume Lestrelin, Jean-Christophe Castella, Olle Mertz, Rikke Brandt Broegaard, and Sithong Thongmanivong. "REDD+ On Hold: Lessons from an Emerging Institutional Setup in Laos." *Asia Pacific Viewpoint* 57, no. 3 (2016): 393–405.

Walker, Gordon. *Environmental Justice: Concepts, Evidence and Politics*. New York: Routledge, 2011.

Westholm, Lisa, and Seema Arora-Jonsson. "Defining Solutions, Finding Problems: Deforestation, Gender, and REDD+ in Burkina Faso." *Conservation and Society* 13, no. 2 (2015): 189–99.

White, Daniel. "A Perfect Storm? Indigenous Rights within a National REDD+ Readiness Process in Peru." *Mitigation and Adaptation Strategies for Global Change* 19, no. 6 (2014): 657–767.

Wilson, Kalpana. "Towards a Radical Re-appropriation: Gender, Development and Neoliberal Feminism." *Development and Change* 46, no. 4 (2015): 803–32.

World Bank. "Dedicated Grant Mechanism for Local Communities Project." Accessed July 12, 2019. http://projects.worldbank.org/P145316?lang=en.

Part 3

Territorial Rights, Ecocosmology, and the Quest for Environmental Justice

After exploring the historical context and the inequities perpetuated by contemporary development projects, this section examines efforts by indigenous communities to map a more equitable future through consolidation of their territorial rights. The chapters highlight the promise and shortcomings of these efforts by focusing on the difficulties of protecting the rights of isolated and uncontacted groups in Brazil, analyzing advances in and challenges for territorial rights for indigenous populations in Peru, and providing a multicountry comparison of advances in and challenges for the process of gaining the free, prior, and informed consent of indigenous people about development projects that affect their collective rights.

6 Indigenism, Isolation, and Socioenvironmental Conflicts in the Javari River Valley

BARBARA ARISI AND FELIPE MILANEZ

The Javari Valley (Vale do Javari)

The 8.5-million-hectare Javari Valley Indigenous Land, the second largest in Brazil, is a continuous high forest area that is home to the largest number of isolated, or "uncontacted," Indians in Latin America, and probably in the world. Although this vast territory is fairly well protected from massive deforestation, there are still threats to the forest and its inhabitants from illegal fishermen, hunters, loggers, miners, and exotic fish and drug traffickers. The only means of transportation to and within the territory is by river or air. More than five thousand indigenous people live in the vast Javari territory. They include eight indigenous groups that are considered to be "contacted" or in "recent contact," as well as groups considered by the Brazilian state to be "isolated," who generally avoid contact, such as the Korubo group we discuss in this chapter. Within the territory there probably are still groups of Marubo and Mayoruna/Matsés people living in isolation from the other groups. The region close to the Jaquirana and Jandiatuba rivers is also believed to be home to other isolated groups, who probably speak Panoan and Katukina languages. Altogether there are sixteen reports of the existence of different groups. The last contact of an isolated group in the region happened in March 2019, during an expedition led by the National Indian Foundation (Fundação Nacional do Índio, FUNAI).[1] This chapter describes the demands made by the Matis themselves in an effort to make public their questions and their autonomy.

The concept of isolation comes from the policy of the Indian Protection Service (Serviço de Proteção ao Índio, SPI), which was created in 1910 to intervene in situations of conflict between indigenous peoples and colonizers. The SPI was succeeded in 1967 by the, FUNAI. In *The Indians and the Civilization*, published in Brazil in 1970, the Brazilian anthropologist Darcy Ribeiro outlined four classifications, which initially came from SPI (where he had worked), of indigenous peoples living in "isolation," with "intermittent contact," with "permanent contact," or "integrated," considering integration or isolation as being in relation to "civilization." The Indian Statute, Law 6001 of 1973, defines isolated people as "groups that are still unknown or when there are just a few and vague reports through casual contacts with the national society."[2] This remains the official definition, even after the creation in 1987 of the Department of Isolated Indians, within FUNAI, to address issues related to isolated people and those in recent contact, in order to enforce respect for isolation over forced contact led by the state. In the past, isolated people were referred to in Portuguese as *arredios* (antisocial) or *bravos* (fierce or hostile people), meaning that they avoided contact with other groups. They can also be understood as people who "have not yet submitted to the present forms of conquest and state action toward populations and territories," following Antonio Carlos de Souza Lima.[3] Although the term "uncontacted" is often used to describe these peoples, Beatriz Huertas Castillo, who studies such groups in Peru, instead proposes the terms "isolated indigenous people" or indigenous people "in isolation," arguing that they are groups that chose isolation from national society because of previous traumatic experiences.[4] Since Ribeiro, scholars use these terms to describe people who live in an "anomalous" situation. Although they took refuge in remote regions to escape colonial exploitation, these indigenous people do not live in a timeless bubble, nor are they unaware of other groups that inhabit their forests. Most have maintained some contact with other people, although in a way that is mediated by objects that circulate in huge networks of trade or war.

Pots and pans and metal tools such as axes and machetes were—and still are—used to promote what the Brazilian government's indigenist agency calls a "front of attraction." An example of this approach is seen in a recent "first contact" event involving gift giving, which occurred in 2014 on the Envira River in the western state of Acre. A video of the event of contact with the "Igarapé Xinane isolated people" shows Ashaninka people in the village of Simpatia, on the Envira River, offering a bunch of bananas to three young men from the isolated group who were crossing the river, while a young FUNAI employee, wearing only underpants, shouts in the background. The scene reflects his lack of experience and his disregard for the indigenous way of reestablishing contact among indigenous groups that have lived in these areas since time immemorial. The video, which went viral on the internet, also displays the lack of preparedness of the younger generation of FUNAI employees. Older employees practiced a style of indigenism rooted in a tradition of learning by doing things the "indigenous way," imitating as closely as possible the behaviors of indigenous people who are local to the area. Such an effort is totally absent from this "contact event."

In the past, SPI and FUNAI agents used such objects from the industrialized world as bait to attract these people. The bait was placed in *tapiris* (tents in the sites where indigenous people circulated) so the agents could try to establish nonviolent contact, hence the term used for years in Brazilian indigenist policy, of "pacifying" indigenous people. These contacts, forced by the seduction of material goods, would serve the pacifying purpose for which the agents of the state were paid. They sought to be at the forefront of economic or religious development, the frontiers of "civilization" as described by Ribeiro in 1970, whether advanced by loggers, *seringalists* (rubber tappers), *garimpeiros* (gold miners), or missionaries (in many cases, this was done with the best of intentions, of protecting the indigenous people from the colonial agents).[5]

On occasion SPI and FUNAI agents clashed with indigenous people and used the strategy of pacifying those who resisted development in their territories. The contradiction is that such development projects were promoted by the state itself, associated with private interests, such as the emblematically tragic cases of the construction of the Transamazonian Highway (BR 230), the Cuiabá-Santarém road (BR 163), and both the Tucuruí and Balbina hydroelectric dams, all of which were planned and constructed during the military dictatorship. They all resulted in tragic confrontations and forced contact with and displacement of indigenous peoples, some of whom were decimated by epidemics and government negligence. Pacification for economic enterprises and to facilitate the expansion of extraction of natural resources was also conducted by private agents, as in the cases of the catechist and *seringalist* Felizardo Cerqueira during the rubber boom in Acre, and the removal and exile of the Xavante in Maraiwatsède, Mato Grosso, in the 1960s by the landowner Ariosto da Riva.[6]

All of these terms—"front of attraction," "isolated peoples," and "surrounding society"—are part of official FUNAI discourse. Since 1987, with the agency's policy on isolation, some new words have entered the official lexicon, such as "ethno-environmental protection," "system of protection and promotion of rights," the prerogative of "noncontact" and Indians in "recent contact." The last term is currently used by the General Coordination of Indians in Isolation and Recent Contact (CGIIRC) to refer to the Matis from the Javari Valley. This categorization is part of a FUNAI strategy to try to control these people and prevent them from living again in areas where some of the older Matis were born and lived before contact.

In the most recent expedition, which took place in the Javari on March 2019, FUNAI used high technology to monitor the area. Finding the contacted Korubo who were reencountering their isolated relatives took no more than a week of walking through the forest; thus the expedition was characterized as a success.[7] FUNAI's new vocabulary includes "distension," referring to the conflict between Korubo and Matis. But both FUNAI and the press made no mention of the conflict between FUNAI and the Matis in 2015. Next we present a brief narrative of what led to this recent conflict and (re)contact. Since the inauguration of the conservative and authoritarian government of President Jair Bolsonaro, the entire public policy to assist indigenous peoples, including protection, health and education programs, have undergone a huge deconstruction, leading to a high risk of genocide among uncontacted and recently contacted indigenous peoples.[8] The Amazon has burned and been deforested at very high levels since Bolsonaro took office.

Contact of the Korubo and Isolation of the Matis

The first contact with a group of Korubo Indians by FUNAI agents occurred in 1996, during an expedition led by the *sertanista* (FUNAI backwoods contact agent) Sydney Possuelo. The Matis played a leading role in this contact, because they were among the Indians invited by FUNAI to serve as translators during the process of contact. The language spoken by the Matis proved more comprehensible to the Korubo than those of the Mayoruna/Matsés and Marubo. After that, the Matis worked for years in the Javari Valley Ethno-Environmental Protection Front at the confluence of the Ituí and Itacoaí rivers. They recognized themselves as descendants of two Korubo women who had been kidnapped by the Matis as children, probably in the 1930s.[9]

In many cases, there is a long history that brings together the people considered to live in isolation and those who are considered contacted. This relationship, which exists before official state contact, weighs heavily in the relations the two indigenous peoples establish between themselves during and after this new encounter. The relationship becomes even more complicated because of the presence of new, non-Indian agents. During the period of isolation, a people or small groups of a people avoid contact with the nonindigenous world and with other Indians who have already been contacted and who remain in the forest. Sometimes they watch the movements of others along rivers, in canoes first powered by paddles and then engines, or flying in the sky with their helicopters or light aircraft. After contact, these people return to live in the same area, continuing their previous relationships.[10]

The Brazilian government's contact with the Matis was motivated by its interest in oil prospecting in the region by Petrobras, the state oil company, between 1976 and 1978, the years of the Matis contact. It is estimated that two-thirds of the Matis population died at that time. The trauma of having buried virtually all his relatives never left Binan Tuku, one of the four Matis men who participated in FUNAI's front of attraction to contact the Korubo in 1996. The Matis and Korubo have a relationship that stems from a common ancestral past.[11] In his studies of the northern Panoan Indians, Philippe Erikson determined that the Matis participated in a "Pan Mayoruna system" and that the ethnic discontinuities seen today are the result of displacement, because the Matis' reaction to the arrival of rubber explorers was the opposite of that of their Mayoruna/Matsés neighbors. While the Mayoruna/Matsés exterminated or incorporated most of their neighbors as a way of escaping extinction, the Matis opted to isolate themselves to avoid confrontation: "The existence of isolated units is new empirical data to which the traditional representations of the self do not seem yet resigned."[12]

In the nineteenth century there were still groups of people who participated in huge networks of interchange and were active in matrimonial, material, and political alliances. In Erikson's view, the fragmentation of the past consisted of unions and relations of interdependence that have now disappeared.[13] We argue that past relationships still play a major role in present-day relationships and still serve as guidelines for defining the new situations of indigenous people when they leave isolation in search of contact. We believe they appear in recent events in the Javari, particularly the violent clashes that resulted in the deaths of Korubo and Matis people near the Coari River. Kinship between the Matis and Korubo peoples is also recent, as shown by the abduction of two Korubo girls by Matis in the early twentieth century.

Since 2009, the Matis had planned to return to the Coari River region, from which they had been displaced after the tragic contact perpetrated by FUNAI. Txami, one of the men leading a *fratria*, had been born in that area of the Coari River, as he informed one of the authors during her fieldwork in 2006.[14] That same year Txami and his brothers had planted cassava and banana crops and cleared a *varadouro* (a footpath in the woods) to connect the banks of the Coari River to the old village of Aurelio, where Txami lived.[15] The group of brothers created a new village, which they called Todowak. A serious conflict occurred in this area of the Coari River when a group of Korubo attacked and killed two of Txami's younger brothers, Damë, an indigenous health agent, and Iva Xukurutá, on December 5, 2014.

According to the former coordinator of the Matis Indigenous Association, Marcelo Makë Turu Matis, the Korubo attacked his relatives because the group contacted in 1996, known as the "Mayá group," did not have enough women. In other words, the need for more women prompted the

Korubo group, which had been in contact with FUNAI and the Matis since 1996, to approach the other group of the Korubo people, who lived in the area of the Coari River. According to Marcelo Makë Matis, the conflict may have begun when the group of Korubo still living in isolation in the Coari approached the other Korubo who now live near the Matis village. In the following excerpt from an article published in *Amazônia Real*, Marcelo Makë Turu Matis explains the possible motivation for the attack that resulted in the deaths of the Txami brothers: "That is why they [the Mayá group] began to seek a woman where they have other Korubo relatives: This group from Maiá [Mayá] was in the village of Korubo, on the Coari River, and attacked the isolated people. The Korubo of the Coari began to spread out near the Matis villages, near Todowak, which is also on the Coari River, where my uncles [Txami, Damë and Iva Xukurutá] lived."[16]

Conflicts between isolated and contacted indigenous groups in the Javari Valley have increased in recent years, as the groups have come under pressure to allow the extraction of natural resources from their territories. There has also been an escalation of contact processes without adequate preparation of FUNAI's agents and government health teams and without the budget allocations necessary to meet these challenges. FUNAI's employees considered these encounters between Indians who had been living in isolation and Indians already in contact with national society to have been "indigenous initiatives." But the violent conflicts merit closer examination. First, a group of six Korubo natives were contacted by the Kanamari near the Massapê village in September 2014. Subsequently another fifteen Korubo from this group were contacted and FUNAI took these Indians to the region where the "Mayá group" lives. They became known as the Korubo living at "Tronqueira creek" and the "Visa and Pino group." They were found to have contracted diseases before contact, such as influenza and malaria. Although the Brazilian government considered them isolated Indians, they were already being contaminated by the surrounding societies.

The Javari Valley may be affected by a broad forest contamination effect, as described by Ramenofsky and Galloway in relation to the tragic expedition of Hernando de Soto in the southeastern United States in the sixteenth century, as well as the effects of different depositories, reservoirs, and mechanisms for the transmission of diseases and infections in the current epidemic.[17] Isolation that is political, and thus voluntary, does not necessarily function as epidemiological isolation, a thesis developed by Arisi, Francisco, and Cesarino in a health assessment specifically focusing on health care in the Vale do Javari Indigenous Land.[18]

During this same period, at the end of 2014, the conflict involving the Korubo, Matis, and FUNAI intensified. FUNAI news described the deaths of Damë and Iva Xukurutá as follows: "A meeting occurred in November 2014, when isolated Korubo Indians from the Coari River approached the plantations of the village Mavis Todowak and a conflict ensued between them, resulting in the death of two Matis Indians."[19] The Union of Indigenous Peoples of the Javari Valley (União dos Povos Indígenas do Vale do Javari, UNIVAJA) reported on December 8, 2014, that the two Matis men were "killed by four Korubo isolated Indians in the village of Todowak on the Coari River, on December 5." The coordinator of FUNAI's CGIIRC, Carlos Travassos, described the conflict as "a breach of protocol" because he considered that previous attacks had "always" come from the Matis side: "The Matis have always stolen the Korubo women, but this [had] never occurred before."[20] Under attack by public opinion, the Matis made an effort to make public their questions and their autonomy.

The conflicts between the Matis and Korubo intensified, culminating in new deaths and new contacts. They prompted political protests by UNIVAJA and by the Matis, in particular, against FUNAI. At some point a retaliation by the Matis occurred, resulting, according to FUNAI, in the deaths of "at least" nine Korubo. In a "response" to an article by Felipe Milanez, FUNAI provided previously unpublished information about the seriousness of the conflict, which the agency considered an "interethnic" affair.[21] FUNAI officials initially stated that the agency was informed on September 26 about a meeting between some Matis and ten Korubo in the village of Matis Tawaya. The health team would have arrived in the village on September 29. Then, on October 7, a second group, composed of eleven Korubo, was also contacted by the Matis.

According to Travassos, FUNAI's CGIIRC coordinator at the time, "contact with the Korubo group resulted from the Matis' coercion"; he claimed that the Matis "captured Korubo children during the first contact with the Korubo" and carried out an "invasion of the Korubo camp," while a government employee "was detained by the Matis." In the same interview, he further stated that the "Matis intimidated personnel from FUNAI and the Secretaria Especial de Saúde Indígena [Special Secretariat for Indigenous Health], making the atmosphere even more tense." He concluded, "The autonomy of one people should never be superimposed on the autonomy of another."[22] That information was published by FUNAI in the press and as a "note" on the agency's website but was contested by Matis and other indigenous leaders in the Javari Valley. There was a protest, with letters sent to the president of FUNAI and public communiqués, culminating in the occupation of FUNAI's headquarters in Atalaia do Norte (Amazonas) in early 2016, resulting in the expulsion of the local FUNAI coordinator. The Indians demanded a dialogue with the government that would enable them to participate in policies that affect their territory and their people, including the process of contact with their Korubo neighbors, because they considered themselves ignored by the official indigenist policy.

FUNAI usually prides itself on implementing an innovative indigenist policy with "rigid protocols and procedures," which it considers pioneering and an example for the world. In the Javari Valley, however, this position has been publicly contested by the indigenous peoples whom FUNAI considers contacted or in recent contact. In this debate leaders of UNIVAJA accused FUNAI of omission, arguing that the official indigenist policy had not been effective in protecting the isolated Indians and that it had only succeeded in ignoring them and sweeping the problem of their survival and autonomy under the rug. In response, FUNAI began to criminalize the natives who were contacted or in recent contact, issuing a letter of repudiation against the Matis and promoting a new program to protect the Korubo against the Matis, who were forced to resign from FUNAI.

FUNAI's CGIIRC accused the Matis of errors and of possible negligence that we believe were the responsibility of the agency itself. However, it is obvious that the conflict currently unfolding in the region is also due to the state's involvement, by either its actions or omissions, and is not theoretically distant from what Brian Ferguson and Neil Whitehead classify as "war in tribal territories" to describe the expansion of state interests into indigenous territories.[23] After all, if FUNAI accuses an indigenous people (the Matis) of disrespecting the official rules created by the Brazilian state, through FUNAI, inside traditional Matis territory, this conflict is also related to the expansion of state territorial control, to FUNAI's implementation of a protectionist policy associated with financial investments of an indigenist NGO, and to extractive interests over natural resources such as oil interests in the region.

The Isolation of the Policy

We argue is that this may be more a case of isolation of Brazilian policy rather than of the indigenous groups living in the Amazon. Currently all indigenous peoples living in isolation in the Brazilian Amazon are in a situation of extreme vulnerability and risk, in a forest that is being devastated by large projects financed by public and private money. In the Javari Valley this occurs on both the Brazilian and Peruvian sides of the border; the indigenous populations on both sides are also exposed to incursions by drug and arms traffickers. The governments of both countries have no policies that defend indigenous interests by placing them on equal footing with economic interests with regard to rights to the natural resources of the territories they occupy. In general, the Brazilian government has only promoted greater dependence on the state by indigenous peoples, in a kind of welfare policy, sometimes outsourcing state tutelage, in what is essentially the privatization of welfare, by contracting NGOs to carry out work that should be the government's responsibility.

In the Javari Valley, a place of extreme natural beauty, where several documentaries have been filmed for international channels such as National Geographic, MBC, and BBC, a war has now been fought between the Matis and the Korubo, two peoples who have a long history in common. Both are known internationally, and their image has been popularized by various films and reports about them.

The Matis have been in contact with the Brazilian government since 1976. The front of attraction at the time was so ill-prepared that even the boat's small engine did not work. The contact was motivated by developmental and extractivist interests, as FUNAI was helping Petrobras drill exploratory wells to determine whether petroleum existed in the region, and studies were being conducted for the construction of the North Perimetral Highway, which was never completed. At that time contacting meant pacifying the Indians to favor the power of economic interests over their territories in a domesticated and controlled environment.

It is estimated that two-thirds of the Matis population died as a result of that contact. Marcelo Makë Turu Matis recently informed the authors that at least eighty-seven Matis died because of "governmental contact." However, the National Truth Commission's December 2014 report on atrocities and crimes committed by the Brazilian dictatorship contains nothing about the institutional violence perpetrated by the Brazilian state against the Matis. Twenty years later, with the historical experience of having been attracted and pacified in a kind of "epidemiological genocide," the Matis agreed to participate in another attraction front, this time in 1996, to contact the Korubo Indians who lived near the place where FUNAI had installed a surveillance base at the confluence of the Ituí and Itacoaí rivers to prevent incursions by hunters and loggers.

In 2006, ten years after contact with the Korubo, the author Barbara Arisi conducted a study to understand why the Matis participated in the team created to contact the Korubo.[24] The question was why they would help the government make contact with an isolated group when they knew, from their own experience, that the Korubo were at risk of very high mortality because they lacked antibodies that would give them resistance even to common diseases like the flu.

The "protection" policies, which gave rise to the idea of pacification, had their roots in the creation of the Indian Protection Service in 1910. The idea of pacification had a humanitarian sense and, according to Ribeiro, implied a guarantee of the right to be different. During the dictatorship, and with the encroachment of development projects on indigenous territories, FUNAI's institutional development was also marked by the creation of fronts of attraction and attraction strategies as part of pacification processes.

The policy for the protection of isolated Indians and the prerogative of noncontact was implemented in 1987, with the creation of the Department of Isolated Indians and has undergone several transformations over the past three decades, directly affecting the peoples living in the

Javari Valley (both those considered to be isolated and those who already had ties with wider Brazilian society). These changes culminated in the most recent "system of protection and promotion of rights," which, in the opinion of one indigenist expert, is currently in the process of collapse without ever having been implemented and which made the prerogative of noncontact "a rhetorical fiction."[25]

With the strengthening of indigenous protagonism and mobilizations by the indigenous movement, it became clear that at no point in history was the contact process unilateral, under the control of those who previously fulfilled the function of the sertanista. There was always a response from the victims, and indigenous protagonism has grown and become even more vigorous in recent years. Examples of indigenous protagonism in situations of contact with national society were studied by Albert and Ramos in a collection of essays; indigenous people reported that they were the ones who chose to go out to encounter nonindigenous people in an effort to make them more peaceful and less aggressive.[26]

In the case of policies for isolated peoples, indigenous people always responded with many challenges, tensions, and contradictions, constructing different possibilities to resist state-imposed control. In the case of the conflicts analyzed here, isolation and the Matis' request to play a greater role in political processes concerning them show the gap between the reality of the forest and the policies developed in a distant office, as the Matis struggle for the nonisolation of the group of isolated Indians. If historical experience shows the indigenous agency in the process of "pacification of the whites" (nonindigenous people), the isolation (or alienation) of groups from the possibility of interceding in diplomatic processes in their own territories indicates a contradiction in the idea of no contact (the policy of keeping indigenous people outside of decision-making processes concerning their own lives and land).[27] Finally, this isolation policy presents itself as a new problem with regard to consolidation of what is meant by the protagonism of subaltern groups. In fact the distance from contact with state or private agents is not only voluntary but is also a political decision by indigenous peoples, and it can also be understood as environmental resistance to dispossession. What is at stake, ultimately, is their possibility of living in autonomy inside their own territory.

The Tragedy of Contact with the Matis

The story recounted earlier has a tragic history. Beginning in 1969 FUNAI officials documented the complexity of the Javari Valley territory. The sertanista Raimundo Pio Carvalho Lima proposed the creation of four indigenous posts in the region that includes the Javari, Curuçá, Itaquaí, and Ituí rivers. Meanwhile FUNAI's work in the region in 1971 was mainly to support the construction of the North Perimetral Highway, which would cut across the Javari Valley in several places inhabited by isolated Indians, including the Jandiatuba and Jutaí rivers in the area being studied for the creation of the Javari Indigenous Park, as it was called at the time. The Brazilian government began the process of attracting isolated groups that were in areas that FUNAI considered critical in the Itaquaí watershed. This front was deactivated after the official Jaime Pimentel was killed by Indians who were reported to have hair cut in the shape of a gourd and who used clubs. From the description of an official reporter, it is likely that the killers may have been Korubo.

The indigenist Wellington Figueiredo participated in the first contact with the Matis in the Javari Valley, along with the sertanista Sydney Possuelo, between 1975 and 1977. He described his feelings upon meeting the Matis for the first time as a mixture of fascination and tragedy:

> I was present at the seventh meeting with the Matis. It was my first experience with isolated Indians. From a traditional perspective, there are two ways of promoting contact with an isolated indigenous group. The first uses the technique of putting gifts out for the Indians and expecting them to be accepted. The second would be to meet the Indians and promote contact more abruptly, if the people were less aloof, or aggressive, or if there were the possibility of having an interpreter. This is the traditional model, which for many years was the basic rule of thumb in Brazil. The work of contact with the Matis followed this traditional model: there was a post and we made paths through the forest, and we distributed gifts from time to time and waited for them to come to us. One thing that struck me a lot about the Matis is that it was the first time in my life that I saw naked Indians, like those in the stories I had heard, like those of Ismael Leitão about the

> Xavante. This was when the Matis maintained their seventh contact with the Funai Itui Attraction Front, a visit they made to the base. This brief encounter was so impressive that I had a hard time sleeping for the next I two days, so great was the shock of seeing people appear that way. I was struck by the concern that those people, whose living conditions were so good at that moment, would later become like the Ticuna: dependent, dressed in rags, with diseases. This disturbed me greatly. It marked this first encounter with the Indians and then guided all my work with isolated Indians for the rest of my life. This shock I had when I saw the Matis for the first time.[28]

Just as Figueiredo has always kept the tragedy in mind, for the Matis the experience of contact persists in their bodies and in their own life histories. In 1974 FUNAI created an encampment to try to attract the indigenous groups still living in isolation on the left bank of the Ituí River, just above the stream called Jacurapá. It is believed that before the contact in 1976, the Matis lived in about five villages in the region of the Jacurapá, in the center of the Javari Valley Indigenous Land. Several Matis marked the birthplaces of their relatives on a map for Arisi: the left bank of the Branco River (where Iva Xukurutá, one of the two Matis killed in the Korubo attack in 2014, was born) and the right bank of the Coari River, the head of the Jacurapá stream (the place where his brother, Tumi Tuku, was born).

After the postcontact epidemiological genocide, which killed around two-thirds of the Matis, they regrouped into two villages in order to try to control disease and death. They probably did so because of the need for access to remedies provided, albeit insufficiently, by FUNAI, and because of pressure from the indigenous agency, which preferred to provide care to the Indians in a single place. They may also have been motivated by curiosity and access to many objects that were previously scarce or even unknown to younger members of the group, such as dogs. The FUNAI front of attraction (*posto indígena de atração*) was moved in 1982 to the stream called Boeiro, on the left bank of the Ituí River. At that time the Matis lived in two villages, one at the mouth of the stream and another deeper in the forest, far from the river. In 1987 they returned to area near the bank of the Ituí River.

Conflicts between members of the surrounding society, especially fishermen, loggers, and people living along the river, and the "hostile people" were also mentioned to justify the need for renewed government action in the area. Besides plans for construction of the North Perimetral Highway, Petrobras was interested in prospecting for oil in the Javari Valley. In Itaquaí the company would carry out seismic testing using explosives in lands inhabited by the people now known as Korubo but which were recorded in documents at the time as Marubo or Marubão. The 1981 document already foresaw conflicts between the indigenous group and Petrobras workers and a possible "involvement of the group bordering on Matis territory."[29] That is exactly what happened. In 1984 the Korubo killed three workers involved in Petrobras activities (one from FUNAI and two from the oil company).

Change in Attraction and Contact Policy

The decision to establish a FUNAI control base at the confluence of rivers that were key for access was a standard indigenist strategy for territorial protection in areas inhabited by isolated people, as travel in the Javari Valley is mainly by river. For Possuelo, the main purpose of contact with the Korubo was to pacify through state intervention amid a cycle of violence. By 1996 there was no pressure from development activities, as there had been in the 1980s, but there was also no effective protection of the territory; conflicts with an environmentally predatory local economy, such as logging, were still very intense.

The 8.5-million-hectare area was recognized by the government as indigenous land in 1999, demarcated in 2000, and approved by Brazil's president in May 2001. In 2002 Possuelo organized a large expedition called Alípio Bandeira in honor of a Rondonian sertanista, with the goal of obtaining information about Indians known as the *flecheiros*, or "arrow people." Accompanied by journalists, Possuelo and his work also helped draw media attention to the land, reinforcing the myth of wild nature and the pristine world that could be reached only by a special expedition of explorers, reminiscent of the myth of the British adventurer Colonel Fawcett and others colonialists. However, proving the area to be pristine and occupied by wild Indians, reinforcing the ideology of wilderness, could be an effective way to respond to questions from the military, which dislikes indigenous territory on international frontiers, and agribusiness sectors, which seek to invade and occupy indigenous land; these interests argue that the land is too large to house such a small indigenous population. Even an ex-president of FUNAI, Mércio Gomes, stated once, "There is too much land for such [a] small population."[30]

We would like to emphasize that we consider this a very narrow vision of the importance of indigenous territories to the planet Earth and humankind as whole, because forests are home to many different species of animals and plants, and they do help to keep clean territories that generate oxygen (that is why the Amazon is called "the lungs of the planet") and keep rivers and reservoirs with good quality, unpolluted water. The indigenous people are like gardeners and keepers of the complex Amazon forest ecosystem. The Amazon region also keeps the balance of the rain system in South America and the planet a place where living beings can still breathe, making Earth cleaner and kept at a healthy temperature. Descriptions of wild, pristine nature and the beauty of Indians who have no technological industrial goods reinforce the idea of remote places, exoticism, and other myths of colonialism, trying to hide the socioenvironmental conflicts that, unfortunately, exist all over the Amazon region.

The perception that interventions by the sertanistas were meant to protect the indigenous groups is related to a guarantee of the region's territorial integrity. After the promulgation of the 1988 Constitution, land demarcation became a priority. Provisions for demarcation of indigenous lands were included in the Indian Statute of 1973 but had never been fully implemented. The Department of Isolated Indians was created in 1987 in response to the state-sanctioned violence against indigenous peoples that had occurred during the dictatorship, especially related to development projects. The creation of the department was promoted by FUNAI's agents and was motivated by their previous tragic experiences, culminating in the formulation of the new public policy to protect those defined as isolated. Possuelo recalls that the idea of a new policy stemmed from the experience of the Arara Front in 1980, "one of the best experiences, from the standpoint of organization, that FUNAI had with a Front of Attraction and Contact," and it was grounded in "technical" criteria of organization and methodology rather than being motivated by political or economic interests.[31]

The Arara Front was created for the purpose of attracting and establishing contact with and pacifying the Arara to clear the way for construction of the Trans-Amazonian Highway. The proposed route of the highway would cross their territory and would directly affect one village. Under the authoritarian government, there was no public debate about strategic military projects; pacification served the interests of territorial state projects and national defense. In the case of the Arara Front, dialogue between sertanistas and the military officers in Altamira led to a plan for "controlled" contact, with more resources, a health team and a team with field training, and a guarantee that the territory would be protected against invasion by outsiders. The army supported FUNAI's work by expelling squatters and settlers who were living inside the indigenous territory, including those who had been settled by the Brazilian state.

According to Figueiredo, the "methodology of contact" with the Arara was based on the idea that it was not a matter of "refusing" to make contact but of "avoiding" it. The change in procedures was as follows: "Our proposal was to monitor the area, preventing the entry of people who put the Indians 'to flight,' on the run. Contact would be a possibility and not a certainty to be pursued." Instead of having a small, mobile team follow signs indicating the presence of indigenous people and approach the villages with interpreters to make contact, as was previously done in the region of Altamira (Pará), FUNAI's new strategy would be to have a physically structured base, distribute gifts through the forest, and expect the indigenous people to take the initiative to make contact. This was the approach taken with the Arara Indians. Although the contact occurred sooner than the military expected and was therefore initially considered successful, in the following months it turned into a disaster and a tragedy, with many deaths as a result of an influenza epidemic. "There was a feeling of despair," Figueiredo recalled. "We drew great lessons from this tragedy, which were later reflected in the contact with the Parakanã. There were 119 Indians involved in the contact, and only one died, of snakebite. Because we had already learned how to handle the situation. We learned from the Arara."[32]

The challenge of avoiding epidemics was the main concern. In the case of isolated people, delimiting the territory implies defining something without dialogue or negotiation with those who are directly affected. It is clearly an extreme situation to ensure survival, but it raises the question of representation: Who represents those who do not want contact? The state was responsible for representing the interests of indigenous peoples living in voluntary isolation, or isolated people, as defined by the Brazilian state. This representation occurred without consultation, however. And that created a problem, because those who do not want to be represented (who have chosen to live in voluntary isolation) cannot be represented by the state or by FUNAI. This problem of representation and consultation is still an issue and is directly related to the case discussed here, because CGIIRC, despite its claim that it "does not represent," has essentially made itself the representative of the isolated Korubo people while attacking the Matis in interviews supposedly to protect the Korubo.

Legally the state clearly must defend the rights of people living in isolation, but it does not have to represent the interests of isolated people. There is even a conflict of interest when it comes to development projects, because they are almost always funded directly or indirectly by the state. On top of this inherent contradiction, FUNAI decided to attack and directly accuse another indigenous group, which it also has the responsibility of defending. As noted earlier, in early 2016 FUNAI posted a note on its website, "Repudiation Note against Intimidation and Violence in Javari." It was addressed to the Matis, "repudiating" their behavior and, ironically, criticizing an indigenous people the agency considered to be in "recent contact."[33] For the first time, FUNAI's official note used the term "to repudiate" in the title of an official message and deployed it to criticize an action of an indigenous people. Taking a position of "solidarity" with the Korubo in the face of aggression suffered by a government official, FUNAI said that the Matis' "attitude demonstrates disrespect for the institution [FUNAI] and a threat to the Korubo group"; in other words, "disrespect" toward Funai is regarded as equivalent to "disrespect" for the Korubo. By comparing itself to an isolated group, FUNAI attempted to confront the Matis and to insert itself internally in what the agency itself, based on an anthropological report on the conflict that it had commissioned, classified as an "internal conflict." According to the sertanista Possuelo, who symbolically represented FUNAI in the Javari for several years, "FUNAI has always defended the interests of its master, the Brazilian state."[34]

Relations between the Matis and the Korubo

The earliest of the various accounts of contact between the Matis and Korubo recorded by Arisi deals with the abduction of two Korubo girls, recounted by several Matis men, including Binan Tuku, Txema, Iva Rapa, Txami, and others. The Matis considered themselves to be descendants of these Korubo women. Txami Matis remembers listening to people tell how, long ago—a time the Matis express with the words "Ënden kimon"—four men and one woman went "to get a woman, because there were no more women to marry the Matis." They came back with two Korubo girls, both called Mënke; one of them would become the mother of Binan Tukun. Of his mother, he says, "My father told it like this. A long time ago, they took my mother when she was a little girl. They brought her here and raised her among the Matis. That's why I'm here, the son of Korubo. I saw my mother when I was a little boy. I only remember when they buried my mother. From my father, yes, I remember well."[35]

Txema Matis, who is older than Txami and Binan Tuku, preferred not to mention his memories of the kidnapped Korubo girls and answered Arisi's questions with a phrase that reveals the still-present fear of retaliation: "I do not want to talk about it. That was a long time ago. The Korubo are many, they are growing. And we [Matis] are few. The Korubo can come here in our village to seek revenge for the Matis stealing their women."[36]

The abduction of the Korubo girls is recounted by Iva Rapa. Makë Wassá, an unmarried man in the village of Aurelio, says that the Korubo are his relatives because his father was Korubo. In 2006 Makë Wassá recalled that Baritsika and Binan Tuku had already met members of the large Korubo group that lived on the Coari River and that those Korubo did not want to "capture" the Matis. That is the group that provoked the attacks in 2014. Binan Tuku remembers when two FUNAI agents were killed on the Itaquaí River, near the Marubo Indigenous Attraction Post. The encampment to promote contact was given that name because the

Korubo were also called Marubo at that time. This station was established to attract the isolated people at the confluence of the Ituí and Itaquai rivers.

Nowadays the Matis pursue complex diplomacy with the Korubo. Unfortunately this was marked by two recent tragedies, the deaths of Matis and of Korubo. Attributing these events to a "historical conflict" is a way for state agents to avoid their responsibilities and ignore the Matis' calls for dialogue. The Matis had formally demanded meetings with FUNAI to address the issue of the isolated Korubo, increasing tensions in the area. They were ignored by FUNAI or were simply told that the current rule is the noncontact paradigm, even without paying attention to the imminent conflict.

This is especially true considering the various transformations and disruptions that have occurred in recent years, with the serious problem of contamination of the forest and the incorporation of new technologies and tools (pots, firearms, axes, etc.) into everyday life. There were also changes in FUNAI, with the ending of local policies in Javari following the departure of Possuelo in 2006, a complete administrative overhaul of the organization in 2010, and the more intense collaboration of the NGO Centro de Trabalho Indigenista with FUNAI in recent years. Although Centro de Trabalho Indigenista had served FUNAI's Department of Isolated Indians before, it was after 2013 that the relationship between the two in the Javari deepened, mainly after contradictory investments were made by the Brazilian National Development Bank, which also funded development megaprojects in the Amazon. After 2013 FUNAI financed projects in the areas of indigenous education and hired employees for surveillance under the responsibility of the NGO. Among other political and social impacts, these changes reconfigured the protagonism of indigenous peoples and increased pressure concerning nonindigenous use of and access to the indigenous territories.

After the two contacts in September and October 2015, the Korubo cases were treated as a health emergency, and a contingency plan was implemented. Initial field reports indicated that respiratory diseases had been controlled. Nevertheless, while immediate care was efficient at controlling an epidemic, it raised expectations and showed that the challenge of protecting the Korubo from an epidemiological tragedy is long term. At that point of initial contact, it became clear that FUNAI did not send to the field a team that was able to ensure the safety of the Indians and the agency's employees; that situation, combined with the strategy that FUNAI implemented of politically isolating the Matis, resulted in a prolonged local protest by the indigenous movement, which led to the resignation of the coordinator of the Department of Isolated Indians, Carlos Travassos from CGIIRC.

Protection or Subalternation Policy?

On March 8, 2014, FUNAI issued the note repudiating "intimidation and violence" that the state agency attributed to the Matis. This official statement coupled with other aggressive messages from FUNAI agents directed at Matis leaders provoked an immediate reaction among young leaders from the social movement in Javari. Expressing indignation at what they considered criminalization, the young leaders also began to review the history of their people's relationship with the Brazilian state. The conflict between the Matis and FUNAI, which accompanied the conflict with the Korubo, prompted a reassessment of previously invisible historical problems, particularly regarding the violence committed by the Brazilian state during the last four decades of contact and postcontact. A letter from the Matis Indigenous Association, dated February 12, 2016, stated that the contact was a tragedy that had never been investigated and for which government agents had never been held accountable. The Matis had turned the marginalization they were suffering from FUNAI against the state itself. Although the Matis describe the deaths of eighty-seven people and rape and other forms of physical aggression, the National Truth Commission report contains not a single mention of these atrocities, which almost led to genocide. For the Matis, this impunity, injustice, and lack of reparation reflect the state's denial of its past and of the direct impacts of the current process of inferiorization and criminalization by FUNAI.

Turning directly to public opinion to expose their differences with FUNAI has become a Matis strategy for building their protagonism, which, despite inclusion in FUNAI's official policy since the 2010 reform and in the Federal Constitution of 1988, does not find space for exercise. Te Matis had written public letters harshly criticizing FUNAI and recounting his people's version of the process of contact and the conflicts with the Korubo, revealing that the president of FUNAI had systematically refused to receive the Matis for dialogue.

For the first time in the case of contact with isolated Indians, indigenous protagonism took the form of public insurgency against the state and was documented. Because of the complexity of the interventions, this was reflected in the construction of a critique of indigenist policy of isolation.

In the conflict between the Matis and an isolated Korubo group, two contradictory narratives emerged regarding the role of the state and the history of conflicts. On the one hand, the Matis, in the process of decolonization of their relationship with the state, have taken action, exposing the omissions of state agents and demanding effective protagonism with participation in decision making and sharing of resources, positions, and posts, whether within the state or outsourced through NGOs. The state, represented by FUNAI, has tried to assert itself as a humanitarian organization defending groups subalternized by other indigenous groups within so-called internal relations, delegating to itself the ability to represent those who do not accept representation due their chosen isolation—those who are, in the strict sense of the concept of will, in voluntary isolation. In this narrative of protection and representation of isolated people, the Matis were criminalized and accused of violence and intimidation against state agents.

The Matis recount that the conflict with the Korubo stems from ill-considered interventions by FUNAI and various omissions by the state, a narrative contextualized, in a historical perspective, by the first contact with FUNAI agents in 1976, accompanied by a history of specific policies formulated for isolated groups. In statements made to state agents and public communiqués, the Matis, through the Matis Indigenous Association, attribute responsibility for the conflict—as well as the idea of maintaining the isolation of the Korubo at all costs—to FUNAI and its ill-conceived interventions, as well as its failure to act when it should. It is a deep sociohistorical argument that FUNAI has countered by calling the events an "interethnic" and "historical conflict" that has occurred independently of the relationship with and the "presence" or "absence" of the state. FUNAI's former coordinator has even accused the Matis of "territorial disputes," a charge the Matis leader Make Turu promptly called an "invention" of FUNAI.[37]

Contradictions about which historical events should define the socio-environmental conflict we analyze in this chapter, either in the past or internal to an indigenous culture, as claimed by the Brazilian state, are used against indigenous people in order to advance control by the state. When considering that the relationship between the Matis and the Korubo is also shaped by state interests in infrastructure and development projects, such as oil exploration, land demarcation, logging, and invasion by fishermen and hunters, due to lack of state oversight, as well as to the intensified dispute over materials and objects made available by FUNAI, suddenly the presence of state agents disappears from the scene when the Matis are accused of a territorial dispute. Although the federal government has not given the requested support to FUNAI, as it should have done to fulfill its official role in protecting indigenous rights, and instead has reduced FUNAI's staff and resources and weakened the agency, there has also been a change in the situation involving the peoples who are considered to be isolated and in the political practices of the agency charged with protecting them. This constraint was exposed during the four processes of contact in 2014 and 2015 with indigenous peoples who had so far been living in voluntary isolation; all were unplanned contacts, and FUNAI and health agents in the field had not been prepared or lacked the material support they needed.

The Javari Valley has been a key region in the Amazon for innovative and challenging experiences for FUNAI and the Brazilian state for the past half-century: on the one hand, managing and implanting developmentalism, and on the other, guaranteeing the rights of indigenous peoples whose relationships with the state differ in intensity. Recent controversy over official policies toward indigenous peoples can therefore be seen not as rhetorical criticism addressing specific interests but rather as a paradigm shift, as a broader crisis that is also occurring in other regions of the Amazon.

Tribal Warfare or Institutional Disrespect?

Throughout the years prior to contact, the Matis had been telling FUNAI about the presence of isolated Indians in the forest and on river banks, where hunters occasionally encountered them. The Matis worried that there could be a conflict. These communications were mainly transmitted by radio and to the agency's local coordinator, as well as in meetings with FUNAI indigenist agents, who tried to dissuade the Matis from reoccupying a territory from which they had been displaced after contact in the 1970s. Imminent tension erupted on December 5, 2014, when a group of six Korubo attacked the Matis near the village of Todawak on the Coari River, killing Iva Xukurutá and Damë. The indigenous social movement and the UNIVAJA formally requested FUNAI's support in mediation and requested a meeting in Brasília to outline a peacemaking strategy.

The Matis and other indigenous peoples work under precarious conditions marked by a lack of administrative resources; the same is true of government officials living on the protection fronts, where health care is also provided to a small group of Korubo Indians who were contacted in 1996 (the "Maya group"). As we have discussed in this chapter, FUNAI officials referred to the conflict between the Matis and the Korubo as an interethnic dispute, which could have occurred regardless of the state presence in the area and the pressure from the extraction of natural resources. But Matis leaders attributed the killings of the Matis and of the Korubo to the negligence of FUNAI officials, not to interethnic relations—a conflict that is difficult to define: is it between ethnic groups, or is it a "tribal" war? Just before two of their key leaders were killed by Korubo warriors, the Matis had warned the agents of an imminent conflict, but nothing was done; nor was any response provided to prevent the massacre of isolated Korubo people, which became publicly known in late September 2015. The Matis engaged in contact with the Korubo near the village of Tawaya on September 26, initiating contact with ten people; on October 10, eleven more previously isolated Korubo arrived. A preliminary FUNAI report prepared after the contact estimated that "at least" eight and possibly as many as fifteen, or even more, Korubo people died. If twenty-one people were contacted and if possibly fifteen others were killed, this would reflect a very violent process. There is no evidence of such violence previously in the Javari in the past century, or at least since violent attacks from rubber tappers. This means that something very serious has changed in the Javari, which should not be attributed to a so-called tradition but is directly related to state interventions and economic pressures surrounding the territory.

In early 2016 the indigenous movement of the Javari Valley occupied the FUNAI headquarters in Atalaia do Norte and demanded the resignation of the official in charge of that office to pressure for dialogue between the federal agency and the Indians. As usual, the employees closest to the point where services are provided are the ones who experience the consequences of a poorly planned public policy managed from Brasília and far from the grassroots. The Matis demanded to be heard by FUNAI, especially because two of their people had been killed.

The Matis, as well as the Marubo, Matsés/Mayoruna, Kanamari, Kulina, and Korubo, all want to take part in decision making about their uncontacted or isolated neighbors because they understand that all of FUNAI's political interventions affect their use of the territory and their access to resources that they need for survival. Autonomy and self-government are part of the vocabulary. Dialogue seems to be the first step toward the end of a colonial policy of "ignored Indians." An ideal of isolation based on a romantic notion of Indians living wholly outside of the historical processes, as if they did not suffer from the pressures of development promoted by the state and from the presence of expansionist fronts such as logging or dam construction, cannot take precedence over the self-determination of indigenous peoples.

NOTES

1. Phillips, "Brazil."
2. Government of Brazil, *Estatuto*.
3. Lima, *O Grande Cerco*.
4. Huertas, *Los Pueblos Indígenas*, 22.
5. Ribeiro, *Os índios*.
6. On Cerqueira, see Iglesias, *Os Kaxinawá*.
7. Phillips, "Brazil."
8. Funes, "Brazil's New President."
9. Arisi, "Matis e Korubo"; Arisi, "A Dádiva."
10. Arisi, "Matis e Korubo," 16.
11. Arisi, "Matis e Korubo"; Arisi, "Matis y Korubo."
12. Erikson, *El Sello*, 80.

13. Erikson, "Bibliografia."
14. "Fratria" is an anthropological term used to refer to a group of brothers.
15. Arisi, "Matis e Korubo."
16. Marcelo Make Turu Matis, interviewed in Brasil, "Vale do Javari."
17. Ramenofsky and Galloway, "Disease," 261.
18. Arisi, Francisco, and Cesarino, *Saúde*.
19. Fundação Nacional do Índio, "Resposta."
20. Fundação Nacional do Índio, "Resposta."
21. "Omissão da Funai."
22. Brasil and Farias, "Vale do Javari."
23. Ferguson and Whitehead, *War*, 127.
24. Arisi, "Matis e Korubo"; Arisi, "Matis y Korubo."
25. Vaz, "A que será," 29.
26. Albert and Ramos, *Pacificando*.
27. Albert and Ramos, *Pacificando*.
28. Figueiredo, "Depoimento," 257.
29. Fundação Nacional do Índio, *Informação No. 331/DID/DGPI/81*, 5.
30. Instituto Socioambiental, "Coiab."
31. Possuelo, "Depoimento," 28.
32. Figueiredo, "Depoimento," 271, 278.
33. Fundação Nacional do Índio, "Nota da Funai."
34. Possuelo, "Depoimento," 232.
35. Arisi, "Matis e Korubo," 76; Arisi, "La No-Frontera Pano," 23, 24.
36. Arisi, "Matis e Korubo."
37. Brasil, "Vale do Javari."

BIBLIOGRAPHY

Associação Indígena Matis. *Carta no. 11/AIMA/2015*. Made public November 17, 2015. In authors' possession.

———. *Informe 001/AIMA/2015*. Made public December 2, 2015. In authors' possession.

———. *Informe 001/AIMA/2016*. Made public January 15, 2016. In authors' possession.

———. *Nota de Repúdio*. Made public February 8, 2016. In authors' possession.

———. *Nota Pública*. Made public January 27, 2016. In authors' possession.

Albert, Bruce, and Alcida R. Ramos. *Pacificando o Branco: Cosmologias do contato no norte-amazônico*. São Paulo: UNESP, 2002.

Arisi, Barbara M. "A Dádiva, a Sovinice e a Beleza: Economia da cultura Matis, Vale do Javari, Amazônia." PhD dissertation, Florianópolis: Universidade Federal de Santa Catarina, 2011.

———. "La No-Frontera Pano: Etnónimos como categorías alternativas y múltiples entre Matis y Korubo." *Tipití: Journal of the Society for the Anthropology of Lowland South America* 10, no. 1 (2012): 19–36. http://digitalcommons.trinity.edu/tipiti/vol10/iss1/2.

———. "Matis e Korubo: Contato e índios isolados no Vale do Javari, Amazônia." Master's thesis. Florianópolis: Universidade Federal de Santa Catarina, 2007.

———. "Matis y Korubo, contacto y pueblos aislados: Narrativas y etnografia en la Amazonía brasilera." *Mundo Amazónico* 1 (2010): 41–64. doi:10.5113/ma.1.10291.

Arisi, Barbara M., Deise Francisco, and Pedro Cesarino. *Saúde na Terra Indígena Vale do Javari. Diagnóstico médico-antropológico: Subsídios e recomendações para uma política de assistência.* São Paulo: ISA/CTI, 2011. https://www.socioambiental.org/banco_imagens/pdfs/saudev02baixa.pdf.

Arisi, Barbara, and Felipe Milanez. "De Índios isolados a ignorados: Conflitos no Vale do Javari, AM." In *Fazendo Antropologia no Alto Solimões: Diversidade étnica e fronteira*, edited by Gilse Elisa Rodrigues and Michel Justamandm Tharcísio Santiago Cruz. Embu: Alexa Cultural, 2016.

Brasil, Kátia. "Vale do Javari: 'Não brigamos por terra com os Korubo, isso é uma invenção da Funai,' diz Marke Matís." *Amazônia Real*, February 16, 2016. http://amazoniareal.com.br/vale-do-javari-nao-brigamos-por-terra-com-os-korubo-isso-e-uma-invencao-da-funai-diz-marke-matis/.

Brasil, Kátia, and Elaíze Farias. "Vale do Javari: Doenças e mudança de aldeia podem ter causado um 'choque' entre Korubo e Matís." *Amazonia Real*, February 22, 2016. http://amazoniareal.com.br/vale-do-javari-doencas-e-mudanca-de-aldeia-podem-ter-causado-um-choque-entre-korubo-e-matis/.

Erikson, Philippe. "Bibliografia anotada de fuentes con interés para la Etnologia y Etnohistória de los Pano septentrionales (Matses, Matis, Korubo . . .)." *Amazonía Peruana* 14, no. 27 (2000): 231–87.

———. *El Sello de los Antepasados*. Quito: Abya yala/IFEA, 1999.

Ferguson, R. Brian, and Neil L. Whitehead. *War in the Tribal Zone: Expanding States and Indigenous Warfare*. Santa Fe NM: School of American Research Press, 1992.

Figueiredo, Wellington. "Depoimento." In *Memórias sertanistas: Cem anos de indigenismo no Brasil*, edited by Felipe Milanez. São Paulo: Edições Sesc São Paulo, 2015.

Fundação Nacional do Índio. *Informação No. 331/DID/DGPI/81. Ref: Proc. FUNAI/BSB/1074/80. Ass: Eleição do Parque Indígena do Javari. Brasília*, 1981. Archived by FUNAI.

———. *Relatório de Identificação e Delimitação da Terra Indígena Vale do Javari, GT Portarias No. 174/95 and 158/96. Brasília*, 1995. Archived by FUNAI.

———. *Relatório do Centro de Documentação Etnológica, solicitado pelo radiograma no. 45/DGPC, de 22/01/1980-controle 3902-sobre os grupos indígenas Maruno, Canamari, Culina, Mayoruna, Matses e maih [sic]-Amazônia Ocidental, Brasília*, 1980. Archived by FUNAI.

———. "Resposta à Carta Capital." November 27, 2015. http://www.funai.gov.br/index.php/comunicacao/noticias/3553-resposta-a-carta-capital.

———. "Nota da Funai em repúdio à intimidação e violência no Vale do Javari." Amazonia, March 10, 2016. https://amazonia.org.br/2016/03/nota-da-funai-em-repudio-a-intimidacao-e-violencia-no-vale-do-javari/.

Funes, Yessenia. "Brazil's New President Moves to Kick Indigenous People Off Their Land Just Hours after Taking Office." *Gizmodo*, January 4, 2019. https://earther.gizmodo.com/brazil-s-new-president-moves-to-kick-indigenous-people-1831466539.

Government of Brazil. *Estatuto do Índio. Law 6.001. Brasília*, December 19, 1973. http://www.planalto.gov.br/ccivil_03/Leis/L6001.htm.

Huertas Castillo, Beatriz. *Los Pueblos Indígenas en Aislamiento: Su lucha por su sobrevivéncia y la libertad*. Copenhagen: International Working Group on Indigenous Affairs, 2002.

Iglesias, Marcelo Piedrafita. *Os Kaxinawá de Felizardo: Correrias, trabalho e civilização no Alto Juruá*. Brasília: Ed. Paralelo, 2010.

Instituto Socioambiental. "Coiab pede exoneração do presidente da Funai." October 8, 2006. https://pib.socioambiental.org/es/Not%C3%ADcias?id=44770.

Lima, Antônio Carlos de Souza. *O Grande Cerco de Paz: Poder tutelar, indianidade e formação do Estado no Brasil*. Rio de Janeiro: Ed. Vozes, 1995.

Milanez, Felipe, ed. *Memórias sertanistas: Cem anos de indigenismo no Brasil*. São Paulo: Edições Sesc São Paulo, 2015.

Milanez, Felipe, and Glenn Shepard. "Esqueça mitos coloniais: Contato dos Xatanawa põe fim a resistência centenária." *Carta Capital*, August 7, 2014. https://acervo.socioambiental.org/acervo/noticias/esqueca-mitos-coloniais-o-contato-dos-xatanawa-no-acre-poe-fim-uma-resistencia.

Milanez, Felipe, and Glenn Shepard. "A Tale of Two Contacts: The Government Approach to Isolated Peoples." *Indian Country Today*, September 19, 2014. http://indiancountrytodaymedianetwork.com/2014/09/19/tale-two-contacts-government-approach-isolated-peoples-156935.

"Omissão da Funai pode ter levado a conflito violento no Vale do Javari." *Carta Capital*, November 19, 2015. https://www.cartacapital.com.br/sociedade/omissao-da-funai-pode-ter-levado-a-conflito-violento-no-vale-do-javari-6376.html.

Phillips, Dom. "Brazil: High-Risk Expedition to Contact Isolated Tribe Declared Success." *The Guardian*, April 5, 2019. https://www.theguardian.com/world/2019/apr/05/brazil-high-risk-expedition-indigenous-tribe-success.

Possuelo, Sydney. "Depoimento." in *Memórias sertanistas: Cem anos de indigenismo no Brasil*, edited by Felipe Milanez. São Paulo: Edições Sesc São Paulo, 2015.

Ramenofsky, Ann, and Patricia Galloway. "Disease and the Soto Entrada." In *The Hernando de Soto Expedition: History, Historiography and "Discovery" in the Southeast*, edited by Patricia Galloway. Lincoln: University of Nebraska Press, 2006.

Ribeiro, Darcy. *Os índios e a civilização*. São Paulo: Civilização Brasileira, 1970.

Vaz, Antenor. "A que será que se destinam?" *Le Monde Diplomatique* (São Paulo), August 4, 2014.

7 We Are Here
The State of Community-Based Landscapes in Peru

RICHARD CHASE SMITH

Indigenous Communities and Their Rural Landscapes in Peru
The Tragic Confrontation in Bagua-Amazonas

On June 5, 2009, the front pages of Lima's newspapers showed a picture of the thousands of Amazon indigenous peoples, faces painted red and armed with wooden lances, who delivered a hard blow to President Alan García's government. The result of the confrontation with more than a thousand police left thirty-three dead on the outskirts of the city of Bagua, the majority of whom were members of the National Police sent by the central government to put an end to the protest.[1]

In August of the previous year, an already frustrated and distrustful indigenous people's organization, the Interethnic Association for the Development of the Peruvian Amazon (Asociación Interétnica de Desarrollo de la Selva Peruana, AIDESEP), had convened a national strike of the Amazon's indigenous peoples, demanding the immediate repeal of eight legislative decrees they considered harmful to their territorial rights. After eight frustrating months filled with plenty of promises, committees, reports, and dialogues, AIDESEP called for a second national strike over the very same demands. This time, facing the blockade of an important road near the city of Bagua by thousands of Amazonian and Andean indigenous peoples, the government sent over a thousand members of the National Police to forcefully break the roadblock. Given the previous treatment of indigenous and mestizo, rural populations by the National Police, and the distrust and frustration these groups harbor, confrontation was inevitable.[2]

These two Amazon and Andean indigenous mobilizations centered around three key issues: (1) the threat to their culture that President García's anticommunities discourse represented, blaming their poverty on the lack of productivity of their natural resources and their lands and on a purported ignorance of production matters; (2) the transparent infringement of fundamental aspects of the relationship of indigenous peoples with their communal territories demonstrated in the content of eight of the ninety-nine legislative decrees approved by García; and (3) the lack of prior consultation of indigenous peoples about these legislative decrees, in violation of ILO Convention 169 concerning indigenous and tribal peoples, which was ratified by Peru.

Eighty-four protestors were arrested and jailed on the day of the confrontation.[3] Of these, fifty-four indigenous participants of the protest were charged with eight different crimes, including first-degree murder of the twenty-four members of the police force who died in the confrontation. Among them was Alberto Pizango Chota, the president of AIDESEP, who did not participate directly in the protest and later fled to Nicaragua. Four years passed before the cases were brought before a judge in Bagua, and it took three more years, until September 2016, for the judge to make public his ruling to acquit all of the accused on all counts due to the lack of any clear evidence that linked them individually to the crimes of which they were accused. There was no parallel investigation carried out nor any charges made against anyone for the death of nine indigenous protesters.[4]

A detailed analysis of the events that transpired between August 2008 and June 2009 suggests that—just as the governments of Alberto Fujimori, Ollanta Humala, and apparently the short-lived government of Pedro Pablo Kusczynski all tried to do with similar policies and legislation—the García government sought to weaken the social fabric of rural indigenous communities, facilitate the parceling of communal lands, strengthen the marketability of these parcels, and facilitate easement rights over communal lands for mining and oil companies, all without prior consultation. To argue, as some do, that indigenous peoples are not capable of understanding these public policies and to insist that they represent some larger conspiracy not only delegitimizes these social actors but legitimizes irresponsible state policies.

The Origin of Peru's Legalized Communities

The institution of an "indigenous" community has precolonial (indigenous land tenure and production systems) and colonial (the "reduction of indigenous peoples" of the sixteenth and seventeenth centuries) roots. Between 1550 and 1575 the Spanish Viceroyalty in Peru forced indigenous peoples to resettle in "Indian reductions," each one based around a small church or "doctrine" and put under the care of a high-ranking Spaniard called the *encomendero*.[5] Eventually the Crown recognized and protected the lands associated with each indigenous reduction, in many cases with titles extended by the Crown in order to ensure the production of food and goods for the annual payment of tribute to the Spanish colonial regime and to facilitate the collection of these tributes and regulate the flow of indigenous labor to compulsory service in the mines.[6]

In 1821, soon after declaring the Independence of Peru, José de San Martin proclaimed the incorporation of indigenous peoples into the new Peruvian state through the elimination of tribute and forced service and labor in the mines. Three years later, in 1824, Simón Bolívar ratified San Martín's proclamations and, in the same spirit of liberalism, declared the redistribution of indigenous peoples' communal lands into individual lots and the abolition of the indigenous community as a corporate entity.[7]

After more than one hundred years of conflicts between the Republic of Peru and the indigenous peoples concerning the subdivision of their communal lands and other abuses, the 1920 Constitution recognized the "Indigenous Race" and their right to collective and private property of their territories in the *comunidad indígena* (indigenous community). The community titling process began a few years later. Subsequently, in 1969, the military government, in the framework of the Agrarian Reform, changed the name of this legal figure into a *comunidad campesina* (peasant community).

Five years later the same military government recognized a second type of indigenous community, the *comunidad nativa* (native community), and the right of indigenous settlements in the Peruvian Amazon to be recognized as such and to private property rights for their community lands. The Amazonian indigenous population, descendent of more than fifty-five peoples with different languages and cultures, was deeply affected by the rubber boom that took place between 1880 and 1920.[8] Numerous groups fled to more remote zones to hide from the forced recruitment of slaves, or *correrías*. While the great majority of them have had their lands recognized and titled as native communities, a few indigenous groups remain isolated from the dangers that Peruvian society represents to them. Five of these groups are protected within five territorial reserves, created between 1988 and 2000, and are currently under consideration for a more stringent protection regime, denominated "indigenous reserves."

It is because of these historic reasons that native and peasant communities are today the concrete, juridical, and historic expression of indigenous peoples in Peru. However, and despite this legal precedent, the Peruvian state, with few exceptions, has preferred to cover them with the mantle of invisibility and neglect. It is important to ask ourselves why.

The Roots of Modern Racism in Peru

The response to this question lies in the history of the indigenous peoples of Peru. Unfortunately this history is widely ignored by most Peruvians, including the indigenous population itself. However, it is fundamental for understanding modern racism and the discriminatory attitudes that still dominate social relationships and power in Peru.

The Colonial Roots of Racism

We have already briefly reviewed some of the colonial institutions imposed on the indigenous population during the three centuries of Spanish colonial rule. These include the forced resettlements of Indians, indoctrination and evangelization centers under the Catholic Church, the payment of tribute (at first in goods and later in silver and gold), and the forced conscription of young males for long periods of labor in the mines and of women to the same in the homes of the colonial elites.

We know that by 1575, dissatisfied with the inefficiency of the governance under the system of distribution of Indians among the *encomenderos*, the Crown gave the order to put them under the jurisdiction and governance of *corregidores de indios*, assigned by the Crown to new districts composed of rural Indians. According to orders given by several viceroys in Lima, the first task of these local governors was to "keep the Indians in a state of peace and justice" by treating them with the Christian values of "love and humility" and "not with contempt nor as dogs."[9] However, this system became one of abuse and corruption that lined the pockets of the *corregidores* and complicit indigenous *caciques*, subjecting indigenous peoples to their arbitrariness. In the words of Solórzano de Pereira, "No thief or pirate would be as greedy with strangers as is the bad and iniquitous Corregidor with his [Indians]; having once arrived in the Indias [Americas] they degenerate their obligations and bring out their vices and desires, especially guided by that of accumulating gold and silver, in order to return quickly to Spain as a rich person, running rampant over all respect for reason and justice."[10]

All of these institutions and others were powerful symbols that strongly separated the Indians and marked them with the status of vanquished and dominated peoples and, at the same time, underlined the status of the Spanish and later the *criollo* as the conqueror with an inherent right to dominate the vanquished.[11] But they were also institutions that over time became the basis for an institutionalized system of discrimination and abuse based on origins and "race."

The attempts of San Martín and Bolívar to proclaim the liberation of the Indians from these institutions of domination were not long-lasting. In the first place, they were not well received by the Peruvian *criollos*, who resisted the loss of their colonial-based privileges and status as well as the attempts of these liberal outsiders to incorporate "free" Indians as citizens of the new Republic of Peru. Second, with the treasury of the new republic in bankruptcy after years of war with Spain, and the interest on the external debt rapidly accumulating, the new Peruvian state began to show its colonial face in August 1826 when the new Congress approved a bill reinstating the institution of tribute for the indigenous population, this time euphemistically labeling it an "indigenous contribution." In fact the highest rate on the scale for this "tax" was to be paid by the "Indians of origin" who had colonial titles for their community property.[12] According to the same source, data on annual income for the Peruvian state demonstrate that the annual tax collected under this new form of indigenous tribute was more than twice that from sources of nonindigenous income. Later, in the 1840s, the Peruvian government reinstated forced labor by enacting a law that allowed the conscription (called *pongaje*) of adult males living in communities to construct and maintain public infrastructure.

During the remainder of the nineteenth century and the first two decades of the twentieth there was a series of upheavals in indigenous zones in the country in protest of the loss of community lands, the "indigenous contribution," pongaje, and the increasingly abusive exploitation of indigenous people by a rising class of criollo landowners and merchants.[13] Paralleling this was the growth of urban and intellectual cultural movements and a debate on laws that would prohibit the more common abuses associated with these practices. It was only with the Constitution of 1920 that the Peruvian state recognized the existence of the "indigenous race" and their right to communal property.

This brief history shows in broad strokes one of the roots of the modern racism and discrimination that populations with indigenous origins face today in Peru and explains, in part, the mantle of invisibility and neglect that covers them. Since the beginning of the colony, in the collective worldview of the Spanish and their criollo descendants, indigenous peoples were the defeated-dominated and thus owed payment of tribute and obligatory service as a signal of their inferior status. Neither San Martín nor Bolívar, with their good intentions and liberal ideas, was able to change this collective image in criollo minds. To this day this colonial artifact sustains common notions about the inferior position (second class) of indigenous people in the social hierarchy of Peru.

The Pseudo-Scientific Roots of Racism

Toward the end of the nineteenth century, Father Gabriel Sala, a Franciscan missionary and a researcher and official explorer associated with the prestigious Geographic Society of Lima, wrote in his diary, "Among our Indians, whether in the highlands or in the Amazon . . . it is necessary to bend and subjugate their will, even if it means beating the hell out of them, so that eventually they will receive the light and open their understanding. That is how it was done in the time of the Viceroys and how it is still done along the Ucayali River; in other words, there is no other solution for the Indian who does not attend the Mass or indoctrination classes than to lower his pants and give him another kind of lesson."[14] These words reveal another root of racism and discrimination in Peru. Father Sala was considered an educated, illustrious, modern, and enterprising man by many of his time, one who sought development and modernization for Peru. His words reflect the great industrial boom in Europe and the colonial dominion it held over a large portion of the world's population. Like many Europeans of his time, he was influenced by the new "scientific" theories that arose in the mid-nineteenth century through the cross-fertilization of ideas of biological evolution as introduced by Charles Darwin and social evolution, most notably espoused by Herbert Spencer. Some of these theories argued that a small number of peoples or nations had evolved to a superior state of intelligence, capacity, and modernity. By the early twentieth century such theories were being used to explain and justify Europe's dominant position and colonial obligation in the world.

When the pseudo-scientific concept of race, defined as a human group with certain physical (and social) characteristics, became incorporated into this doctrine, a new collective worldview arose that asserted that the "more evolved" white or European race had a duty within the social evolutionary hierarchy to dominate and make races with darker skin, considered inferior or backward, to produce, even if it meant "beating the hell out of them."

Sala was a spokesman for these ideas among the enterprising and modern sectors of Peru at the end of the nineteenth century. But the line between this pseudo-scientific perspective and raw racism was often not very clear. Sala also wrote in his diary, "*Chuncho* means the same as a person who is false, a traitor, ungrateful, lazy, a drunkard, vengeful and unpredictable. And what should we do with people like this? Just what they do in the rest of the world: assuming they do not want to live like human beings, treat them like animals. And shoot them when they unjustly oppose the life and well-being of the rest."[15]

It was at the beginning of the twentieth century that this perspective began to infiltrate and serve the dominant sectors of Peruvian society to justify their abuse of and contempt for the indigenous race and the social exclusion and invisibility to which they were subjected. The words used by Article 58 of the Constitution of 1920 are no surprise: "The State will protect the indigenous race and dictate special laws for their development and culture in harmony with their needs." Words like those of Sala can easily fall out of the mouths of ladies and gentlemen from modern high society in Lima today. One still hears them talk about the "indigenous or copper-colored race."

Urban Pressure to Dismantle Rural Communities: 1991–2018

The Fujimori Government (1991–2001): The Beginning of the Modern Dismantlement of Indigenous Communities

With the weight of the World Bank and the International Monetary Fund behind them, the ideological winds changed sharply after the Peruvian military regime of the 1970s. In 1981, I participated in a meeting between the new minister of agriculture and a delegation of Yánesha and Asháninca communities from the Selva Central. They expressed their concern about the impact of the large Pichis-Palcazu Special Project announced by the second Fernando Belaúnde government that threatened to settle their territories with people from other parts of Peru. The minister responded that the concept of indigenous community was obsolete, that it impeded the people's entry into the market economy and the benefits it offered, and then he encouraged them to parcel and individually title their community lands with government support. It was a portent of what was to come.

Outside of Peru the international legal framework associated with the rights of indigenous peoples had been growing stronger for three decades. Through the United Nations Working Group on Indigenous Populations, new concepts and rights were introduced, including the right to an identity as an indigenous people, the right of an indigenous people to a territory, and the right to self-determination.[16] At the same time, new instruments favoring indigenous communities were introduced, such as the International Labor Organization Convention 169 (1989), the

United Nations Declaration on the Rights of Indigenous Peoples (2007), and various historic rulings in the Inter-American Court.

Despite these advances and Peru's being a signatory of both instruments, the country experienced a notorious rollback over the collective rights of indigenous peoples during the same period. Between the Constitution of 1979 and that of 1993, two fundamental guarantees for indigenous territories from the Constitution of 1920 were excised: immunity from seizure, and inalienability. All that remained was imprescriptibility, land rights not subject to being taken away by prescription or by the passage of time. After the constitution promoted by President Fujimori in 1993, the Peruvian state began to harden its public discourse regarding communal property of indigenous peoples.

Thus 175 years after Bolívar's declaration on the subdivision of indigenous communities, the 1995 Law of Investment in Agricultural Lands (Law 26505) put in place a new neoliberal strategy to accomplish the same purpose.[17] Starting with this law, three strategies were set in motion to achieve this goal: (1) ensuri ng that the Special Project for the Titling of Rural Lands prioritize breaking up indigenous communities into individual parcels over titling them collectively; (2) changing the internal processes by which a community may approve the parceling up of its territory; and (3) redefining the state's treatment of "unproductive" and "abandoned" lands in order to expropriate sectors of communal territories.[18]

Two years later a new law was enacted for coastal communities, legalizing the procedures needed to implement these three strategies.[19] The introductory article states the following: "Article 1. In the granting of property and the registry inscription of lands of the Peasant Communities of the Coast, it is of national interest and [the interest] of its members to adopt the decision to proceed to the allocation of individual lots . . . in order to guarantee . . . the rights of members to individual property and inheritance, and the exercise of private initiative."[20]

This law made four important changes to facilitate the breaking up of the communities in the coast. It specified that members with the right to vote in the Communal Assembly were limited to those who habitually live within the community and could prove their status as a "possessor" of community lands; in other words, those who regularly undertook agricultural activities in the community could vote. This eliminated from the voting lists those members who lived outside the community yet maintained a relation with it through inheritance and ancestral rights. It lowered the number of votes needed to approve the separation and titling of individual lots for these possessors from 67 percent of the Assembly quorum to 50 percent. It also reduced the votes needed to approve the separation and titling of individual lots to illegal settlers (land invaders) within communities to 30 percent of the Assembly quorum. If the community does not approve the titling of lots for illegal landholders and residents, the law empowers the Peruvian state to declare the lands occupied by invaders as abandoned and expropriate them. Once the abandoned land is expropriated, the state can sell it back to the former squatters.

Even though Peru had already signed ILO Convention 169, there was no consultation with communities about any of these laws despite their direct impact on the communities' internal decision-making mechanisms and the state of their territories. The justification given by the state for this law was the need to "strengthen" the rural land market.[21]

The García Government (2006–2011) on the Attack: Indigenous Lands for Investors

Toward the end of 2007 President García published an op-ed in Lima's newspaper *El Comercio*, announcing his intention to dismantle indigenous and peasant communities: "Here we find . . . millions of hectares that communities and associations have not cultivated and will not cultivate . . . unused lands because their owner does not have the necessary education or economic resources; and as such their property is only apparent. This same land, sold off in big parcels, would bring technology that would also benefit the community members."[22]

Just as Fujimori had done before him, García approved a block of ninety-nine legislative decrees, eight of which sought to facilitate and further the process of parceling indigenous communities, not only in the coastal region but throughout the entire country. This political process began with a campaign that included a series of articles and editorials published in *El Comercio*, seeking to justify these legislative decrees by accusing communities of being "antidevelopment," calling them a "dog in the manger," and accusing them of blocking the modernization and development of Peru.[23]

In 2008 Legislative Decree 1073 extended the same rules that applied to coastal communities since 1997 to indigenous and peasant communities in the rest of the country. The government sought to justify it as a "modernizing and democratic" improvement. Legislative Decree 1064 established exceptions to the security of communal property for all the country's communities. These included "community lands that are declared [by the state] as abandoned" and "lands that are occupied by centers of human settlement." Left intentionally vague, these exceptions allowed the state to demarcate and change the status of property in virtually any part of a community, exclude it from communal property, and transfer it to a third party, without prior consultation or vote by the Communal Assembly.[24]

It is also important to note that there was a simultaneous slowdown of community titling beginning in 1997, a trend that persisted until 2008, when community land-titling simply stopped. It is no surprise that during the five years of the García government (2006–11) title was granted to only eight Amazonian communities out of more than six hundred awaiting legal security of their lands. Beginning in 2008, and in less than three years, this government dismantled the Special Program for the Titling of Rural Lands and passed its functions to the Office for the Formalization of Informal Property (Organismo de Formalización de la Propiedad Informal, COFOPRI), the state agency of the Ministry of Housing, Construction and Sanitation in charge of titling urban lots. Two years later, also by legislative decree, the responsibility for titling rural lands, and related documentation, was transferred to the regional governments. These steps sowed chaos and confusion and eviscerated the state's capacity for legalizing and titling rural lands.

The Humala Government (2011–2016): The Security of Indigenous Lands in Limbo

Despite electoral promises, the government of President Ollanta Humala brought little or no change for the indigenous populations of Peru. This was despite the fact that the Congress, then controlled by the president's party, approved the Law of Prior Consultation for Indigenous and Original Peoples, which Humala immediately signed into law. Four months later he changed his mind, and terminology, by publicly declaring that Peru's more than seven thousand peasant communities were, in fact, *not* indigenous communities but rather groups of farmers dedicated to agricultural activities and thus not entitled to the right to prior consultation enshrined by the new law. This impasse remains unresolved well into 2019.

In 2013 this government began an aggressive campaign to strengthen the rural land market through a series of laws designed to facilitate and attract foreign investment. These "packages" consisted of sets of norms, laws, legislative decrees, supreme decrees, and ministerial resolutions that, under the justification of reactivating the economy and boosting private and public investment, introduced more flexibility into the legal framework of social, environmental, territorial, and indigenous rights and standards. Between May 2013 and December 2015 five packages of eight norms were issued for this purpose.[25]

The first package came out in May 2013 and provided an easy way out of the restrictions for lands containing archaeological remains by limiting the period for the official registration of such remains to a couple of months, while also making it easier to certify that archaeological remains did not exist (D.S. 054-2013 PCM). This first package also included a decree loosening up the standards for the approval of environmental impact studies for all large-scale projects in infrastructure, mining, and hydrocarbons extraction (D.S 060-2013-PCM).

A year later a second package contained a pair of executive orders giving extraordinary powers to a few government agencies for the expropriation of land and real estate. Most insidious was Law 30230, granting special powers to the Ministry of Housing, Construction and Sanitation to seize rural property, even land with proper titles or with crops already planted, via unspecified "special" procedures, and hand it over to a state entity or private company in order to favor any investment project declared to be of national interest.[26]

Thanks to advocacy work by several civil society institutions, the Ministry of Housing, Construction and Sanitation publicly committed to incorporating an exclusionary clause protecting communal lands in the law's regulations.[27] In mid-2015 a group of civil society institutions presented to the Constitutional Court of Peru a petition to declare Law 30230 unconstitutional, citing many different legal grievances. In January 2016 the Constitutional Court formally admitted the petition, which was still being processed in early 2019.

During the debates regarding Law 30230, the Ministry of Energy and Mines enacted Decree 001-2015-EM, granting to the Executive Committee of each community the authority to put communal lands in third-party hands, ignoring the role assigned by law to the General Assembly of the community, through the vote of two-thirds of its members. Many experts have declared this decree to be unconstitutional as well.

Throughout 2015 the Humala government approved several additional norms that sought, through other means, to facilitate transferring rural lands to the hands of investment projects with foreign capital. In some cases, civil society institutions were successful in incorporating exclusions for communal lands. Most of these norm changes were driven by the Ministry of Economy and Finances, and in various cases were based on previous studies commissioned by the National Council of Competitiveness, a body attached to the ministry.[28] Two other key sectors driving the issuance of these norms were the Ministry of Energy and Mines and the Ministry of Housing, Construction and Sanitation. It should be noted that the development of these norms did not entail public debate, which could have helped identify the contradictions contained within these norms with the Constitution and with various organic laws and international treaties, such as ILO Convention 169.

The Kuczynski Government (2016–2018): Taking the Same Path

In June 2016, before being elected president, Pedro Pablo Kuczynski signed a pact with a group of indigenous leaders. It included a list of nine points he committed to fulfill should he be victorious in the elections. One of these points was a promise not to break up community lands and to grant collective title to communities that lacked them, as established by the Constitution's communal property regime. In a televised interview soon after Kuczynski won the election, his minister of economy and finance declared that the only way to address "the problem of communal lands" in Peru was to parcel them up and grant title to individual owners. Judging by what has taken place since, the voice of this minister defined the policies followed by this government.

As with the previous four governments, Congress delegated to Kuczynski's executive branch the powers to legislate for a period of ninety days, with the express mandate to create the necessary norms for economic reactivation, address citizen security, fight against corruption, improve water and sanitation, and reorganize Petroperú, the national oil company. The result at the end of the ninety-day period was an unprecedented 112 legislative decrees, 65 of which were related to "reactivation of the economy."

We must remember that the governments of Fujimori, García, and Humala also used these extraordinary powers to enact a large number of norms, many seeking to weaken environmental safeguards and the communal regime of indigenous lands, always under the guise of reenergizing the economy and boosting private and public investment. We have shown that many of these norms had the intent of weakening the legal framework protecting indigenous people's communal territories. Up to the beginning of 2018, many of the legislative decrees and new proposals approved by President Kuczynski followed the same pattern and espoused the same justification: the need to "liberate" communal lands and those of other owners to promote foreign investment and reactivate the country's economy.

Three of the legislative decrees (1251, 1330, 1333), like those of Fujimori, García, and Humala, including Law 30230, change the rules of the game for the transfer of rural property—be it private, titled, public, communal, or individual—and hand it over for implementation of investment projects in infrastructure, mining, hydrocarbons, and agroindustry through expropriation processes. This time around, however, there is no mention of COFOPRI from the Ministry of Housing,

Construction and Sanitation. Instead the key role is given to ProInversión (Pro-Investment), an agency of the Ministry of Economy and Finance, through the creation of a Special Project for Access to Land for Priority Investment Projects (APIP). Once again, by giving the APIP Special Project the powers to expropriate, clear hurdles, and grant title to rural lands and property, confusion and administrative conflicts are generated with the Office for Reorganization of Rural Property and Land Registry of the Ministry of Agriculture.[29]

Through continuous advocacy efforts by civil society, including indigenous organizations, to alert members of the Congress of the Republic about the threat these decrees represent to indigenous lands, it was possible to send DL 1333 into a legal limbo with an "observation" from Congress. In mid-2017 Kuczynski presented a new legislative decree, DL 1718, to replace the stalled DL1333 and to further strengthen APIP's role in the management of rural lands. Unlike the previous legislative decree, this new proposal contemplates exclusion for the lands of indigenous peoples. However, exclusion for the lands of indigenous peoples has no legal weight in Peru, as there is no norm that defines and recognizes the legal figure of "indigenous peoples." By law and by the Constitution, the legal owners of the lands of indigenous peoples in Peru are either native communities or peasant communities. Therefore the exclusion would have to be for native and peasant communities.

Perhaps that is a moot point now because this back-and-forth debate over rural lands between Kuczynski and the Congress got lost in the storm of accusations between Fujimoristas in the Congress and the president's party regarding mutual accusations of corruption and bribes made over the past two decades by the Brazilian construction behemoth, Odebrecht. And this has come to an end with the sudden renunciation of Kuczynski as president on March 21, 2018. At that point in time the vice president, Martín Vizcarra, has assumed the presidency, and he has already been through a change in prime minister and in many of his cabinet ministers. However, during the time that has passed since Vizcarra assumed the presidency, most political attention has been focused on two different corruption scandals, that involving past presidents and ministers with very large bribes made by Odebrecht and a new scandal within the Peruvian justice system involving systematic bribes taken by a group of high-level prosecutors and judges. So far Vizcarra has not pushed, at least not openly, the approval of any legislation that would threaten community lands. But neither has he given any clear statement regarding greater protection for indigenous lands or creation of government programs to work with the indigenous communities to raise production levels or alleviate the very high indices of poverty found there.

The Big Data Cover-Up: The Continuing Effort to Keep Community-Based Landscapes Invisible

One Hundred Years Later and Still No Official Numbers

Today, nearly a hundred years since their recognition by the Constitution, the Peruvian state cannot officially inform the public of how many indigenous communities exist, where they are located, or what lands they occupy. In the middle of the digital era, at a time when the Ministry of Energy and Mines has almost fifty-five thousand mining concessions perfectly recorded in a modern registry system, there is no similar database for the ten thousand existing indigenous communities.

Data about communities are available unofficially, scattered throughout the archives of regional governments, COFOPRI, the archives of the Public Registry, and other state entities. Since the publication of the last Official Directory of Communities (Native and Peasant) by the Special Project of the Titling of Rural Lands (Proyecto Especial de Titulación de Tierras, PETT) in 1998–99, there has been no further official information published on the subject.[30] Between 2008, when PETT was dismantled, and 2016 there was no government entity in Peru responsible for the systematization and publication of reliable and official figures on the status and number of recognized and recorded indigenous communities. The official figures about the number of acres titled or to be titled to communities are in the same situation. Beginning in 2017 this responsibility was assigned to the newly created General Agency for the Legalization of Rural Property and Cadastral System of the Ministry of Agriculture (Dirección General de Saneamiento de la Propiedad Agraria y Catastro Rural, DIGESPARC).

Because of this void and apparent lack of state will, in 1998 the Information System for Native Communities (Sistema de Informacíon sobre Comunidades Nativas, SICNA) was established as a private initiative of the Institute for the Commons (Instituto del Bien Común, IBC).[31] SICNA has the objective of collecting information about every native community, titled or not, and making it accessible to the public. To this end, for twenty years SICNA staff has visited every native community and searched the archives of the governments of eleven regions and the national records of the public registry of real estate property to collect copies of the documentation and property titles as well as perimeter maps. During the visit to each native community, staff recorded the coordinates of the accessible landmarks that delineate their boundaries. Later, back in the central office, all this information was entered into a database linked to a geographical information system (GIS) and a documental archive.

According to the most recent information in SICNA, there are a total of 2,496 registered native communities in the eleven Amazonian regions in the country, of which 1,689 have communal property titles (see table 4).[32] There are 30.72 million acres titled in total. SICNA has also collected data about legally recognized but untitled communities, as well as communities still to be recognized. The smallest native community is the Yánesha Alto Churumazú community (Pasco) with 280 acres, and the largest is the Matsés community (Loreto), with 1.119 million acres.

TABLE 4. Number of native communities in Peru, April 2019

Native Communities	Total number of communities	Number recognized and titled	Number recognized without titles	Number not recognized without titles
TOTAL	2,496	1,689	636	171

Source: IBC-SICNA, Lima, Peru.

Toward the end of 2018 the government agency in charge of rural property at the national level made public an online GIS-based platform claiming that it contained all the official data regarding lands held by native communities. While we applaud the effort, unfortunately the data turn out to be a 2008 version of that produced by the SICNA initiative of the IBC that was put in the hands of the COFOPRI agency in charge of rural lands at that time. IBC has offered its latest dataset to replace the

2008 version, but the agency has not accepted the offer, stating that the updated IBC dataset is not official!

The same lack of clear and up-to-date information on peasant communities drove the Institute of the Commons and the Center for Social Studies in Peru to create an Information System for Peasant Communities (Sistema de Información sobre Comunidades Campesinas, SICCAM), starting in January 2015. This process involved the organizations of communities and indigenous populations that are members of the Unity Agreement (Pacto de Unidad) that brings together five peasant/indigenous organizations in Peru. The objective is similar to SICNA's: to collect and systematize data and numbers about the peasant communities of Peru and make them public. The first step was to compare the available databases regarding peasant communities in all the regions of Peru, collected by four different national agencies (PETT 1998, COFOPRI 2010, MINAGRI-DIGESPARC 2015 and 2016, and the Ministry of Culture 2015), plus additional information from fourteen regional governments. The purpose of this first step was to construct a single, clean, and accurate database.

Data for 1,122 riverine communities located in the flood-prone areas (*varzea*) of the primary rivers of the Peruvian Amazon collected by SICNA was added to this database. These communities have their origins in both local indigenous groups and indigenous populations from the northern mountains that migrated to the Amazon during the rubber boom in the nineteenth and early twentieth centuries. By law they have the right to recognition and title of their lands as peasant communities. To date only 143 have been recognized as peasant communities, of which 60 have received their communal property titles.

As of December 2018 SICCAM has registered a total of 7,282 peasant communities. Of these, 5,297 are recognized and have property titles covering a total of 61.12 million acres; 1,006 have been recognized but still lack property titles, and 979 have yet to be recognized and given title (see table 5).[33] SICCAM continues gathering and updating information from the regional governments and comparing these data with the information available in the public registry.

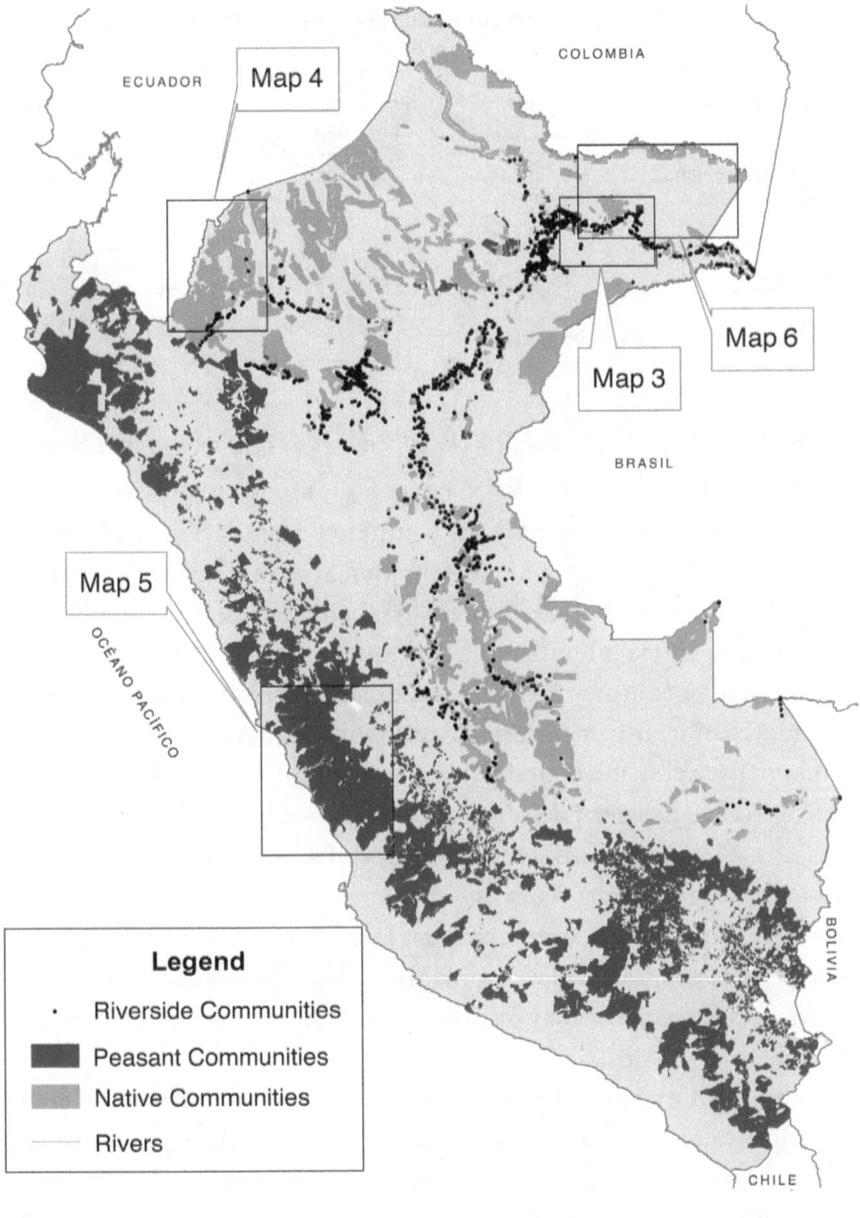

2. Peasant and native communities in Peru, 2018. Courtesy of SICNA and SICCAM Information Systems, IBC; Richard Chase Smith.

TABLE 5. Number of peasant communities in Peru, December 2018

Peasant Communities	Total number of communities	Number recognized and titled	Number recognized without titles	Number not recognized without titles
Original (highlands and coast)	6,160	5,237	923	0
River Bank (Amazon)	1,122	60	83	979
TOTAL	7,282	5,297	1,006	979

Source: IBC-SICCAM, Lima, Peru.

One Hundred Years Later and Still No Official Maps of Community Lands

The state of spatial information about communities—that is, the georeferencing of polygons that mark the communal boundaries based on GPS and processed by a geographic information system to incorporate them in the land registry map—is even further behind and more complicated. According to information from COFOPRI (2010), fewer than 7 percent of titled native communities and 38 percent of titled peasant communities have official georeferencing of their boundaries.[34] The majority of maps that show the boundaries of each community and accompany their technical dossier for its titling were drawn by hand at the desk in the offices and before PETT incorporated the use of modern GPS and GIS technology twenty years ago. In the case of native communities, these old maps do not even contain referential coordinates; the only sign of their geographic location is their possible adjacency to a river or stream whose name may or may not be identified in the most recent national map of Peru.

Both SICNA and SICCAM have been making native and peasant communities visible on maps as a second objective. In the case of SICNA, during each visit to a native community, accessible landmarks within the community's boundary perimeter were identified and their GPS coordinates taken. Later these georeferenced points were combined with the information of official maps (for example, the distance and angle of each segment, or the natural element that serves as a boundary) to project a polygon of property boundaries for each community (see map 2).

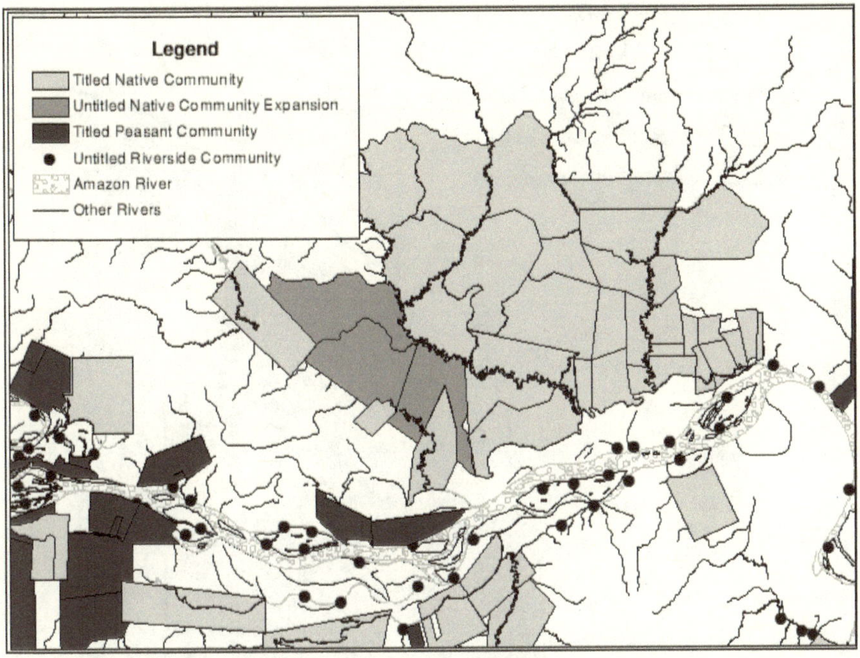

3. Indigenous communities along the Amazon River, Loreto Region, Peru. Courtesy of SICNA and SICCAM Information Systems, IBC; Richard Chase Smith.

Today SICNA has georeferenced polygons elaborated in this manner for 100 percent of titled native communities. The gray polygons in map 3 represent some of the 1,343 titled native communities that have been georeferenced in the field by SICNA staff. In the case of untitled native or riverine communities, SICNA personnel took points for each community at the school or football field. In those cases it is not possible to establish a polygon of property boundaries as these have yet to be formalized.

Until early 2011 it was impossible to similarly visualize peasant communities. In 2011 SICNA gained access to a shapefile with approximately 1,900 georeferenced peasant communities, or 38 percent of those titled.[35] Since SICCAM began working in 2015, the georeferenced polygons for 1,250 more communities have been found in regional governments' files. With this information SICCAM's database now has polygons for approximately 69 percent of peasant communities.[36] Further, during fieldwork SICNA's staff were able to georeference one point in the schools of more

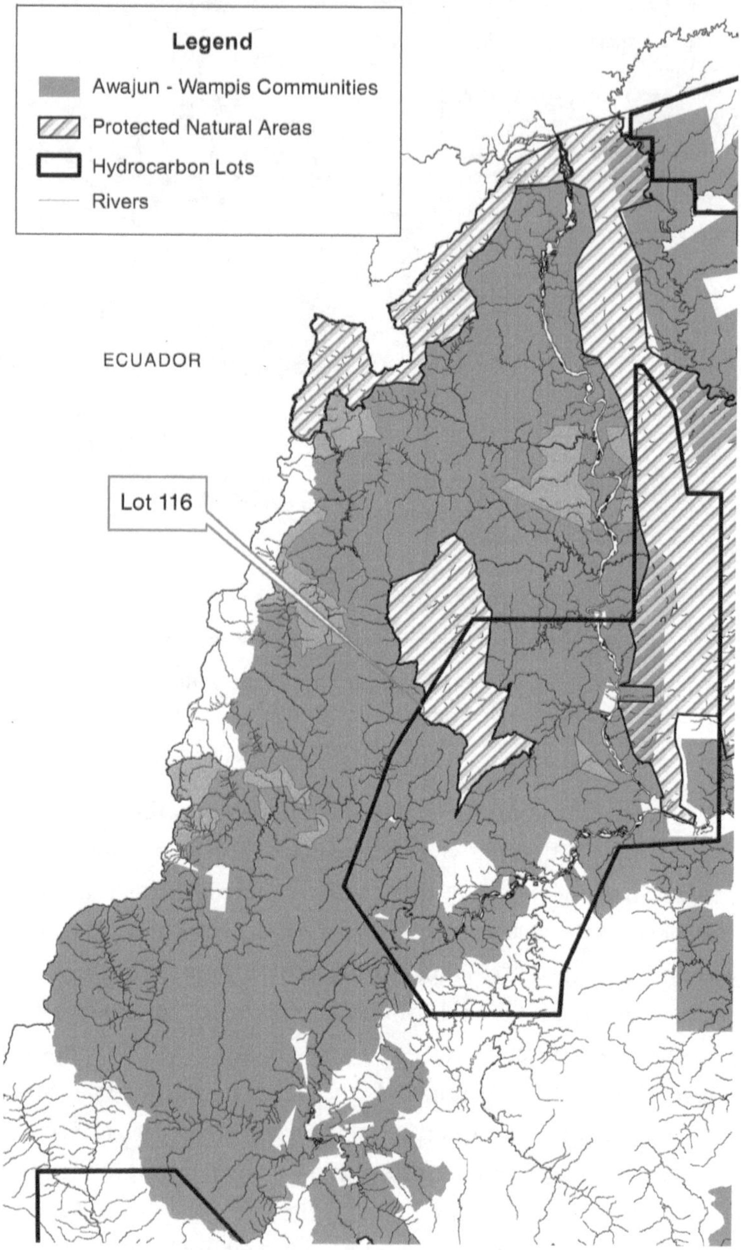

4. Landscapes of Awajun and Wampis community lands with hydrocarbon concessions, Amazonas Region, Peru. Courtesy of SICNA and SICCAM Information Systems, IBC; Richard Chase Smith.

than a thousand riverine communities (see map 3). It is worth noting that this georeferenced information is more complete for certain parts of the country, such as the regions of Piura, Lima, Cuzco, and Puno, but is very scarce for other parts of the highlands, where large numbers of peasant communities are located (Apurímac, Ayacucho, Junín, Huánuco).

The official invisibility of communities is not explained by their being few, or small, nor by their remote location. In 2016 IBC published an estimate of the total area in possession of indigenous communities in Peru (155 million acres), which represents 49.1 percent of the country's surface.[37] Similarly IBC calculated that 47.3 percent of the area of Amazon forests in Peru is within the hands of indigenous communities.[38] This official information gap isn't due to a lack of funds, lack of technology, or lack of information sources. It is related to a lack of political will by the ruling political parties—be they progressive, neoliberal, or conservative—who do not consider indigenous issues worth the official effort to correct this vacuum. Behind this lack of political will are the prejudices that historically come from the two roots of modern racism and that create the mantle of invisibility and neglect.

We Are Here: Efforts to Reconstruct Community-Based Landscapes in Peru

The possibility of visualizing indigenous communities on maps opens a wide range of striking topics for analysis. In the Peruvian Amazon we can now assess how hydrocarbon lot 116 overlaps many Wampis-Awajun communities and with the protected natural areas of the Santiago River (see map 4).[39] Between 2013 and 2018, the total number of hydrocarbon concessions has decreased significantly, but those set aside for exploration and exploitation still cover more than 37 million acres, or 20 percent of the Peruvian Amazon.[40] The ability to see these overlaps between hydrocarbon concessions and native communities helps one understand the more common sources of pressure on the collective indigenous property and conflict in rural Peru.

In map 5 we can visualize the area of the peasant communities of the Lima region and the overlapping of more than 3,350 mining titles that cover approximately 43 percent of the region. There are 289 peasant

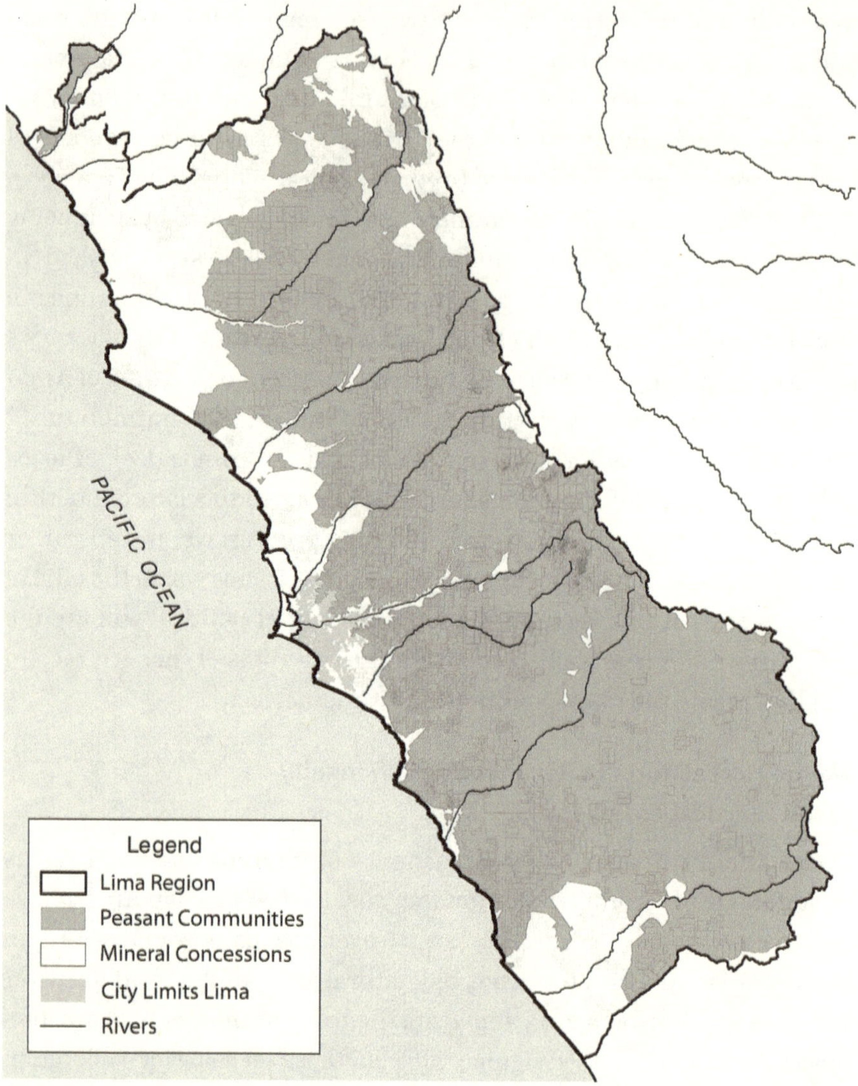

5. Landscapes of Andean peasant community lands with mining concessions, Lima Region, Peru. Courtesy of SICNA and SICCAM Information Systems, IBC; Richard Chase Smith.

communities with official recognition in this region, 234 of them with title. SICCAM has a georeferenced polygon for 222 of these, covering a total of 5.6 million acres; 48 percent of their area is overlapped by mining concessions. Of these, only 54 are in production and 34 others are in exploration.[41] But the important point here is that we can now visualize this situation on a map, calculate the overlaps, and better understand the growing resistance to the mining industry in Andean Peru.

In both of these cases, if we look underneath mineral or hydrocarbon concessions, we can also see that there are still some very large and intact indigenous landscapes, many of which are mosaics of community-owned land plus lands under some form of protection within the national or regional parks service. Looking again at map 5, we can now see 222 out of 289 registered peasant communities in the political jurisdiction of the Lima region that form an enormous and contiguous indigenous landscape of 5.4 million acres. In other words, even without the 67 communities yet to receive land titles or to have their titled lands georeferenced, 63 percent of the Lima region is already property of indigenous communities. In this case, the very early formation of indigenous reductions with colonial land titles played an important role in creating and protecting this landscape. And we will certainly find similar landscapes in Piura, Puno, Cuzco, Junín, and likely Ayacucho for the same reasons.

In the Amazonian part of the country, we find similar large indigenous landscapes that are the result of conscious strategic thinking and planning by indigenous peoples, their organizations, and others who have been supporting them over the past fifty years. One of the important strategies used is to ensure that, when there are multiple communities in the same area being titled at the same time, their boundaries should be as contiguous as possible, leaving no space in between for invasions by outsiders or land traffickers. A well-known example is the case of the Wampis and Awajun communities of the Upper Marañón. If we look behind the hydrocarbon concession on map 4, we see a very extensive community-based landscape titled to over eighty individual communities. In this case, some spaces occupied by colonists and small urban centers have been respected, while other, larger, and more remote spaces were declared protected natural areas under agreements between the local indigenous organizations and the national park service after lengthy, and sometimes acrimonious, negotiations.

A similar case can be found in the middle of the Urubamba watershed, where a land-titling process was carried out by a sympathetic civil society organization on behalf of the Machiguenga and Ashaninka communities.[42] A strip of land along the Urubamba River, now occupied by the Camisea Gas Project headquarters and installations, was left to the small number of colonists that had settled there. On the eastern side of this community-based landscape, in the headwaters of the Camisea River, a large area was formally reserved for two different populations of isolated indigenous peoples. Over time, and also through negotiations with conservation organizations and the national park service, this indigenous landscape was bordered on its western flank by a communal reserve and a national park and on its northern flank by a national sanctuary. Altogether this community-protected area covers more than four million acres.[43]

A second strategy is a mosaic approach to establish such community-based landscapes in large areas with relatively few indigenous communities and large unclaimed spaces. An example of this is an area now called the Putumayo-Amazonas Indigenous Landscape, located between the middle and lower Putumayo River corridor and the Pebas plateau area of the lower Amazon River in Peru. In this case, the first steps, after a government-led community land-titling process in the late 1970s, was the collaborative effort made in 1998 by the local indigenous organization of the Ampiyacu River basin, Federation of Native Communities of the Ampiyacu [River] (Federación de Comunidades Nativas del Ampiyacu, FECONA), and the IBC-SICNA team to georeference and map the boundaries of the fourteen native communities found there.[44] Two years later, in an attempt to dissuade the Fujimori government from granting a 617,000-acre land-forest concession to a Korean company in the headwaters of this and neighboring basins, the IBC-SICNA team worked with FECONA, a second indigenous peoples organization in the neighboring Apayacu basin, and a third in the Putumayo River basin to research and map their use of forest resources in an area of over 2.4 million acres. The study and the map served as the basis for a proposal to create a communal reserve over the entire area for future use by the communities.

For the second stage of this process, the conservation biology and social teams from the Field Museum of Chicago and the Special Project for Integral Development of the Putumayo Watershed (Proyecto Especial de

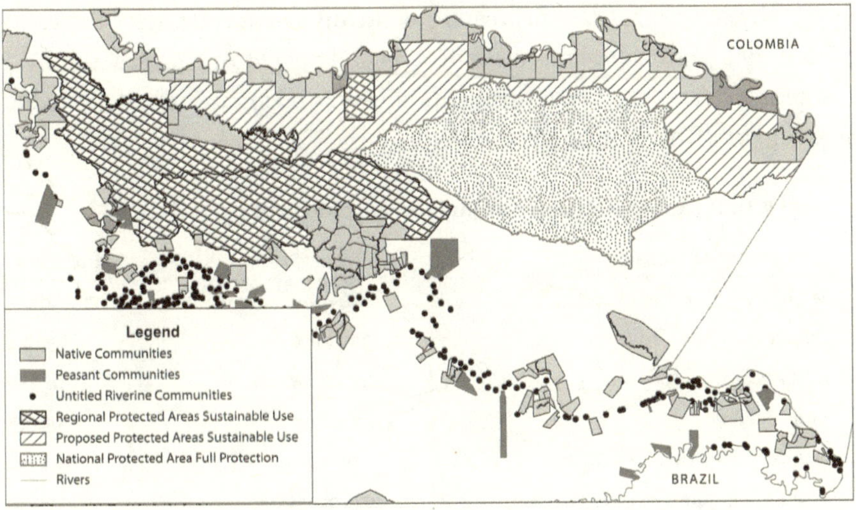

6. Putumayo-Amazonas indigenous landscape, Loreto Region, Peru. Courtesy of SICNA and SICCAM Information Systems, IBC; Richard Chase Smith.

Dearrollo Integral de la Cuenca del río Putumayo, PEDICP), a binational government development project in the Putumayo, joined the effort. By 2006 a long-term plan had been mapped out and agreed upon by the indigenous communities and their organizations to establish a 6.7-million-acre mosaic combining enlarged spaces with land titles for forty-four communities on the periphery of the area, a very large and strictly protected core area covering two-thirds of the Yaguas River watershed, and four protected areas as buffer zones between the communities and the core area for managed use by the indigenous peoples (see map 6 and table 6).

In 2018, following four biodiversity and social inventories, a government-led consultation process involving twenty-nine communities, along with some difficult conflicts with groups of illegal loggers and gold miners, 73 percent of the area planned for the Putumayo-Amazonas Indigenous Landscape is now secure (see table 6).[45] The Yaguas National Park, established in January 2018, was the latest component of the mosaic put in place. Two protected buffer areas for community management and use with a total area of 1.8 million acres are in process, but, depending on the political will of the government to move forward with them, it could take another three years or more to complete them. The two major challenges for the future are creating greater capacity for local governance of this enormous landscape and developing productive and marketing strategies that assure economically and environmentally viable livelihoods within the landscape.

TABLE 6. Putumayo-Amazonas indigenous landscape

Component of Mosaic	Current State	Area Acres
Territories of 44 communities	95% complete	1,201,274
Core area: Yaguas National Park	Complete	2,093,614
ACR Ampiyacu-Apayacu	Complete	1,070,188
ACR Maijuna	Complete	555,728
ACR Algodón-Medio Putumayo	In process	1,083,155
ACR Bajo Putumayo-Boca Yaguas	In process	729,558
TOTAL AREA		6,733,517

Source: IBC, Lima, Peru.

A third strategy has been developing in the middle and upper reaches of the Amazonian portions of the Marañón River over the past decade, initiated by the Regional Coordinating Group of Indigenous Peoples based in San Lorenzo (Loreto). Some have called it a self-demarcation of integrated territories of indigenous peoples with the declaration of "autonomous governments," as in the case of the Awajún-Wampis communities of the Santiago River.[46] According to Surrallés and Tuesta, advisors to this movement, this strategy is not a negation of the value of land titles for individual communities but rather should be seen as a step forward in the exercise of the right to self-determination in accordance with international law.[47]

As defined by this movement, an integrated territory of an indigenous people is not just the surface of a polygon of land needed to feed a group of people. It is a special concept that includes "the social, political and legal conditions, both individual and collective of the occupation of a particular space, along with the processes, expectations, transformations and conflicts that arise from those conditions,"[48] Since the first effort to self-demarcate an integrated territory surrounding the titled communities of the Candoshi peoples in 2007, groups of communities titled to Achuar, Shapra, Shiwilo, Shawi, Cocama, Awajún, and Wampis peoples have carried out a similar process of territorial self-demarcation.[49] This process has not been without conflict among indigenous groups over boundaries, leading, in some cases, to the declaration of interethnic territories between boundaries as well.

Other than ILO Convention 169, there is no specific law or lesser ordinance within the Peruvian legal framework that recognizes and defines at the national level a legal status for what we are calling here an "indigenous people," and there is no recognition of the right of any human group to self-demarcate an "integrated territory" for themselves. In the case of the Achuar and other groups living within the Datem del Marañón Province (Loreto), there is an ordinance passed in 2008 by the municipal government recognizing the right of indigenous peoples to govern their territories that is used as the legal reinforcement for this process. To date, however, no integrated territory has received recognition of its legal existence. The short-term value of this process is the raising of awareness among these peoples of their complex relationship with a territorial unit and their right to lobby for it. It is too early in the process to be able to judge the success of this territorial strategy in the long term.

Despite the devastating effects of many centuries of colonial domination and deep-rooted racism, there are still an important number of community-based landscapes under the local governance of indigenous peoples in Peru. In fact we have shown that about half of the terrestrial surface of Peru is still possessed and/or owned by indigenous communities. At the same time, there continues to be enormous government and private-sector pressure on these communities and their lands to parcel them out in individual lots that would be subject to the commercial rules of a rural land market. However, at the same time there continues to be widespread pushback by these rural communities and their organizations to protect what is rightfully theirs.

NOTES

1. Smith, *Bagua.*
2. International Federation for Human Rights, *Bagua.*
3. International Federation for Human Rights, *Bagua.*
4. Coordinadora Nacional de Derechos Humanos, "CNDDHH"; La Mula, *Bagua*; Zabarburu S., Cabrero B., and Montenegro G., *Sentencia.*
5. The term *indio* (Indian), introduced by the Spanish when referring to the indigenous population, was used during the colonial and republican era until the 1970s and 1980s, when the terms *indígena* (indigenous person) and *campesino* (peasant) started being used.
6. Varallanos, *Historia*, 239–46. See also Guaman Poma, *El Primer*, 489–95.
7. Bonilla, *El Perú*; Bonilla and Spalding, "La independencia"; Dobyns and Doughty, *Peru*; Smith, *Liberal Ideology*.
8. See Barclay and Santos, *La Frontera*; Chirif, *Después del Caucho*.
9. This began a slow change in the colonial governance of Indians from the "repartimientos de indios encomendados" to the "corregimientos de indios." See Varallanos, *Historia*, 185.
10. Solórzano de Pereira, *Politica Indiana*, 26–30, as cited in Varallanos, *Historiia*, 192. See also Guaman Poma, *El Primer*, 454–78.
11. *Criollos* were descendants of the Spanish but born in the colonies.
12. Dobyns and Doughty, *Peru*; Smith, *Liberal Ideology*, 79.
13. Smith, *Liberal Ideology*, 80.
14. Sala, "Diario," 68. In 1974, due to special circumstances, I had in my possession the original handwritten manuscript of Father Sala's travel diary and was able to take copious notes on it. The diary was published for the first time by the Peruvian government's Ministry of Development in 1897 (Sala, *Apuntes de Viaje*), and then as parts of different works in Peru, including Izaguirre's great work on the history of the Franciscan Order in Peru (1920–29). In comparing the original manuscript with the various published versions, I discovered they did not contain many of the phrases, such as the ones quoted here, that reflect Sala's racist ideas.
15. Sala, "Diario," 565. *Chuncho* is a term of disrespect used to this day for the indigenous inhabitants of the Peruvian Amazon.
16. Berman, *International Labour Organization*; Berman, *Indigenous Peoples*; Chirif, García, and Smith, *El Indígena*, 19–30.
17. Law 26505: Law of Private Investment in the Development of Economic Activities in the Lands of National Territory and Those of Peasant and Native Communities, enacted on July 17, 1995. This law and the new strategy to strengthen the market of rural lands issued from the advice given by Hernando de Soto and the Institute of Liberty and Democracy to the Fujimori government.
18. Smith and Salazar, *La Seguridad*, 24–25.
19. This law was limited to the communities on the coast for two reasons: (1) fear of the negative reaction of communities in the Andes and Amazon and (2) the interest in expanding agro-industry along the coast.

20. Peruvian Law 26845, Titling of Lands of Peasant Communities in the Coast (2007).
21. These and other arguments were provided by Hernando de Soto and other consultants at the Institute of Liberty and Democracy who advised the Fujimori government on land issues (see Institute for Liberty and Democracy, *Mystery*; de Soto, "Peruvian Amazon"). Rosser, "Anticipating," warns us about the negative impact that this objective could have on the indigenous territories of the United States and Canada.
22. García, "El síndrome."
23. Smith, *Bagua*, 50. *The Dog in the Manger* is the title of a sixteenth-century palatine comedy by Lope de Vega, that, in popular parlance, refers with contempt to a person who "neither eats nor allows others to eat."
24. Smith and Salazar, *La Seguridad*, 26.
25. Smith and Salazar, *La Tierras*, 56.
26. In early 2013 the Ministry of Agriculture and Irrigation had already been declared by law the governing body at the national level for the Reorganization of Rural Property and Rural Land Registry. Law 30230 assigned similar decisive power over rural lands and property to an agency in a different ministry in charge of urban properties and construction. In addition, the law did not even include the Ministry of Agriculture and Irrigation among the members of the Multisector Committee charged with establishing the law's regulations. See Mongylardi, "Superposición."
27. Ruiz, "La constitucionalidad"; Smith and Salazar, *Las Tierras*, 33.
28. Smith and Salazar, *Las Tierras*, 34.
29. Mongylardi, "Superposición."
30. Proyecto Especial de Titulación de Tierras y Catastro Rural, *Directorio de Comunidades Campesinas*; Proyecto Especial de Titulación de Tierras y Catastro Rural, *Directorio de Comunidades Nativas*.
31. See the SICNA website: www.ibcperu.org/mapas/sicna.
32. In comparison with the data from December 2016 published by SICNA, the overall number of registered native communities has increased by 262 and the number of titled communities by 523 (Sistema de Información sobre Comunidades Nativas, *El directorio*). This is an important and encouraging increase during 2017–18.
33. In comparison with the data from December 2016 published by SICCAM, the overall number of registered peasant communities has increased by eighty-one and the number of titled communities also by eighty-one (Sistema de Información sobre Comunidades Campesinas, *El directorio*).
34. del Castillo, "La formalización."
35. Shapefile is a specific GIS data file format for storing geographic and tabular data in a single file, used by the ARCGis system software.
36. Although all the georeferenced polygons for peasant communities come from government sources, SICCAM cannot guarantee their accuracy nor the thoroughness of the methodology used to establish them.

37. Smith and Salazar, *La Tierras*, 53.

38. Smith and Salazar, *La Tierras*, 53.

39. This concession was granted to Maurel et Prom Peru SAC (France) and Pacific Stratus Energy SA, Peru Branch in 2006, and has a total area of 1.62 million acres. It is in the exploration phase.

40. Figures and the map of plots of contracts, sedimentary basins and protected natural areas, July 2014, were taken from the PerúPetro website, http://www.perupetro.com.pe/relaciondecontratos/relacion.jsp?token=70.

41. Mining projects map; Main production units and exploration projects, Ministry of Energy and Mines, Lima, 2014.

42. Smith, *Can David*.

43. Data from SICNA show the following: 23 communities with 1,154,798 acres; Kogapakori-Nahua Territorial Reserve with 1,128,462 acres; Megantoni National Sanctuary with 533,423 acres; Machiguenga Communal Reserve with 540,927 acres; and the Otishi National Park with 756,075 acres; for a total of 4,113,685 acres.

44. Smith, *El gran mosaico*, 86.

45. See Pitman et al., *Perú: Ampiyacu*; Pitman et al., *Perú: Yaguas-Cothue*.

46. Surrallés and Tuesta, "Los territorios."

47. Surrallés and Tuesta, "Los territorios," 54. See also García and Surrallés, "Introducción."

48. Surrallés and Tuesta, "Los territorios," 54; see also Chirif, García, and Smith, *El Indígena*; Surrallés, "Entre derecho y realidad."

49. Surrallés and Tuesta, "Los territorios."

BIBLIOGRAPHY

Barclay, Frederica, and Fernando Santos G. *La Frontera Domesticada; Historia económica y social de Loreto, 1850–2000*. Lima: Pontificia Universidad Católica del Perú. 2002.

Berman Howard R. "Indigenous Peoples and the Right to Self-Determination." *In Proceedings of the American Society of International Law* 190 (1994): 190–204.

———. "The International Labour Organization and Indigenous Peoples: Revision of ILO Convention 107." 75th Session of the ILO Conference, International Commission of Jurists Review 48. Geneva, 1988.

Bonilla, Heraclio. "El Perú entre la Independencia y la guerra con Chile." In *Historia del Perú*, vol. 6, edited by Juan Mejía Baca. Lima: Editorial Juan Mejía Baca, 1980.

Bonilla, Heraclio, and Karen Spalding. "La independencia en el Perú: Las palabras y los hechos." In *La Independencia en el Perú: Perú Problema No. 7*. Lima: Instituto de Estudios Peruanos, 1972.

Chirif, Alberto. *Después del Caucho*. Lima: Lluvia Editores, CAAAP, IWGIA, IBC, 2017.

Chirif, Alberto, Pedro García H., and Richard Ch. Smith. *El Indígena y su territorio son uno sólo: Estrategias para la defensa de los pueblos y territorios indígenas de la Cuenca Amazónica*. Lima: Oxfam América, 1991.

Coordinadora Nacional de Derechos Humanos. "CNDDHH sobre la Sentencia del Caso Bagua: Comunicado." September 23, 2016. http://derechoshumanos.pe/2016/09/cnddhh-sobre-la-sentencia-del-caso-bagua/2016.

del Castillo, Laureano. "La formalización de las comunidades y de sus territorios." *Debate Agrario* 46 (2014): 119–140.

Delran, Guido C. *Historia Rural del Perú*. Cuzco: Centro de Estudios Rurales Andinos Bartolome de las Casas, 1981.

de Soto, Hernando. "The Peruvian Amazon Is Not Avatar." Center for International Private Enterprise. 2010. https://www.cipe.org/legacy/publication-docs/The%20Peruvian%20Amazon%20is%20not%20Avatar.pdf.

Dobyns, Henry F., and Paul L. Doughty. *Peru: A Cultural History*. New York: Oxford University Press, 1976.

García, Alan. "El síndrome del perro del hortelano." *El Comercio*, October 27, 2007.

García H., Pedro, and Alexandre Surrallés. "Introducción." In *Tierra adentro: Territorio indígena y percepción del entorno*, edited by A. Surrallés and P. García H. Copenhagen: International Work Group for Indigenous Affairs, 2004.

Guaman Poma de Ayala, Felipe. *El Primer Nueva Corónica y Buen Gobierno*. Edited by John V. Murra and Rolena Adorno. México DF: Siglo Veintiuno Editores, 1980.

Institute for Liberty and Democracy. *The Mystery of Capital among the Indigenous Peoples of the Amazon: Part One*. YouTube, February 16, 2017. https://www.youtube.com/watch?v=o86bXOG5GcU.

International Federation for Human Rights. "Bagua: Bloodshed in the Context of Amazon Protest—Urgent Need for Good Faith Dialogue." October 26, 2009. https://www.fidh.org/en/region/americas/peru/Bagua-Bloodshed-in-the-context-of.

La Mula. "Bagua: La sentencia y la polémica." https://reflexionesperuanas.lamula.pe/2016/09/26/bagua-la-sentencia-y-la-polemica/reflexionesperuanas/.

Mongylardi C., Cinthia. "Superposición de competencias derivada de la aplicación del Título III de la Ley 30230: COFOPRI, Direccion de Saneamiento de la Propiedad Agraria y Catastro Rural (MINAGRI) y gobiernos regionales." In *Estrategias del Despojo: La propiedad comunal en el Perú. Análisis jurídico de la Ley 30230*, edited by Miluska Carhuavilca. Lima: Instituto del Bien Común, 2015.

Pitman, Nigel, Richard Chase Smith, Corine Vriesendorp, Debra Moskovits, Renzo Piana, Guillermo Knell, and Tyana Wachter, eds. *Perú: Ampiyacu, Apayacu, Yaguas, Medio Putumayo (Rapid Biological Inventories 12)*. Chicago: Field Museum, 2004.

Pitman, Nigel, Corine Vriesendorp, Debra K. Moskovits, Rudolf von May, Diana Alvira, Tatzyana Wachter, Douglas F. Stotz, and Álvaro del Campo. *Perú: Yaguas-Cothue (Rapid Biological Inventories 23)*. Chicago: Field Museum, 2011.

Proyecto Especial de Titulación de Tierras y Catastro Rural PETT. *Directorio de Comunidades Campesinas del Perú—1998*. Lima: Presidencia de la Republica, 1998.

———. *Directorio de Comunidades Nativas del Perú—1999*. Lima: Presidencia de la Republica. 1999.

Rosser, Ezra. "Anticipating de Soto: Allotment of Indian Reservations and the Dangers of Land-Titling." In *Hernando de Soto and Property in a Market Economy*, edited by D. Benjamin Barros. London: Ashgate, 2010.

Ruiz, Juan Carlos. "La constitucionalidad de la Ley 30230 en materia de saneamiento de predios." In *Estrategias del Despojo: La propiedad comunal en el Perú. Análisis jurídico de la Ley 30230*, edited by Miluska Carhuavilca. Lima: Instituto del Bien Común, 2015.

Sala, Gabriel. *Apuntes de Viaje: Exploración de los ríos Pichis, Pachitea y alto Ucayali*. Lima: Ministerio de Fomento, 1897.

———. "Diario." Unpublished manuscript. 1895.

Sistema de Información sobre Comunidades Campesinas. *El directorio de Comunidades Campesinas del Perú 2016*. Lima: Instituto del Bien Común, 2016.

Sistema de Información sobre Comunidades Nativas. *El directorio de Comunidades Nativas del Perú 2016*. Lima: Instituto del Bien Común, 2016.

Smith, Richard Chase. "Bagua: La verdadera amenaza." *Poder* (Lima), July 2009, 48–53.

———. "Can David and Goliath Have a Happy Marriage? The Machiguenga People and the Camisea Gas Project in the Peruvian Amazon." In *Representing Communities: Histories and Politics of Community-Based Resource Management*, edited by Peter Brosius, Anna Ling, and Charles Zehner. Lanham MD: Altamira Press, 2004.

———. "El gran mosaico de territorio indígena, de conservación y de uso sostenible entre los ríos Putumayo, Yaguas, Apayacu y Ampiyacu." In *Atlas de comunidades nativas y áreas naturales protegidas del nordeste de la amazonia peruana*, edited by Margarita Benavides. Lima: Instituto del Bien Común, 2010.

———. "Liberal Ideology and Indigenous Communities in Post-Independence Peru." *Journal of International Affairs* 36, no. 1 (1982): 23–31.

Smith, Richard Chase, and Milagros Salazar. *La Seguridad Territorial en el Limbo: El estado de las comunidades indígenas en el Perú*. Lima: Instituto del Bien Común, 2014.

———. *La Tierras Comunales: Más que preservar el pasado es asegurar el futuro. El estado de las comunidades indígenas en el Perú*. Lima: Instituto del Bien Común, 2016.

Solórzano Pereira, Juan de. *Política Indiana*. Madrid: Cía. Iberoamericana de Publicaciones, 1930.

Surrallés, Alexandre. "Entre derecho y realidad: Antropología de los territorios indígenas amazónicos en un futuro próximo." *Bulletin de l'Institut Français d'Études Andines* 38, no. 1 (2009): 29–45.

Surrallés, Alexandre, and Ermeto Tuesta. "Los territorios integrales y la libre determinación de los pueblos en el noroeste de la Amazonía peruana." In *Atlas de comunidades nativas y áreas naturales protegidas del noroeste de la Amazonía Peruana*, edited by Richard Ch. Smith and Carla Soria. Lima: Instituto del Bien Común, 2017.

Varallanos, José. *Historia de Huanuco: De la era prehistórica a nuetros días*. Buenos Aires: Imprenta López, 1959.

Zabarburu Saavedra, Gonzalo, Norberto Cabrero Barrantes, and Enrique Montenegro Guimaraez. *Sentencia*. Expediente No. 00194-2009 [0163-2013]. Corte Superior de Justicia de Amazonas. http://derechoshumanos.pe/2017/06/la-sentencia-del-caso-baguazo/.

8 In Search of Justice and Power

Contentious Experiences of Free, Prior, and Informed Consent in Latin America

ROGER MERINO

The right to free, prior, and informed consent (FPIC) is an international standard first recognized in 1989 in the International Labor Organization Convention 169 regarding indigenous and tribal peoples, then developed by the United Nations Declaration on the Rights of Indigenous Peoples (UNDRIP) of 2007 and international jurisprudence on human rights.[1] After centuries of exclusion and exploitation, indigenous peoples would benefit from this right, which would overcome incorporation policies of the mid-twentieth century and result in a shift toward policies of self-determination.[2]

FPIC is usually understood as a procedural right similar to the right of citizen participation in state policies and investment projects, but more intense, since FPIC can help to overcome historical and structural injustices against indigenous peoples. This is particularly relevant in Latin America, which is home to between 42 million and 45 million indigenous people, representing an estimated of 10 to 12 percent of the total population.[3] FPIC would allow these segments of the society access to substantial rights, such as the protection of their territories and livelihoods and respect for their ancestral culture, as state policies would have to be reframed depending on the outcomes of consultation processes.

This, however, is a rather romantic view of the realities and possibilities of FPIC in deeply unequal contexts, with historically racialized state structures and exclusionary national elites. Since the 1990s Latin

American countries have approved multicultural policies with provisions for social inclusion of indigenous peoples, including this approach in constitutional reforms and new constitutions of this decade.[4] These constitutions recognized collective rights, such as land rights, customary law, and language rights, and some degree of autonomy (limited territorial control and the recognition of indigenous systems of justice).[5] Beyond the debates on the real impact of multicultural policies, indigenous self-determination has never been fully recognized.[6] When Latin American multicultural constitutions recognized the right to self-determination or autonomy, what they protected was the right of each community to govern itself within the boundaries of limited territorial units. They did not recognize the self-determination of a whole indigenous nation.[7]

The lack of strong recognition of self-determination is also reflected in the incorporation of FPIC into national laws. While the international standard requires the actual consent of indigenous peoples in several situations, governments have implemented a weaker FPIC standard called "prior consultation," whereby consent is a general purpose of the processes but not necessarily a requirement. Moreover, as consultation does not refer to whole indigenous nations but to specific affected communities, the political power of decisions of indigenous peoples is often limited and fragmented. Institutions then are designed not to increase indigenous decision-making power but to limit it.

Consultation thus expresses a continuing struggle for justice and power. Many indigenous peoples appropriate consultation to challenge structural injustices unaddressed by the state, and in this way they attempt to go beyond mere participation in specific projects or policies. Consultation becomes a means to occupy deliberative spaces for claiming broader institutional transformations, such as territorial zoning, dominium of natural resources, or the reframing of extractive policies.

On the basis of a documentary analysis of laws, policies, reports, and case studies, this chapter compares and discusses the implementation and application of consultation in the four countries with the most extensive experience in these processes in Latin America: Colombia, Bolivia, Ecuador, and Peru. Although they have developed different institutional designs for consultation, they share similar colonial legacies of exclusion of indigenous peoples and a political economy deeply dependent on resource extraction, despite the different ideological orientations of their governments.[8]

Free, Prior, and Informed Consent: International Sources, Definition, and Scope

In the international arena, a unique standard for FPIC does not exist.[9] ILO Convention 169 incorporated the right of indigenous peoples to be consulted before the approval of any norm or measure that could affect their collective rights and required the state to obtain consent if people would be displaced. The UNDRIP, a declaratory, nonbinding instrument according to international law, establishes that all consultation processes should try to obtain consent, and adds that states must obtain consent in cases of displacement, storage, or disposal of hazardous materials in indigenous territory and the use or occupation of indigenous material and immaterial property. The Inter-American Court of Human Rights asserted that consent is also necessary in cases of "large-scale development or investment projects" that would have a "major impact" on "a large part of their territory" (*Saramaka v. Suriname*).[10] Meanwhile multilateral organizations such as the World Bank, the International Finance Corporation, and the Inter-American Development Bank and private global organizations such as the International Council on Mining and Metals have approved guidelines on consultation as a kind of participatory right instead of an expression of self-determination.[11]

There is therefore a *consent standard*, which includes the obligation to obtain consent in cases recognized by the UNDRIP and international jurisprudence; the *consultation standard*, a limited right of participation recognized in the binding ILO Convention 169 and most guidelines of multilateral organizations (where actual consent is an exception); and *consent as an expression of self-determination*, as a total power of decision of indigenous peoples (where actual consent should be the rule), not fully recognized by national and international standards.[12]

The main holders of rights to FPIC are indigenous peoples, who are defined by ILO Convention 169 as descendant from precolonial populations, maintaining cultural and sociopolitical patterns and identifying themselves as members of ancestral peoples.[13] As the ILO Convention also applies to tribal peoples, they also should hold the right of FPIC; nevertheless new international instruments, such as the UNDRIP and the American Declaration on the Rights of Indigenous Peoples, focus exclusively on indigenous rights.[14]

Despite the high number of adherents to the UNDRIP and ratifications of the ILO Convention in Latin America, actual implementation of FPIC has been resisted or has entailed the legalization of a weaker standard by emphasizing the need for consultation rather than consent.[15] Most mechanisms implemented are related more to participation than to self-determination, being trapped within bureaucratic provisions.[16] This situation is deepened by the fact that most peoples occupy a marginal political and economic status in a context of a growing global demand for exploitation of natural resources.[17]

Implementation of consultation has also raised several critiques related to procedural issues (such as the short time for deliberation and non-inclusion of all communities involved) and more structural concerns (asymmetric relations, tensions between local visions of development and the objectives of state development, the transformation of consultation into a simple social responsibility label, among others).[18] Despite these critiques, most scholars agree that consultation has an intrinsic inclusionary nature by deepening indigenous participation in policymaking, and it would need only institutional adjustments aimed at giving more decision-making power to indigenous peoples.[19] Others argue that FPIC is inherently undemocratic because it limits participation to specific communities, excluding the majority of the poor and contradicting fundamental ideas of representation, legitimacy, and inclusiveness.[20]

It is important to differentiate the aims of the policy from its actual implementation. The policy design conceives consultation as a participatory tool for social inclusion, justified as a special right because of the structural and historical injustices that indigenous peoples suffered as a result of state actions and omissions. (The critique of consultation as a privilege is therefore weak.) At the moment of implementation, however, it becomes a political technology, a tool for action and resistance by indigenous peoples, and a technique for governance and domination by the state.[21] In this sense, consultation is a space for political negotiations and struggles, whose tensions are deepened by the expansion of extractive industries.

Policy and Legal Implementation in Bolivia, Ecuador, Colombia, and Peru

The four countries studied ratified ILO Convention 169 in the 1990s, have a history of exclusion of indigenous people, and have experience with the development and implementation of the policy of consultation. They also exemplify how consultation processes are associated with social conflicts related to extractive activities, which are exacerbated by the lack of protection of indigenous ancestral territories and the aggressive granting of rights for the exploitation of natural resources.[22]

Bolivia

Bolivia ratified ILO Convention 169 in November 1991 (Law 1257). The 1995 Constitution included the concept of multiculturalism and recognized a number of indigenous rights but did not legalize consultation. It was only expressly recognized in the current Constitution of 2009 (article 30; article 403) as a right of indigenous peoples and a duty of the state regarding the exploitation of natural resources on indigenous territories. Only two sectors have enacted specific regulations based on this constitutional provision: hydrocarbons (Law 3058 of May 2005 and enabling legislation approved by Supreme Decree 29033) and mining (Law 535 of May 2014).

There is no centralized institution in charge of consultation governance. In hydrocarbons the responsible entity is the Ministry of Hydrocarbons and Energy, whereas in mining it is the Administrative-Jurisdictional Authority of Mining. Both are in charge of identifying indigenous peoples to be consulted, and organizing and leading deliberative meetings. The Intercultural Service for Democratic Strengthening of the Supreme Electoral Tribunal (an entity ascribed to the Organo Electoral Plurinacional) supervises consultations and formulates a report on the results (Law 26/2010, article 41). There is no clarity about responsibilities and procedural rules for consultations on other issues. For example, in the consultation on construction of a highway into the ecological area known as TIPNIS (Isiboro Sécure National Park and Indigenous Territory), the ministries of environment and water, public works, services, and housing participated in a 120-day consultation process under Law 222, which was especially enacted to establish the responsibilities and time frames for this process; this law, however, is not a framework regulation for other similar processes.

The subjects of consultation are those "nations" and "first indigenous peasant peoples" in whose territory natural resources are exploited (Constitution, article 352). These peoples are organized as communities and are represented by their traditional authorities (Law 3058, article 118). They also can create indigenous First Peoples' peasant autonomies (*autonomías indígena originaria campesinas*) in ancestral territories (Constitution, article 290). There is, however, no unique terminology for naming indigenous populations. The hydrocarbon regulation refers to "indigenous peoples, native and peasant communities" (SD 29033), and mining legislation refers to "nations and first indigenous peoples, peasants, intercultural communities and Afro-Bolivian people" (Law 535). In the latter regulation, they are defined as groups whose existence predated the colonial era and which maintain cultural patterns, ways of life, and institutions, self-identify, and have collective access to and management of their lands and territories (Law 535, article 209). This definition is inspired by the criteria defined in ILO Convention 169.

Consultation is an administrative procedure. In hydrocarbons it has four phases: (1) coordination and information, (2) organization and planning, (3) execution of consultation, and (4) agreement (SD 29033, articles 9–13). The process lasts forty-five days, though it can be extended for an additional period of up to one month (SD 29574, article 2). The administrative procedure in mining has seven phases: (1) preparatory phase, (2) notifications and meetings for intercultural dialogue, (3) representation, (4) deliberation and additional meetings, (5) mediation and decision, (6) final resolution, and (7) costs (Law 535/2014, articles 210–16). The entire process has a maximum period of four months.

Consultation in hydrocarbons occurs at two different moments: first, in the bidding, authorization, contracting, call, and approval of measures, works, or hydrocarbon projects (SD 29033, articles 14, 15); and second, prior to approval of the environmental impact assessment. In mining, consultation applies to any request for the signing of a mining contract that could directly affect the collective rights of indigenous peoples or Afro-Bolivian peoples (Law 535, article 207).

The Organo Electoral Plurinacional reported that between October 2015 and October 2016, the Administrative-Jurisdictional Authority of Mining began a total of 165 consultations, of which 96 were completed (84 met the standards, while 12 did not). In the hydrocarbon sector, the General Directorate of Socioenvironmental Management of the Ministry of Hydrocarbons reported 70 consultations by 2017.[23]

Indigenous organizations have raised several critiques of the consultation processes. They criticize the lack of an institution that centralizes consultations and standards for them, as well as the fact that only hydrocarbons and mining have regulations, leaving aside other social and economic sectors.[24] In addition, they claim that current norms do not reflect the interests of indigenous populations. In 2013 the Bolivian state discussed with indigenous organizations the contents of the draft legislation for the Framework of the Right to Prior Consultation, which was presented in 2014 by President Evo Morales and was still awaiting approval in the Plurinational Legislative Assembly as of early 2018. The government asserts that this project is the result of an intercultural dialogue, but indigenous organizations say their participation was limited, and in some cases some federations were not invited. (In their place, mining cooperatives participated.)[25] Meanwhile organizations have drafted their own proposed legislation. In July 2011 the National Council of Ayllus and Markas of Qullasuyu delivered a proposal to the Plurinational Legislative Assembly; in 2012 CONAMAQ and the Confederation of Indigenous Peoples of Bolivia also delivered proposals.[26] More recently the Guaraní people of the Bolivian Chaco have been preparing a proposal based on their experience on consultations about hydrocarbon activities.

Another important critique is the lack of protection of indigenous territories due to the promotion of infrastructure and extractive projects.[27] Yet another contentious issue is the definition of who should be consulted; some people maintain that only indigenous nations and first indigenous peasant peoples with communitarian lands should be consulted, while intercultural communities and Afro-Bolivians support a less strict right of generic participation.[28]

Ecuador

Ecuador ratified ILO Convention 169 in 1998 (Executive Decree 1387), and consultation is now expressly recognized in the 2008 Constitution (articles 57, 398). A general prior consultation law does not exist; it is only generally mentioned in sectoral norms, such as the Organic Law of Water Resources, Use and Exploitation (Law of 2014, article 68), the Organic Law of Citizen Participation (Law of 2010, article 81), and the Regulations for the Execution of Prior Consultation in Bidding Processes and Allocation of Hydrocarbon Areas and Blocks (Decree 1247, August 2012). Nonetheless Ecuador is the only country that has incorporated prelegislative consultation under the responsibility of the National Assembly and the corresponding Specialized Commission (Instructive for Prelegislative Consultation, RO 733, June 2012, article 4). In an important decision, the Constitutional Court has defined standards of good faith for implementing this right. It also has established that prelegislative consultation must be substantial rather than formal and that consultation is required for any mining activity intended to be conducted in ancestral territories (Constitutional Court 001-10-SIN-CC, March 18, 2010). The Confederation of Indigenous Nationalities of Ecuador has questioned this decision, however, on the grounds that it does not establish that the outcomes of consultation are binding for the state.[29]

Governance of consultation is not centralized, but sectoral. In hydrocarbons, the responsible authority is the Hydrocarbons Secretariat of the Ministry of Nonrenewable Natural Resources, with the support of the Ministry of Environment and the Secretariat of Peoples, Social Movements, and Citizen Participation (Decree 1247, article 6). Other activities are governed by the general standards indicated in the Constitution and Executive Decree 1040/2008, which develops participatory mechanisms established in the Environmental Management Law and regulates consultation as a general participation right dependent on the Ministry of Environment. Over the years some sectors have been setting standards for consultation, distancing from those who understand it as a mere mechanism for participation. An example is the Organic Law of Water Resources (Law of 2014, article 68), which establishes that the Single Water Authority (Autoridad Única del Agua) is responsible for consultations on collective rights to water.

The subjects of consultation are not only indigenous *comunas*, communities, peoples, and nationalities (Constitution, article 57) but also the Afro-Ecuadorians and the *montubios* (Constitution, article 57).[30] All of these peoples organize through communities and designate their representatives in consultations, according to their uses and customs. Companies do not play a specific role in consultation; they are only informed of it through oversight committees ascribed to the Hydrocarbons Secretariat.[31]

Consultation in Ecuador is an administrative procedure. In hydrocarbons, it has five phases: (1) previous activities, (2) public notice by the Hydrocarbons Secretariat, (3) consultation process, (4) systematization, and (5) conclusion.[32] The maximum period for this process is thirty days, counted from the last publication of the public notice (Decree 1247, article 12). Prelegislative consultation has four phases: (1) preparation, (2) public notice, (3) information and development, and (4) analysis of results and closure of consultation (Instructive, articles 7–21). This process has no specific maximum time limit, but counting the partial periods of all stages it would last approximately 150 to 180 days.

The object of consultation in hydrocarbons is the administrative decision that approves plans and programs for prospection, exploitation, and commercialization of nonrenewable resources (Constitution, article 57.7); in prelegislative consultation it refers to any legislative measure that may affect collective rights of comunas, communities, indigenous peoples and nationalities, Afro-Ecuadorian people, and the montubio people (instructive, articles 1–2). In both cases, the aim of the consultation process is consent (Constitutional Court, 001-10-SIN-CC, 52). There are few experiences of prior consultation, mainly because of a lack of political will. Between 2012 and 2015 there were four processes of prior consultation in the hydrocarbons sector.[33] Figures on other sectors were not found. There is no information about any prelegislative consultation.

The lack of implementation of the instructive might derive from its weak legal nature, as it is neither a law nor a decree.[34]

One of the most serious critiques of consultations is that they are not conducted in all relevant cases. In November 2012 an oil round (*ronda petrolera*) to auction thirteen oil concessions in the Ecuadorian Amazon (covering the territory of seven indigenous nationalities in the provinces of Morona, Santiago, and Pastaza) was held without consultation.[35] In general, the state is promoting the rapid development of extractive industries without complying with international standards. In some cases promotion of extractive activities has been accompanied by coercion and the direct involvement of companies, offering social benefits to local people in exchange for approval of their projects.[36] Another critique is the lack of specific legislation for other sectors.[37]

Colombia

Colombia ratified Convention 169 in 1991 (Law 21). There is no mention of consultation in the Colombian Constitution, and the country has no general law on consultation. Nonetheless it has developed significant jurisprudence and administrative procedures, especially for hydrocarbons, mining, and infrastructure.[38]

Colombia has centralized governance for consultation, with the Directorate of Prior Consultation at the Ministry of Internal Affairs as the governing organ (*ente rector*) for these processes (Decree 2893/2011, article 16). This office directs, coordinates, and grants the necessary guarantees for consultations. It also monitors compliance of the commitments assumed by the state. The Directorate of Indigenous Affairs, Rom, and Minorities and the Directorate of Black Communities, Raizales, and Palenqueros (also within the Ministry of Internal Affairs) are in charge of consultations related to legislative and administrative measures of national scope (Decree 2893).

In Colombia companies have an active role in consultations. In the case of administrative measures related to extractive projects, the company interested in the approval of the license or environmental permit must participate in the process, detailing the environmental impacts that could occur and provisions for managing them. These companies accompany the Directorate of Prior Consultation and finance the process.[39]

The subjects of consultation are not only indigenous peoples but the four national ethnic groups recognized in national legislation (1) indigenous, (2) black and Afro-Colombian, (3) Raizales and Palenqueros, and (4) Rom or Gypsies (Presidential Directive 01/2010).[40] National ethnic groups are organized either as communities or *resguardos*.[41] The Directorate of Indigenous Affairs, Rom, and Minorities has a registry of the indigenous and communal authorities who serve as legal representatives and who are designated autonomously by the people (Decree 2164/1995, article 2).

Consultation is an administrative procedure that lasts four to six months.[42] It must include the following phases: (1) preconsultation, (2) opening of the process, (3) workshop for identification of impacts and definition of management measures, (4) pre-agreements, (5) meeting of formalization, (6) systematization and monitoring of agreements, and (7) closure of the process (Presidential Directive 01/2010).

Objects of consultation include resolutions for approving prospecting for and exploitation of resources in territories of the national ethnic groups; public decisions about land transfer; development, increase, or transformation of the road network in ethnic territories; and adoption of health policies, among others (Presidential Directive 01/2010). Decree 2041/2014 states that exploitation of natural resources must incorporate participatory mechanisms, which in practice are equated with consultation processes. As concerns raised in these processes are included in the environmental impact assessment (a requirement for obtaining the environmental license), some organizations question why projects not requiring environmental licenses (and that might affect collective rights, such as mining exploration) do not involve consultations.[43]

Between 2011 and 2014 a total of 4,042 consultations were conducted.[44] Between 1993 and 2012, however, 2,331 environmental licenses were granted in indigenous territories, and during that period only 156 consultations took place (65 for hydrocarbons, 44 for infrastructure, 24 for electricity, and 6 for mining).[45]

The critiques of consultation begin with the policy design: the lack of a general law and the scant importance of the Directorate of Prior

Consultation in conducting the processes.[46] This critique also refers to the lack of civil servants with appropriate training, lack of interinstitutional coordination, and lack of a common database for use by the entities involved, with solid information about projects.[47] Indigenous organizations also question why internal norms for consultation were not the result of a dialogue.

In 2015 the draft legislation on consultation proposed by the Ministry of Internal Affairs was criticized for having been formulated with no participation by communities and for reducing consultation to a participatory mechanism.[48] Some indigenous organizations (such as the National Indigenous Organization of Colombia) consider the internal regulation of consultation unnecessary, since the Constitutional Court has established more protective rules; moreover discussion of a general law in Congress could weaken this right because indigenous peoples have no strong representation in the parliament.

Indeed the Constitutional Tribunal has been crucial for the development of consultation. Whereas the internal regulation emphasizes the need for consultation rather than consent, Decision T-129/2011 established that the purpose of consultation is to obtain consent, recognizing that consent is an actual requirement for activities that represent a high social, cultural, and environmental impact on an ethnic community. Decision T-769/2009 established that in cases of large-scale development plans and investments with major impacts on ancestral territories of indigenous peoples and Afro-descendants, the state must not only consult but must also obtain consent.

Despite these developments, the direction of jurisprudence has changed, depending on the composition of the court. For example, although decisions SU-383/2003 and T-236/2012 asserted that natural resources located inside ancestral territories are part of those territories and that people who inhabit them have the right of collective property, decision C-077/2012 asserted that national government intervention in cases of natural resources located in indigenous resguardos constitutes the exercise of state sovereignty over natural resources. In general, the Constitutional Court has enacted important decisions, but the aggressive expansion of extractive activities has not ended.[49]

Peru

In 2011 Peru approved Law 29785, the Law of the Right of Prior Consultation of Indigenous Peoples, recognized in ILO Convention 169, better known as the Prior Consultation Law, sixteen years after the country ratified ILO Convention 169 (Legislative Resolution 26253). This law was one of the first measures approved by President Ollanta Humala in response to the indigenous uprising known as the Baguazo, which occurred in the northern Peruvian Amazon in 2009.[50]

Consultation is a 120-day administrative procedure organized by any public entity that enacts a norm or decision that may affect indigenous peoples' collective rights. The stages of the process include identification of the indigenous population affected by the norm or decision; provision, by the state, of information about the implications of the norm or decision; and internal evaluation of the information provided. If by this point there is no agreement, a new phase begins: intercultural dialogue between the indigenous people and the state. The final stage is the decision, which occurs whether or not the indigenous peoples agree with the measures proposed by the state. Thus the entire process is designed to inform and convince the people of a public decision that has already been made.

The governance of consultation is sectoral; it depends on those specific entities (promoter entities, according to the law) that enact norms or decisions that could affect indigenous peoples' collective rights. The role of the Vice Ministry of Intercultural Affairs (through the Directorate of Prior Consultation) is just to provide technical assistance to the process and help identify the indigenous population. In Peru companies have no specific role in the process. As of early 2018 the state had reported thirty-two consultations: in hydrocarbons (eleven), mining (ten), conservation (five), infrastructure (two), and policies (four).

Although the provisions of the law regarding the object of consultation are very general, several discussions have emerged to define which specific measures must be consulted. The only ministry that has defined this is the Ministry of Energy and Mines. Based on Article 6 of the Prior Consultation Law, that ministry enacted two different regulations: Ministerial Resolution no. 209-2015-MEM/DM for energy industries (hydrocarbons and electricity) and Ministerial Resolution no. 362-2015-MEM/DM for mining. In the case of hydrocarbons, consultation applies to the decree that approves the concession contract. In the case of mining, consultation applies to the resolution that authorizes the commencement of operations. In contrast, indigenous organizations and activists are resorting to the judicial system, seeking constitutional injunctions (*demandas de amparo*) to require consultation about concessions and resolutions approving environmental impact assessments for both mining and hydrocarbon activities.[51]

In Peru consultation refers specifically to indigenous peoples. There are fifty-five indigenous peoples registered in the database of the Ministry of Culture (as of 2018); fifty-one are Amazonian and four are Andean. This population is formally organized in 3,029 peasant or *campesino* communities and 2,558 native communities.[52] Nevertheless there is still a struggle over how to define the indigenous population because peasant and native communities may or may not be regarded as indigenous peoples. (Some argue that peasant communities should not have indigenous status.)

Implementation of consultation has been problematic. Indigenous organizations have strongly criticized specific provisions of the law and its enabling legislation (such as the exemption from consultation of infrastructure projects meant to provide "public services" and the limited recognition of the right to provide consent) and consultation practices (such as the lack of technical assistance for the communities).[53] Other critiques emphasize the lack of a specialized entity representing indigenous peoples and the fact that many communities still lack title to their collective property.[54]

Prior Consultation in Comparative Perspective

The experiences of these four countries reflect profound differences in the policy design, the kind of regulation implemented, the governance, the subjects of consultation, the object of consultation, and procedural rules. Whereas Peru has no constitutional support and no strong jurisprudence for consultation, it is the only one with a general law, addressing extractive sectors and general policies. Nonetheless it has no legal provision for prelegislative consultation, and the actual number of consultations is far fewer than in the other countries. In Colombia consultation has no constitutional recognition, but development of jurisprudence and the number of consultations have been more significant than in the other countries. Nevertheless most cases of consultation seem to be a stage in the process of environmental licensing for extractive industries. Ecuador has constitutional recognition of consultation and sectoral regulation in the hydrocarbon sector and is the only country with a specific regulation for prelegislative consultation. But it has not implemented actual consultations on this issue or in other sectors. Bolivia also has constitutional recognition of consultation, but no general law; consultation has been developed basically for hydrocarbon and mining activities.

Regarding governance, in most countries the entities in charge of consultations are specialized sectoral institutions, except in Colombia, where the main responsibilities for consultation are in one specific institution (the Ministry of Internal Affairs). Colombia is also the only country where companies intervene directly in consultations. Although all of the countries use the parameters of ILO Convention 169 for defining indigenous peoples, they differ in the application of consultation to nonindigenous groups. In Peru consultation is a special right of indigenous peoples, whereas in Ecuador, Bolivia, and Colombia consultation extends to other ethnic groups, such as Afro-descendant communities.

These countries have similar procedural rules, although the duration of the process varies from country to country. (Peru has the longest period, with 120 days, whereas the shortest is 30 days in the hydrocarbon sector in Ecuador.) In Peru and Bolivia consultation on hydrocarbon activities refers to the concession contract, while Colombia and Ecuador consult administrative decisions that approve prospecting or exploitation. In Bolivia any request for the signing of a mining contract is consulted; in Peru consultation in mining refers to the authorization of the beginning of activities of exploration or exploitation.

Regarding the number of consultations carried out, there are notable differences among the countries as well as within them. (Most countries focus on the mining and hydrocarbon sectors.) In general, Colombia is the country with the highest number of consultations carried out; between 2011 and 2014, there were 4,042 processes. Peru had only 32 consultation processes concluded as of early 2018; these have not been

TABLE 7. FPIC implementation in Bolivia, Ecuador, Colombia, and Peru

Ratification of Convention 169	Scope of the regulation	Consultation governance	Who is consulted?
Peru			
Legislative Resolution no. 26253 (05/12/93); in force since 02/02/95	Legal and administrative regulation (Law 29785 (07/09/11) Bylaw DS 01–2012-MC (03/04/12) No regulation in the Constitution General scope: applied to all social and economic sectors	In charge of the process: promotor entity Oversight: Directorate of Prior Consultation (Vice Ministry of Intercultural Affairs) Role of companies: no role	Only indigenous peoples; no tribal peoples. Indigenous peoples are organized legally as peasant or native communities. They are consulted through indigenous organizations or directly through communal representatives.
Columbia			
Law no. 21 (07/08/91)	Administrative regulation; no constitutional recognition; not a general law. Jurisprudential decisions have defined standards directly applied by public administration. Limited scope: applied to hydrocarbon, infrastructure, and mining projects.	In charge of the process: Ministry of Interior through the Directorate of Prior Consultation, in coordination with Directorate of Indigenous Affairs, Rom, and Minorities and Directorate of Black Communities, Raizales, and Palenqueros. Oversight: Directorate of Prior Consultation. Role of companies: they must finance and participate in consultations (Presidential Directive 01/2010).	National ethnic groups (indigenous, *negras*, Afro-Colombian, *raizales*, *palenqueras*, and Rom) (Presidential Directive 01/2010). The Directorate of Prior Consultation recognizes the areas where communities are present, as well as the identification of their representatives (Presidential Directive 010/2013).

Legal nature and duration	Object of consultation	Number of consultations
Administrative procedure. Seven stages: (1) identification of the measure, (2) identification of the subjects of consultation, (3) publicity, (4) information, (5) internal evaluation, (6) dialogue, (7) decision. Duration: maximum 120 days (DS 01-2012-MC, art. 24).	(1) Administrative decision that authorizes the start of the activity or project (mining), or authorizes the signing of concession contracts (hydrocarbons); (2) legislative measures, norms with the rank of law, such as laws, legislative resolutions, legislative decrees, regional or municipal ordinances (Regulation, Article 3, j). Consent: displacements, keeping or disposal of dangerous materials on indigenous lands.	32 processes concluded: hydrocarbons (11), mining (10), conservation (5); infrastructure (2); policies and norms (4)
Administrative procedure. Seven stages: (1) preconsultation, (2) opening of the process, (3) workshop for identification of impacts and definition of management measures, (4) pre-agreements, (5) meeting of formalization, (6) systematization and monitoring of agreements, (7) closure of the process (Presidential Directive 01/2010). Duration: approximately, 4 to 6 months.	Legislative and administrative measures that could directly and specifically affect ethnic groups, prospection for exploitation of resources within their territories, decisions related to use of their lands, among others (Presidential Directive 01/2010).	Between 2011 and 2014: 4,042 processes[a]

a. Defensoría del Pueblo del Perú, *Buenas prácticas de las defensorías*, 57.

Ratification of Convention 169	Scope of the regulation	Consultation governance	Who is consulted?
Ecuador			
Executive Decree 1387 (06/05/98)	Constitutional and legal recognition. Jurisprudential decisions have defined standards directly applied by public administration.	In charge of the process: promotor entity: hydrocarbons: Secretariat of Hydrocarbons (Decree 1247). Prelegislative consultation: National Assembly (Prelegislative Consultation Application Instructions, 05/27/12, art. 4). Oversight: Ministry of the Environment and Secretariat of Peoples, Social Movements and Citizen Participation, in their role as coadjutant entities (Decree 1040/2008). Role of companies: Public and private companies do not participate in the process; they are only informed through a Supervisory Committee.	Indigenous communes, communities, peoples and nationalities; Afro-Ecuadorian and Montubio people (Constitution, art. 57).
Bolivia			
Law 1257 (11/06/91)	Constitutional and legal recognition. Only two sectors have regulations on consultations: hydrocarbons (Law 3058) and mining (Law 535).	In charge of the process: hydrocarbons: Ministry of Hydrocarbons and Energy (Supreme Decree 29033) mining: Administrative Jurisdictional Mining Authority (Law 535, art. 40). Oversight: Intercultural Service of Democratic Strengthening of the Supreme Electoral Tribunal (an entity ascribed to the Organo Electoral Plurinacional) (Law 26/2010).	Indigenous nations and first indigenous peasant peoples in whose territory natural resources are exploited (Constitution, art. 352). Hydrocarbons regulation refers to "indigenous peoples, native and peasant communities." Mining regulation refers to "nations and first indigenous peoples, peasants, intercultural communities and Afro-Bolivian people."

Legal nature and duration	Object of consultation	Number of consultations
Administrative procedure. Hydrocarbons consists of five stages: (1) previous activities, (2) public notice by the Secretariat of Hydrocarbons, (3) consultation process, (4) systematization, (5) conclusion. Consultation is no longer than 30 days. Prelegislative consultation has four stages: (1) preparation; (2) public notice; (3) information and development; (4) analysis of results and closure of consultation (Instructive, articles 3, 7–21). Duration: approximately 150 to 180 days.	Administrative decisions that approve plans and programs for prospecting, exploitation, and commercialization of nonrenewable resources (Constitution, art. 57.7) and legislative acts (art. 57.17).	Between 2012 and 2015: 4 consultation processes in the hydrocarbons sector (no information on other sectors and laws were found)
Administrative procedure: hydrocarbons: Four stages: (1) coordination and information, (2) organization and planning, (3) execution of consultation, (4) agreement (SD 29033, arts. 9–13). Duration: 45 days, with possible extension of up to 1 month (SD 29574, art. 2). Mining: seven stages: (1) preparatory phase, (2) notifications and meetings of intercultural dialogue, (3) representation, (4) deliberation and additional meetings, (5) mediation and decision, (6) final resolution, (7) costs. Duration: 4 months (Law 535/2014, arts. 210–216)	In hydrocarbons: decisions aimed at developing hydrocarbon projects, occupying lands, and traditional access by native indigenous peoples and peasant communities. In mining, any request for the signing of mining contracts that directly affects collective rights of indigenous peoples and Afro-Bolivian people (Law 535, art. 207).	Between October 2015 and October 2016, 96 consultations were completed in mining. By 2017 in hydrocarbons, approximately 70 processes of consultation had been concluded.[b]

b. Defensoría del Pueblo del Perú, *Buenas prácticas de las defensorías*, 33.

restricted to extractive or infrastructure activities (they also included policies and norms).

Despite the differences, these four countries share profound similarities (see table 7). All have implemented a weaker standard of FPIC, regulating consultation as a participatory right rather than an expression of self-determination. Indigenous peoples attempt to expand the political dimension of consultation by using it to challenge structural issues related to inequality, environmental impacts, and territorial rights, among others. This situation has become problematic with the aggressive expansion of extractive industries in the region.

In Bolivia, during the negotiations leading to the new Constitution, indigenous peoples accepted state dominium over natural resources in exchange for consultation on extractive industries and participation in the benefits.[55] This trade-off did not work in practice, however, because of pro-extractivist policies and the lack of proper mechanisms for consultation.[56] Indeed despite the repeated assertion in the new Constitution of the need for industrialization to break the country's dependence on extraction (articles 316, 319) and to increase the autonomy of indigenous peoples (articles 1, 2, 289, 290), the state dominates all natural resources (articles 298, 309, 316), allowing access by new transnational capital, particularly foreign public companies from China, Russia, and Venezuela. In practice, the extractive economic model has not been challenged, so indigenous peoples' territories remain under constant threat.[57]

In Ecuador petroleum exploitation generated 40 percent of all export earnings and 33 percent of all tax revenues in 2010.[58] Oil and gas concessions cover vast territories in the Amazon, overlapping protected areas and indigenous lands. Ecuador also expected an investment of US$8 billion in large-scale mining projects by 2018, but indigenous peoples, including the Shuar, Awá, and Saraguro, argue that the granting of concessions does not comply with consultation. Following multiple protests, President Lenín Moreno met with representatives of the Confederation of Indigenous Nationalities of Ecuador. In December 2017 he ordered a halt to the granting of mining and oil concession and an assessment of the conditions of those already operating.[59]

Bolivia and Ecuador are characterized by a type of *progressive neo-extractivism* that uses discourses of nationalization and indigenous rights while implementing redistributive policies, but which in practice affects indigenous territorial rights and environmental sustainability.[60] These contradictions between governmental policies and indigenous politics arise because the emphasis of the latter is the constitution of a true plurinational state, whereas the former focuses on the implementation of a leftist developmental state.[61]

In Peru and Colombia neoliberalism remains in place and indigenous territorial rights are constantly under threat. In Peru by 2013 mining concessions covered 26 million hectares, equivalent to 21.02 percent of the national territory.[62] According to a report from the Munden Project, 96 percent of the concessions granted as of 2014 overlapped spaces inhabited by indigenous and local communities.[63] Peru currently leads the list of Latin American countries exporting zinc, silver, lead, tin, iron, and bismuth, as well as the list of world gold exporters. According to data from the Ministry of Energy and Mines, the country has fifty-one projects in its mining portfolio for the years 2012–18, and extractive industries (mining, gas, and oil) contributed around US$68 billion to gross domestic product in 2016.[64] Much the same is happening with the exploitation of hydrocarbons: 72 percent of the Peruvian Amazon was covered by concessions for the exploration and exploitation of oil and gas as of 2008.[65] For the period 2017–18 Peru had hydrocarbon investment projects totaling US$21 billion.

In Colombia mineral exports totaled US$39 billion between 2000 and 2009; likewise, between 2004 and 2011 the number of mining claims registered in Ingeominas rose from 2,056 to 5,903. All of this has contributed to the boom in the country's primary-export economy, and successive governments have designed policies, plans, and programs aimed at positioning Colombia as a "mining country."[66] In addition, according to the National Hydrocarbons Agency (2017), foreign investment grew by 35 percent between 2016 and 2017 (from US$1.9 billion to approximately US$3 billion). Projections for the current year total between US$4 billion and US$4.5 billion.[67]

Colombia and Peru have not engaged with "post-neoliberal" discourses as Bolivia and Ecuador have, much less implemented plans for indigenizing the state, such as in the plurinational approach. Nonetheless they have deployed discourses of inclusive capitalism and multicultural/pluricultural social relations and have implemented policies designs for these purposes. Ultimately the four countries share the paradox of exalting social inclusion as a major political goal, but as they need revenues to fund social programs and provide social services, they feel compelled to expand extractive activities that often affect the cultural and environmental rights of those communities they are attempting to include. For this reason, indigenous self-determination and the political economy of resource extraction are at the core of the tensions between indigenous communities and state policies. Prior consultation as a legal and political device ultimately channels and reproduces these tensions and conflicts.

Conclusion

The international standard of free, prior, and informed consent—even the weaker standard of prior consultation implemented in Latin America—is an advance in the struggle of indigenous peoples for the historical recognition of their collective rights. Instead of fortifying indigenous self-determination, however, this right has become merely an administrative procedure with formal rules aimed at making policy and investment decisions viable.

Consultation therefore unveils profound tensions around the state's development goals and the environmental and sociocultural concerns of communities. This is the case in the four countries studied, even though they proclaim very different ideological discourses. In those states that could be labeled neoliberal, such as Colombia and Peru, and in post-neoliberal states, such as Ecuador and Bolivia, the policy design similarly limits the scope of consultation and emphasizes exploitation of natural resources as a matter of national interest. This kind of nationalism is located at the very core of the state's political and economic structures, excluding multicultural or even plurinational visions by which indigenous peoples should have the right to make decisions about the management of their livelihoods and territories.

The contentious implementation and application of consultation in these four countries reveal a constant search for justice and power.

Whereas consultation refers to specific projects or legal decisions, indigenous peoples use it as a means to denounce structural injustices related to decades of environmental pollution, displacement, and segregation. Indigenous politics is often so intense that it transcends the design of a consultation and triggers social conflicts. In this sense, consultation also channels demands for more decision-making power. It is particularly important to see how, in these experiences, indigenous organizations criticize the rules of consultation, arguing that they do not sufficiently guarantee indigenous rights, but at the same time, they demand more consultation processes. Because of persistent inequities in justice and power, indigenous peoples use consultation to struggle to expand the limited political space allowed them by the state.

The question, then, is how to transcend a legal and political device that has become a tool for "participation as usual" and a new international label for "inclusive development." More ambitious policy designs are needed to ensure that the state actually considers the concerns and aspirations of indigenous communities and that it recognizes them as proper nations with foundational rights in their territories.

NOTES

I am grateful to Carlos Quispe, research assistant at the Universidad del Pacífico, for their research support in the elaboration of this chapter.

1. Hanna and Vanclay, "Human Rights."
2. Anaya, "International," 14; Gilbert, "Indigenous Rights," 210; Oldham and Frank, "'We the Peoples,'" 7; Tennant, "Indigenous Peoples."
3. International Work Group for Indigenous Affairs, *Indigenous World*; World Bank, *Indigenous*; Economic Commission for Latin American and the Caribbean, *Guaranteeing*.
4. For example, the constitutions of Brazil (1988), Colombia (1991), Paraguay (1992), Peru (1993), Guatemala (1993), Bolivia (1995), Ecuador (1998), and Venezuela (1999). Mexico (1992) and Argentina (1994) made significant constitutional reforms.
5. Horton, "Contesting," 832.
6. Van Cott, *Friendly Liquidation*; Hale, "Neoliberal Multiculturalism," 13; Sieder, *Multiculturalism*.
7. Merino, "Law."
8. The size of the indigenous population and the percentage they represent of the national population vary in these countries: in Peru, it is around 7.6 million people (representing 25 percent of the country); in Colombia, around 1.6 million people (3.4 percent);

in Ecuador, approximately 1.1 million people (7 percent); and in Bolivia, around 6.2 million (approximately 62 percent). See International Work Group for Indigenous Affairs, *Indigenous World*; World Bank, *Indigenous*; Economic Commission for Latin American and the Caribbean, *Guaranteeing*.

9. Owen and Kemp, "'Free,'" 93.
10. Inter-American Court of Human Rights, *Saramaka People v. Surinam*, Decision of the Preliminary Exceptions, Found, Reparations and Costs, November 28, 2007.
11. World Bank, *Indigenous*; International Financial Corporation, "Performance"; Inter-American Development Bank, *Operational Policy*; International Council on Mining and Metals, "Position Statement"; Hanna and Vanclay, "Human Rights"; MacKay, "Draft," 72.
12. Merino, "An Alternative," 274.
13. "Article 1. This Convention applies to":

    (b) peoples in independent countries who are regarded as indigenous on account of *their descent from the populations which inhabited the country*, or a geographical region to which the country belongs, *at the time of conquest or colonisation* or the establishment of present state boundaries and who, irrespective of their legal status, *retain some or all of their own social, economic, cultural and political institutions.*

    . . . *Self-identification* as indigenous or tribal shall be regarded as a *fundamental criterion* for determining the groups to which the provisions of this Convention apply.
14. Tribal peoples are those "whose social, cultural and economic conditions distinguish them from other sections of the national community, and whose status is regulated wholly or partially by their own customs or traditions or by special laws or regulations" (ILO Convention 169, article 1). An example of tribal peoples are Afro-descendants.
15. Martí i Puig, "Emergence," 81. The ILO Convention was ratified by Mexico (1990), Bolivia (1991), Colombia (1991), Paraguay (1993), Costa Rica (1993), Peru (1994), Honduras (1995), Guatemala (1996), Ecuador (1998), Argentina (2000), Dominica (2002), Venezuela (2002), Brazil (2002), Chile (2008), and Nicaragua (2010).
16. Rodríguez-Garavito and Lam, "Addressing," 22.
17. Jackson and Barber, "Recognizing"; Hinojosa et al., "Gas."
18. Schilling-Vacaflor, "Rethinking"; Fontana and Grugel, "Politics"; Leifsen, Sánchez-Vázquez, and Reyes, "Claiming"; Rodríguez-Garavito, *Etnicidad.gov*; Szablowski, "Re-empaquetando."
19. Costanza, "Indigenous," 273; Flemmer and Schilling-Vacaflor, "Unfulfilled Promises," 178; Hernández, "Pueblos," 44; Santamaría, "La consulta," 231.
20. Fontana and Grugel, "Politics."
21. Guzmán-Gallegos, "Between Oil."
22. Tanasescu, "Rights"; Berros, "Constitution."
23. Defensoría del Pueblo del Perú, *Buenas prácticas*, 33.

24. Centro de Estudios Jurídicos e Investigación Social, "Autonomía Indígena."
25. Due Process of Law Foundation, *Right*, 74.
26. Consejo Nacional de Ayllus y Markas del Qullasuyu, "Anteproyecto."
27. International Work Group for Indigenous Affairs, *Indigenous World*.
28. Fontana and Grugel, "The Politics of Indigenous Participation," 256.
29. Confederación de Nacionalidades Indígenas de Ecuador, "Consulta."
30. *Comunas* are urban settlements with colonial legacies where indigenous peoples maintain their own culture and identity. See Andrade, *Las comunas*, 15. The *montubios* are a cultural group located on the coast, organized and self-identified as *montubios*, with their own ancestral characteristics that respond to a complex historical process of adaptation and ethnic, interregional, and sociocultural transformations among Indians, blacks, and whites (Executive Decree 1394/2001).
31. Ministerio de Cultura del Perú, *Memoria*, 18.
32. Defensoría del Pueblo del Perú, *Buenas prácticas*, 72.
33. Defensoría del Pueblo del Perú, *Buenas prácticas*, 73.
34. Centro de Derechos Económicos y Sociales, *La consulta*, 43–44.
35. An oil round is an international call for bids on a group of hydrocarbon blocks. In Ecuador consultation applies to oil rounds, whereas in Peru it applies to each block.
36. Koening, "Community Consent."
37. Altmann, "El Estado Plurinacional."
38. Rodríguez, *De la consulta*.
39. Ministerio de Cultura del Perú, *Memoria*, 18.
40. Law 70/1993 defines the black, mulatto and Afro-Colombian communities as the group of families of Afro-Colombian descent who have their own culture, share a history, traditions, and customs, and conserve awareness of identity that distinguishes them from other ethnic groups. It includes more than 4 million people representing 10.52 percent of the total population. They are located mainly in the municipality of Puerto Tejada, in the department of Valle del Cauca. The Palenqueras and Raizales are considered part of the black or Afro-Colombian population.

    The Rom or Gypsies are an ethnic group that originated in northern India. Decree 2957/10 recognizes that they have their own identity, social organization, and language. There are approximately five thousand Rom people in Colombia (0.012 percent of the total population), mainly in the departments of Atlántico, Bolívar, Valle del Cauca, Norte de Santander, Santander, and Nariño and in the city of Bogotá (Constitutional Court Decision C-359/2013).
41. The indigenous community is the group or group of families of Amerindian descent, who are aware of their identity and share values, traits, uses, or customs of their culture, as well as forms of government, management, social control, or their own regulatory systems that distinguish them from other communities, whether or not they have property titles. The indigenous resguardos are a special sociopolitical and legal institution, made up of one or more indigenous communities that have a property

title, have all or part of their territory, and have an autonomous organization and their own regulatory system. In accordance with articles 63 and 329 of the Colombian Constitution, resguardos are collective properties that are inalienable, imprescriptible, and indefeasible (Constitutional Court, Decision T-387/13).

42. Ministerio de Cultura del Perú, *Memoria*, 18.
43. Due Process of Law Foundation, *Right*, 69.
44. Defensoría del Pueblo del Perú, *Buenas prácticas*, 57.
45. Rodríguez, *De la consulta*, 141.
46. Blanco, "Balance," 16.
47. Due Process of Law Foundation, *Right*, 69–70.
48. Due Process of Law Foundation, *Right*, 58.
49. International Work Group for Indigenous Affairs, *Indigenous World*; Rodríguez, *De la consulta*; Rodríguez-Garavito, *Etnicidad.gov*; Organización Nacional Indígena de Colombia, *El derecho*.
50. Merino, "Law"; Merino, "Politics," 87.
51. By 2016 there were nine requests for constitutional injunctions (*demandas de amparo*) under way. See O'Diana and Vega, *¿Cómo va?*
52. See the website of Base de Datos Oficial de Pueblos Indígenas u Originarios, http://bdpi.cultura.gob.pe/.
53. Reglamento of Law 29785, approved by Supreme Decree no. 001-2012-MC, April 3, 2012. Consent applies only in cases of displacement and keeping of dangerous material in indigenous territories, not in general when state measures might affect a large part of the territory, as ruled by the Inter-American Court of Human Rights in the case *Saramaka v. Suriname* (2007).
54. Due Process of Law Foundation, *Right*, 75; International Work Group for Indigenous Affairs, *Indigenous World*; Flemmer and Schilling-Vacaflor, "Unfulfilled Promises," 179.
55. Valencia and Egido, "Bolivia," 58; Postero, *Indigenous State*.
56. Tockman and Cameron, "Indigenous Autonomy," 57.
57. Merino, "An Alternative."
58. Perreault and Valdivia, "Hydrocarbons."
59. Riofrio, "Ecuador."
60. Gudynas, "Buen Vivir," 446. Laing, "Resource Sovereignties"; Bebbington and Humphreys, "Andean Avatar."
61. Becker, "Correa."
62. Cooperacción, *Noveno informe*; De Echave, "La minería," 71.
63. Fowks, "El 96%."
64. Deustsche Welle, "¿Esperanza?"
65. Finer and Orta-Martinez, "Second Hydrocarbon Boom," 4.
66. Vélez, "Dimensiones," 54.
67. National Hydrocarbons Agency. *Management Report 2017*.

BIBLIOGRAPHY

Altmann, Philip. "El Estado Plurinacional en Ecuador o la esperanza de supervivencia de los pueblos indígenas." Interview with Ileana Almeida. *La línea de fuego*, October 9, 2016.

Anaya, James. "International Human Rights and Indigenous Peoples: The Move toward the Multicultural State." *Arizona Journal of International & Comparative Law* 21, no. 1 (2004): 13–61.

Andrade, Gustavo. *Las comunas ancestrales de Quito: Retos y desafíos en la planificación urbanística*. Quito: Universidad Andina Simón Bolivar, 2016.

Bebbington, Anthony, and Denise Humphreys. "An Andean Avatar: Post-Neoliberal and Neoliberal Strategies for Securing the Unobtainable." *New Political Economy* 16, no. 1 (2011): 131–45.

Becker, Marc. "Correa, Indigenous Movements, and the Writing of a New Constitution in Ecuador." *Latin American Perspectives* 38, no. 1 (2011): 47–62.

Berros, María. "The Constitution of the Republic of Ecuador: Pachamama Has Rights." *Environment & Society Portal, Arcadia*, no. 11 (2015). http://www.environmentandsociety.org/arcadia/constitution-republic-ecuador-pachamama-has-rights.

Blanco, Cristina. "Balance del Perú en el contexto regional: Una mirada comparativa del derecho a la consulta previa con relación a Colombia, Chile y Bolivia." In *La implementación del Derecho a la Consulta Previa en Perú: Aportes para el análisis y la garantía de los derechos colectivos de los pueblos indígenas*, compiled by Vargas Karina. Lima: GIZ, 2016.

Centro de Derechos Económicos y Sociales. *La consulta previa e informada en el Ecuador*. Quito: CDES, 2016.

Centro de Estudios Jurídicos e Investigación Social, "Autonomía Indígena." August 21, 2012. http://cejis.org/autonomia-indigena-originaria/.

Confederación de Nacionalidades Indígenas de Ecuador. "Consulta previa: 17 años en deuda con los pueblos y nacionalidades indígenas del Ecuador." November 26, 2015. https://conaie.org/2015/11/26/consulta-previa-17-anos-en-deuda-con-los-pueblos-y-nacionalidades-indigenas-del-ecuador/.

Consejo Nacional de Ayllus y Markas del Qullasuyu. "Anteproyecto de Ley Marco de Consulta y Consentimiento Previo, Libre e Informado a las Naciones y Pueblos Indígenas Originarios, de cumplimiento obligatorio por el Estado Plurinacional." July 28, 2011. http://www.servindi.org/pdf/PROPUESTA_LEY%20MARCO_CONAMAQ.pdf.

Cooperacción. *Noveno informe cartográfico sobre concesiones mineras en el Perú: Base de datos 2016*. Lima: Cooperacción, 2016.

Costanza, Jennifer. "Indigenous Peoples' Right to Prior Consultation: Transforming Human Rights from the Grassroots in Guatemala." *Journal of Human Rights* 14, no. 2 (2015): 260–85.

De Echave, José. "La minería peruana y los escenarios de transición." In *Transiciones Postextractivismo y alternativas al extractivismo en el Perú*, edited by Alejandra Alayza and Eduardo Gudynas. Lima: CEPES, RedGE, 2011.

Defensoría del Pueblo del Perú. "Adjuntía para la prevención de conflictos sociales y la gobernabilidad." Reporte de Conflictos Sociales, no. 168. 2018.

———. *Buenas prácticas de las defensorías del pueblo de Bolivia, Colombia, Ecuador y Perú en procesos de consulta previa.* Lima: Defensoría del Pueblo del Perú, 2017.

Deustsche Welle. "¿Esperanza de cambio para conflictos mineros en Perú?" January 29, 2016. http://www.dw.com/es/esperanza-de-cambio-para-conflictos-mineros-en-per%c3%BA/a-18976899.

Due Process of Law Foundation. *Right to Free, Prior, and Informed Consultation and Consent in Latin America: Progress and Challenges in Bolivia, Brazil, Chile, Colombia, Guatemala, and Peru.* Washington DC: Oxfam and DPLF, 2015.

Economic Commission for Latin American and the Caribbean. *Guaranteeing Indigenous People's Rights in Latin America: Progress in the past decade and remaining challenges.* Santiago, Chile: ECLAC, 2014.

Federación Internacional de Derechos Humanos. "Bolivia: Informe de verificación de la consulta realizada en el Territorio Indígena Isiboro-Sécure." Report no. 609. 2013.

Finer, Matt, and Martí Orta-Martinez. "A Second Hydrocarbon Boom Threatens the Peruvian Amazon: Trends, Projections, and Policy Implications." *Environmental Research Letters* 5 (2010): 1–10.

Flemmer, Riccarda, and Almut Schilling-Vacaflor. "Unfulfilled Promises of the Consultation Approach: The Limits to Effective Indigenous Participation in Bolivia's and Peru's Extractive Industries." *Third World Quarterly* 37, no. 1 (2016): 172–88.

Fontana, Lorenza, and Jean Grugel. "The Politics of Indigenous Participation through 'Free Prior Informed Consent': Reflections from the Bolivian Case." *World Development* 77 (2016): 249–61.

Fowks, Jacqueline. "El 96% del territorio destinado a la epxlotación en Perú está habitado." *El País, Internacional*, November 5, 2014. https://elpais.com/internacional/2014/11/05/actualidad/1415149053_207294.html.

Gilbert, Jérémie. "Indigenous Rights in the Making: The United Nations Declaration on the Rights of Indigenous Peoples." *International Journal on Minority and Group Rights* 14 (2007): 207–30.

Gudynas, Eduardo. "Buen Vivir: Today's Tomorrow." *Development* 54, no. 4 (2011): 441–47.

Guzmán-Gallegos, María. "Between Oil Contamination and Consultation: Constrained Spaces of Influence in Northern Peruvian Amazonia." *Third World Quarterly* 38, no. 5 (2017): 1110–27.

Hale, Charles. "Neoliberal Multiculturalism: The Remaking of Cultural Rights and Racial Dominance in Central America." *Political and Legal Anthropology Review* 28, no. 1 (2005): 10–28.

Hanna, Philippe, and Frank Vanclay. "Human Rights, Indigenous Peoples and the Concept of Free, Prior and Informed Consent." *Impact Assessment & Project Appraisal* 31, no. 2 (2013): 146–57.

Hernández, Augusto. "Pueblos indígenas y desarrollo: El rol de la consulta previa en los países andinos." *Pensamiento Americano* 7, no. 13 (2014): 39–63.

Hinojosa, Leonith, Anthony Bebbington, Guido Cortez, Juan Chumacero, Denise Bebbington, and Karl Hennermann. "Gas and Development: Rural Territorial Dynamics in Tarija, Bolivia." *World Development* 73 (2015): 105–17.

Horton, Lynn. "Contesting State Multiculturalisms: Indigenous Land Struggles in Eastern Panama." *Journal of Latin American Studies* 38, no. 4 (2006): 829–58.

Inter-American Development Bank. *Operational Policy on Indigenous Peoples and Strategy for Indigenous Development*. Washington DC: IADB, 2006.

International Council on Mining and Metals. "Position Statement on Indigenous Peoples and Mining." May 2013. https://www.icmm.com/website/publications/pdfs/commitments/2013_icmm-ps_indigenous-people.pdf.

International Financial Corporation. "Performance Standard 7: Indigenous Peoples." January 1, 2012. https://www.ifc.org/wps/wcm/connect/c02c2e86-e6cd-4b55-95a2-b3395d204279/IFC_Performance_Standards.pdf?MOD=AJPERES&CVID=kTjHBzk.

International Work Group for Indigenous Affairs. *The Indigenous World 2017*. Copenhagen: IWGIA, 2017.

Jackson, Sue, and M. Barber. "Recognizing Indigenous Water Cultures and Rights in Mine Water Management: The Role of Negotiated Agreements." *Aquatic procedia* 5 (2015): 81–89.

Koening, Kevin. "Community Consent: Business Lessons from the Amazon." *Business & Human Rights Resource Centre, Amazon Watch*, March 2017. http://amazonwatch.org/news/2017/0323-community-consent-business-lessons-from-the-amazon.

Laing, Anna. "Resource Sovereignties in Bolivia: Re-Conceptualising the Relationship between Indigenous Identities and the Environment during the TIPNIS Conflict." *Bulletin of Latin American Research* 34, no. 2 (2015): 149–66.

Leifsen, Esben, Luis Sánchez-Vázquez, and Maleny Reyes. "Claiming Prior Consultation, Monitoring Environmental Impact: Counterwork by the Use of Formal Instruments of Participatory Governance in Ecuador's Emerging Mining Sector." *Third World Quarterly* 38, no. 5 (2017): 1092–109.

MacKay, Fergus. "The Draft World Bank Operational Policy 4.10 on Indigenous Peoples: Progress or More of the Same?" *Arizona Journal of International & Comparative Law* 22, no. 1 (2005): 65–98.

Martí i Puig, Salvador. "The Emergence of Indigenous Movements in Latin America and Their Impact on the Latin American Political Scene: Interpretive Tools at the Local and Global Levels." *Latin American Perspectives* 37, no. 6 (2010): 74–92.

Merino, Roger. "An Alternative to 'Alternative Development'? *Buen vivir* and Human Development in Andean Countries." *Oxford Development Studies* 44, No. 3 (2016): 271–86.

———. "Law and Politics of Indigenous Self-Determination: The Meaning of the Right to Prior Consultation." In *Indigenous Peoples as Subjects of International Law*, edited by Irene Watson. London: Routledge, 2017.

———. "The Politics of Extractive Governance: Indigenous Peoples and Socio-environmental Conflicts in Peru." *Extractives Industries and the Society* 28, no. 1 (2015): 85–92.

Ministerio de Cultura del Perú. *Memoria del Seminario Internacional: Consulta previa en el sector hidrocarburos*. Lima: Mincu, 2014.

National Hydrocarbons Agency. *Management Report 2017*. Bogotá: NHA, 2017.

O'Diana, Richard, and Ismael Vega. *¿Cómo va la aplicación de la Consulta Previa en el Perú? Avances y retos*. Lima: Centro Amazónico de Antropología y Aplicación Práctica, 2016.

Oldham, Paul, and Miriam Frank. "'We the Peoples . . .': The United Nations Declaration on the Rights of Indigenous People." *Anthropology Today* 24, no. 2 (2008): 5–9.

Organización Nacional Indígena de Colombia. *El derecho fundamental a la consulta previa de los pueblos indígenas en Colombia*. Bogotá: ONIC, 2011.

Owen, John, and Deanna Kemp. "'Free Prior and Informed Consent,' Social Complexity and the Mining Industry: Establishing a Knowledge Base." *Resources Policy, The University of Queensland* 41 (2014): 91–100.

Perreault, Tom, and Gabriela Valdivia. "Hydrocarbons, Popular Protest and National Imaginaries: Ecuador and Bolivia in Comparative Context." *Geoforum* 41, no. 5 (2010): 689–99.

Postero, Nancy. *The Indigenous State: Race, Politics, and Performance in Plurinational Bolivia*. Berkeley: University of California Press, 2017.

Quiroga, Catalina. "Hacia un territorio minero-campesino: Propuestas territoriales desde el movimiento socioterritorial en el nordeste de Antioquia, Colombia." In *Extractivismo minero en Colombia y América Latina*, edited by Barbara Göbel and Astrid Ulloa. Bogotá: Universidad Nacional de Colombia & Clacso, 2014.

Riofrio, Isabel. "Ecuador: Lenín Moreno detiene entrega de concesiones mineras y petroleras tras reunirse con líderes indígenas." *Mongabay Latam*, December 15, 2017. https://es.mongabay.com/2017/12/ecuador-lenin-moreno-detiene-entrega-concesiones-mineras-petroleras-tras-reunirse-lideres-indigenas/.

Rodríguez, Gloria A. *De la consulta previa al consentimiento previo, libre e informado a los pueblos indígenas de Colombia*. Bogotá: Universidad del Rosario, GIZ, 2014.

Rodríguez-Garavito, César. *Etnicidad.gov: Los recursos naturales, los pueblos indígenas y el derecho a la consulta previa en los campos sociales minados*. Bogotá: Dejusticia, 2012.

Rodríguez Garavito, César, and Yukyan Lam. "Addressing Violations of Indigenous Peoples' Territory, Land, and Natural Resource Rights during Conflicts and Transitions." In *Strengthening Indigenous Rights through Truth Commissions: A Practitioner's Resource*, edited by International Center for Transitional Justice. New York: ICTJ, 2014.

Santamaría, Alejandro. "La consulta previa desde la perspectiva de la negociación deliberativa." *Revista Derecho del Estado de la Universidad Externado de Colombia*, no. 36 (January–June 2016): 227–47.

Schilling-Vacaflor, Almut. "Rethinking the Link between Consultation and Conflict: Lessons from Bolivia's Gas Sector." *Canadian Journal of Development Studies* 35, no. 4 (2014): 503–21.

———. "Who Controls the Territory and the Resources? Free, Prior and Informed Consent (FPIC) as a Contested Human Rights Practice in Bolivia." *Third World Quarterly* 38, no. 5 (2017): 1058–74.

Sieder, Rachel, ed. *Multiculturalism in Latin America: Indigenous Rights, Diversity, and Democracy*. London: Palgrave Macmillan, 2002.

Szablowski, David. "Re-empaquetando el CLPI: Las conexiones globales y el debate sobre el consentimiento indígena para la extracción industrial de recursos." *Anthopologica* 28, no. 28 (2010): 217–38.

Tanasescu, Mihnea. "The Rights of Nature in Ecuador." In *Environment, Political Representation, and the Challenge of Rights*. London: Palgrave Macmillan, 2016.

Tennant, Chris. "Indigenous Peoples, International Institutions, and the International Legal Literature from 1945–1993." *Human Rights Quarterly* 16, no. 1 (1994): 1–57.

Tockman, Jason, and John Cameron. "Indigenous Autonomy and the Contradictions of Plurinationalism in Bolivia." *Latin American Politics and Society* 56, no. 3 (2014): 46–69.

Valdivia, Gabriela. "Governing Relations between People and Things: Citizenship, Territory, and the Political Economy of Petroleum in Ecuador." *Political Geography* 27, no. 4 (2008): 456–77.

Valencia, María, and Iván Egido. "Bolivia: ¿Estado indio? Reflexiones sobre el Estado Plurinational en el debate constituyente boliviano." *Law and Politics in Africa, Asia and Latin America* 42, no. 1 (2009): 55–72.

Van Cott, Donna. *The Friendly Liquidation of the Past: The Politics of Diversity in Latin America*. Pittsburgh PA: University of Pittsburgh Press, 2000.

Vélez, Irene. "Dimensiones del extractivismo minero en Colombia: Análisis de las racionalidades de gobierno durante la última década." *Análisis Político, Universidad Nacional de Colombia* 27, no. 82 (2014): 45–57.

World Bank. *Indigenous Latin America in the Twenty-First Century: The First Decade*. Washington DC: World Bank, 2015.

9 Indigenous Amazonian Peoples and the Struggle for Environmental Justice in Lowland South America

JONATHAN D. HILL

Across lowland South America indigenous worldviews and cultural practices demonstrate a profound appreciation for the connections between human and nonhuman beings. Natural species and material elements are not relegated to the status of mere things or resources that human beings must confront, control, and consume but are celebrated as a communicative, living being that must be heard and spoken, chanted, sung, danced, and otherwise made into being through the little rituals of everyday life as well as larger collective ceremonial and ritual activities. Indigenous Amazonian peoples place cultural value upon "synchronizing human innovation rates with changes and regenerative cycles in the non-human environment."[1] Indigenous ecocosmologies cannot be reduced to static or unchanging relations between humans and nonhuman environments but must be understood as complex, endlessly changing mythopoetic processes of transforming nonhuman species and materials into fully human persons and communities that are simultaneously differentiated from the nonhuman world of plants, animals, and substances as well as deeply and inextricably connected to these nonhuman environments.

The study of indigenous ecocosmologies can provide a radical critique of the structures and epistemologies of power that fueled colonial and national state expansions in lowland South America and that relied on Western mythic tropes of "natural man" that denied any legitimate concern for the territorial, cultural, and other human rights of indigenous

Amazonian peoples by relegating them to a less than fully human state of nature. To set the stage for a more in-depth analysis of an indigenous ecocosmology based on ethnographic fieldwork with the Arawak-speaking Wakuénai (or Curripaco) of southernmost Venezuela, I begin with an overview of the long-term historical patterns and cultural ideologies underlying the expansion of European empires and Euro-American nation-states across lowland South America. The spread of progressive democracies and new national constitutions across South America in the 1980s and 1990s diverged sharply from the Eurocentric, assimilationist ideologies and policies that had prevailed in earlier periods by giving unprecedented legal recognition to the cultural diversity of indigenous and other minority communities. The new constitutions of the 1980s and 1990s offered legal protection for indigenous lands and cultural rights, and these social advances were coupled with new environmental regulations designed to protect Amazonian rainforests and other sensitive or endangered ecosystems. However, the politically conservative backlash against these newly granted rights has reinscribed older colonial and hypernationalist tropes of indigenous peoples as semi- or subhuman beings whose very existence threatens the sovereignty of the nation-state. Multinational corporations, globalization, and neoliberal economic policies have taken the material and ideological dimensions of structural and actual violence to unprecedented levels that are having irreversibly harmful consequences for the Amazon basin and its indigenous communities. A consideration of hydroelectric dam-building as a socially and environmentally disastrous practice, or a weapon of mass destruction, will serve to illustrate the more general problem of economic development as a mechanism of structural violence.

A detailed analysis of ecocosmology among the Arawak-speaking Wakuénai (Curripaco) of the Venezuelan Amazon will reveal the stark contrast between, on the one hand, indigenous ways of defining human existence as the result of poetic transformations of nonhuman beings and, on the other hand, ecologically destructive projects such as hydroelectric dam-building that are grounded in false objectifications of Amazonian ecosystems and the indigenous peoples who inhabit them. The Wakuénai have a complex cycle of mythic narratives that describes the creation of natural species and objects as the result of the prototypical human being, whose body is said to have consisted of all living species and natural elements and whose powerful voice is said to have opened up the world and musically named into being all the animal and plant species living in the Upper Rio Negro region. These mythic processes of creation are episodically enacted in rites of passage at childbirth and initiation into adulthood through ritually powerful sung, chanted, and spoken discourses (*malikái*). Ritual specialists known as "owners" (*limínali*, keepers or guardians) of malikái are the only individuals who can perform these ritual discourses, which consist of taxonomies of musicalized spirit-names that allow ritual specialists to control movements between stages in the human life cycle, developmental cycles of domestic groups, generations of living humans, spirits of the dead and living humans, and mythic ancestors and their human descendants, both living and dead.[2]

The analysis of Wakuénai ecocosmology would not be complete without an overview of their traditional patterns of settlement and subsistence as well as the ways in which these socioeconomic practices are synchronized with the nonhuman environment of the Upper Rio Negro region through a complex cycle of ceremonial exchanges known as *pudáli*. In addition to celebrating the communicative powers of fish, frogs, and other nonhuman beings, the musical dances of pudáli reintroduce nature into society by creating a naturalized space within human communities that allows people to acknowledge the otherness of the other, including nonhuman species and objects, affines, spirits of the dead, mythic ancestors, and nonindigenous peoples.

The chapter concludes with an exploration of some of the broader implications of indigenous ecocosmologies for righting environmental injustices in lowland South America. The sociocultural and linguistic diversity of indigenous Amazonian communities needs to be appreciated as a tremendous resource for protecting forests, rivers, and grasslands in the interior regions of South America rather than denigrated as an obstacle to economic development or national unity. Indigenous ecocosmologies can provide blueprints for creating truly sustainable human communities that form integral parts of biologically diverse rainforests and other tropical ecosystems. We have reached a point in world history that requires radically different worldviews than the ones based on the false concept of perpetual growth and that replaces such anthropocentric cultural models with ones promoting sustainable human engagement with the plants, animals, and materials making up nonhuman environments.[3]

Historical Roots of Contemporary Identity and Ecopolitics in Lowland South America

After two decades of gaining new legal recognition for their cultural and territorial rights in Brazil, Venezuela, Colombia, and Ecuador, indigenous Amazonian peoples have experienced a steady erosion, or even erasure, of these newly granted legal rights in recent years.[4] This continent-wide deterioration in the relations between South American nation-states and indigenous peoples needs to be understood not only in the broader context of multinational corporations and globalization but also in terms of long-term historical patterns and cultural ideologies that have unfolded over several centuries of colonial and national state expansions in the Amazon basin and adjacent regions. In general terms, these historical processes have resulted in major disruption of indigenous political economies, loss of autonomy, and massive depopulation. Three basic principles that have shaped the long-term history of indigenous Amazonian peoples are (1) biocultural disasters, such as pandemics of Old World diseases; (2) the introduction of firearms, steel tools, and other new technologies; and (3) cultural struggles to impose Christian and/or rationalist concepts of order onto indigenous Amazonian peoples. While each of these three principles has presented tremendous challenges to the long-term survival and persistence of indigenous Amazonian peoples, the combined self-amplifying effects of this threefold historical process are greater than the sum of the three principles taken individually.

The spread of contagious diseases in lowland South America coincided with political and economic changes, such as missionization, enslavement, and ethnic soldiering. However, the massive losses of indigenous population due to Old World diseases must be regarded as a distinct, biosocial principle of change inasmuch as the spread of new diseases preceded the arrival of European populations and followed precontact pathways of alliance and exchange. It is difficult to determine how much of the depopulation in South America resulted from diseases once the colonial projects of missions, slavery, and warfare had become well established. Mission settlements, called *reducciones*, were especially prone to become centers of major epidemics, since they concentrated indigenous populations into reduced areas of heightened interethnic contact. Survivors of such epidemics carried the infections into remote, uncontacted areas, resulting in major losses of population even in areas where no Westerners were present. By the end of the colonial period, indigenous populations of lowland South America declined to only 5 percent of their strengths in 1492 and continued to diminish throughout the nineteenth and early twentieth centuries.[5]

The introduction of steel tools, firearms, and other European trade goods played a major role in the indigenous depopulation of South America by tipping the balance of power in favor of indigenous peoples who had the most direct access to centers of colonial wealth and power and by equipping them with far more lethal technologies for hunting and killing their traditional enemies. Along with the spread of contagious diseases, the most devastating effects of the colonial period were internecine wars aimed at capturing slaves to sell to or barter with Europeans. Although the trade in indigenous American slaves differed in many important ways from the massive African slave trade, both indigenous and Afro-American peoples suffered the same reduction to subhuman, animal-like status in colonial South America. Racist ideologies and practices against indigenous and Afro-American peoples buttressed the founding and expansion of Euro-American states in colonial South America.

The building of colonial American states developed through the reduction of complex, multilingual, culturally diverse patterns of regional alliance and warfare into tribalized, or relatively homogeneous, territorially localized, bounded tribal cultural identities. State expansion and tribalization were interdependent processes, and colonial states flourished to the extent that authorities succeeded in dominating and manipulating these newly created tribalized indigenous peoples.[6] The participation of indigenous peoples in colonial state expansion does not imply their full awareness of or consent to the goals of colonial policymakers. It is important to keep in mind the coercive, unequal quality of social interdependencies between European and indigenous Americans, who had already undergone severe losses of population and autonomy due to diseases, warfare, and cultural domination.

The historical interlocking of colonizing and colonized peoples required the latter to adopt new forms of power based on coercion and military force. These alien forms of power became integral to indigenous social identities in the colonial period. The continued existence of indigenous identities came to be primarily, but not exclusively, conditioned by their geopolitical placements within and between expanding, competing colonial states.

The cultural flows between European and indigenous Americans were not only material and political but also social and meaningful. European historical discourses absorbed many elements of indigenous American mythologies and social practices.[7] And throughout South America the figure of powerful European invaders and their diseases entered into the dynamics of mythic imagination and historical discourses among indigenous peoples.[8] Understanding myth, and culture in general, as dynamic processes of socialization and social reproduction that open up and turn upon history in a multitude of ways has allowed South Americanists to move far beyond the structuralist concept of myth as a "cold," ahistorical structure. Far from working to freeze or deny history, mythic concepts of world-transforming fires and floods form basic elements in an endlessly creative process of shaping specific linguistic and cultural identities in changing social and historical conditions. Throughout colonial South America, such dynamic mythic imagery furnished key ingredients in the formation of historically new social identities in general contexts of domination. Conversely, historical discourses resulting from long-term colonial situations were not simple renderings of past events but complex cultural constructions rooted in the mythic imaginations of specific peoples.

European rationalist theories of the state portrayed indigenous Amazonian peoples as lawless savages living in a state of nature, or a Hobbesian war of all against all, at the same time as colonial and national state expansions redefined their identities as less than fully human beings. Colonial discourses of power tended to cast complex histories into naturalized polarities, such as the civilized, orderly West versus the savage, chaotic Rest.[9]

The rise of independent states in South America failed to create a political space for indigenous peoples to survive and prosper as citizens within the jurisdiction of state sovereignty and as ethic groups who were culturally and linguistically different. From their very inception during the Wars of Independence, the founders of liberal Euroamerican states perceived indigenous peoples as equals only if they could be educated and brought into direct relations with Americans of European descent. The European concept of the nation-state placed indigenous Amazonian peoples into a double bind: on the one hand, the liberal state promised civil and legal equality to all citizens within its borders; on the other hand, the rationalist, assimilationist, and individualist nature of this promised equality was extremely hostile to the older colonial policy of granting collective land rights to indigenous peoples as well as to the persistence of indigenous modes of economic production and social organization that had survived into or emerged during the colonial period. The rise of independent nation-states placed indigenous peoples into contradictory social statuses: full citizens who received equal rights and lawless savages who must be educated and assimilated into the social life of the state.

Throughout the nineteenth and early twentieth centuries, national governments turned to the Amazonian interior as a source of raw materials, such as rubber, oil, minerals, timber, and other forest products, as well as hydroelectric power for industry and land for cattle raising and agriculture. In much the same way as nineteenth-century governments had turned a blind eye to the atrocities committed against indigenous peoples during the rubber boom and embraced cultural ideologies of progress at any cost, South American governments in the late twentieth century expressed faith in development projects as a way of harnessing the natural wealth of the Amazonian interior to national economic growth and relied on assimilationist ideologies that objectified the region as a "breadbasket" for the hungry masses of coastal and highland cities.[10] Development projects often came into direct conflict with indigenous communities, and national governments usually silenced political resistance with brute military force, psychological military operations, and Cold War–style propaganda campaigns using mass media to promote terrifying images of indigenous movements as part of an alleged world communist movement or international terrorism.[11]

As the Cold War drew to a close in the 1980s and 1990s, many South American countries experienced a wave of democratization that came to be enshrined in more inclusive, multicultural (or plurinational) constitutions. In Brazil the Constitution of 1988 became known as the Citizen Constitution and gave legal recognition to indigenous and other minority communities' human rights and cultural diversity. For the first time in its history, Brazil offered full citizenship to its indigenous peoples without the requirement of cultural and linguistic assimilation. The Constitution of 1988 also provided legal protection of the environment and "expressed hope for a regime of justice and democracy."[12] A decade later Venezuelans elected a pro-indigenous socialist president named Hugo Chávez whose administration oversaw a year-long constitutional assembly in 1999 that included three indigenous leaders and that gave strong affirmation to the inclusion of indigenous peoples as parts of the nation-state.[13] Be that as it may, the euphoria generated by the passage of more inclusive constitutions soon ran headlong into the entrenched institutionalized racism that placed indigenous and African-descended peoples at the bottom of racialized social hierarchies.[14] Translation of national legal codes into local political idioms and vice versa proved to be slow-moving and conflict-laden processes that seldom achieved the goal of real legal pluralism.[15] By the time of Dilma Rousseff's election as president of Brazil in January 2011, the Citizen Constitution of 1988 had suffered from "all sorts of distortions: its terms have been violated, as occurs so often with such documents, and, even more seriously, constitutional amendments and decrees have attempted to disfigure it."[16] A centerpiece of the Rousseff administration's new Growth Acceleration Program (Plano de Crescimento Acelerado) was a plan to construct a series of twenty-five mega-hydroelectric dams on the major tributaries of the Amazon, starting at a location called Belo Monte in the lower Xingu River region.

The Erasure of Indigenous Rights in Twenty-First-Century Brazil

In the momentary exhilaration surrounding Rousseff's inauguration on January 1, 2011, many people had great hopes that Brazil might turn away from environmentally and socially destructive forms of development such as the Belo Monte Dam, the world's third largest hydroelectric dam, at a site along the lower Xingu River. In her inaugural speech, Rousseff spoke of continuing the legacy of her predecessor, President Luiz Inácio Lula da Silva, as a rupture from the past, a time when "we Brazilians crossed over to another shore, another time in our history."[17] However, Rousseff was also known as the Iron Lady for her managerial style as Lula da Silva's chief of staff, and as his minister of mines and energy (2003–5) she "had frequent clashes with Environment Minister Marina Silva" at a time when "Brazil was still reeling from electricity shortages, caused by a drought that affected hydroelectric dams."[18]

Rousseff's inaugural address was a masterful interweaving of environmentally friendly language with the kind of instrumentalist "techno-bankster-econo-speak" that has become so prevalent in the early years of the twenty-first century (think of Alan Greenspan's phrase "irrational exuberance"). The Iron Lady's fist concealed brass knuckles beneath layers of velvety soft pronouncements in support of preserving "rich biodiversity" and "valuing cultural diversity." She described Brazil as "the world champions in clean energy, a country that will always know how to grow in a healthy and balanced fashion," then asserted that "ethanol and hydro-energy sources will be greatly encouraged" and concluded that "Brazil will continue to give priority to preserving natural reserves and forests." Rousseff viewed scientific research and technological innovation as "instrument[s] of productivity," ignoring the fact that national and international associations of environmental and social sciences had unanimously condemned the Belo Monte project as an intentional, man-made disaster that would destroy biodiversity as well as sustainable, culturally diverse lifeways whose survival depended on this biodiversity. Her way of valuing cultural diversity reduced cultural value to an economic commodity for sale to the highest bidder on the world market: "We are going to invest in culture, increasing nationwide the production and consumption of our cultural assets and expanding the exportation of our music, cinema and literature, living emblems of our presence in the world."[19]

A myriad of scientific, indigenous, and popular voices spoke out against the government's relentless determination to implement Belo Monte. At the 2011 Conference of the Society for the Anthropology of Lowland South America (SALSA), held in Belém, Pará, Brazil, a number of indigenous Kayapó leaders gave moving speeches, translated into Portuguese by the American anthropologist Terence S. Turner, describing the devastating social and ecological effects that the Belo Monte Dam would have on their communities located on the middle reaches of the Xingu River. By the time of the conference, in late June 2011, the Brazilian government had made it unmistakably clear that it was determined to move ahead with granting a license to the multibillion-dollar corporations to build the enormous hydroelectric dam at Belo Monte. The Inter-American Commission on Human Rights had publicly condemned the Belo Monte project in early June, and the new Rousseff administration had responded by refusing to send its ambassador to the Organization of American States, withdrawing $800,000 of support for the institution, and refusing to take any additional steps to protect indigenous rights. These rights are defined not only in international laws and treaties, such as the UN Declaration on the Rights of Indigenous Peoples and the International Labor Organization Convention 169, but also in the Brazilian Constitution of 1988. Putting an exclamation point on its initial refusal to cooperate with the Organization of American States, the Brazilian government later refused to participate in the October 26, 2011, hearing of the Inter-American Commission on Human Rights that was designed to foster dialogue and resolve potential conflicts.

As president-elect of SALSA and a new member of its executive board, I argued that it was important for our organization to make a public statement of opposition to the building of the hydroelectric dam at Belo Monte in spite of, or perhaps even because of, the government's decision to grant a license allowing the dam's construction to begin. Whether or not the Belo Monte project was a fait accompli, SALSA had a moral obligation to assert its opposition, since our membership includes specialists in the anthropology of lowland South America who are from many different countries of South America and Europe as well as North America and who collectively have broader and deeper knowledge of indigenous Amazonian peoples and their environments than any other scientific association. The board of SALSA composed a detailed letter opposing the Belo Monte hydroelectric project, and the letter was signed by then-President Beth Conklin after translation into Portuguese and sent to President Rousseff in October 2011. The following excerpts provide some of the letter's major points:

> We believe that the project would have disastrous social, cultural, ecological, economic, and political effects on indigenous and other people living along the Xingu River. It would destroy a large segment of tropical rainforest in one of the most important river basins in the world. The increased settlement and deforestation would open the southeastern Amazon to increased degradation from expanding agriculture and ranching, and thereby endanger all of Amazonia. The devastation wreaked by the project is in no way compensated by the returns of the project, which would clearly be insufficient to accomplish the promised rewards. Taking into consideration the relatively short life-expectancy of dams in the Amazon because of silting and acidic erosion of turbine blades, the Belo Monte dam seems likely to prove to be an economic white elephant. In short, SALSA strongly advises the immediate stoppage of all plans toward this project.
>
> Among other reasons for this conclusion are the following highlighted social and environmental problems:
>
> The Xingu River is one of the major tributaries of the Amazon. With its numerous affluents it has created a basin twice the size of the country of Ecuador that remains perhaps the least disturbed and most diverse ecosystem in Brazilian Amazonia. It is unquestionably the most culturally diverse region in the Brazilian Amazon, home to over 24 indigenous peoples of distinct cultures and languages. Over the years, this variegated system of social and cultural groups has evolved a relatively sustainable pattern of coexistence with one another and the even more varied riverine and forest ecosystems of the Xingu basin. All of these systems, however, would be gravely imperiled by the

> federal government's plan to build a series of six giant hydroelectric dams along the Xingu and its largest tributary, the Irirí. . . .
>
> Our Society of anthropologists adds its voice to those of the inhabitants and defenders of the threatened ecosystem, cultures, and social communities of the Xingú Valley in calling for the suspension of the proposed Belo Monte dam and all other hydroelectric dam projects in the Xingú River basin.

The SALSA letter to Rousseff built upon letters and activities initiated by the Brazilian Anthropological Association, a national indigenous movement spearheaded by the Northern Kayapó, and protests organized by Brazilian citizens and NGOs in major urban centers across the country. The Northern Kayapó of Central Brazil, who live in regions along the middle Xingu River and who had led a successful multiethnic resistance to the building of a hydroelectric dam at Altamira on the lower Xingu River in 1989, also played a leading role two decades later in resistance to the building of the hydroelectric dam at Belo Monte. In opposing the project, the Kayapó joined indigenous groups from the Upper Xingu and various other regions of Brazil as members of a national organization of indigenous peoples, Coordenação das Organizações Indígenas da Amazônia Brasileira. The most important and vocal civic organization to oppose the building of the Belo Monte Dam was (and still is) The Xingu Lives Forever (Xingu Vivo para Sempre).[20]

Not only did these collective expressions of concern and opposition fail to stop the Belo Monte Dam from going forward, but the Rousseff administration (2011–16) proceeded to implement its Growth Acceleration Program. This program called for the building of "over 25 large hydroelectric dams in addition to hundreds of small hydroelectric plants in the Brazilian Amazon in the next 30 years . . . with limited or non-existent consultation or participation of affected communities, and at the expense of indigenous and non-indigenous peoples whose livelihoods are tied to Amazonian rivers."[21]

Hydroelectric dams are only one of several types of economic activity that are threatening the future of Amazonian riverine environments and their indigenous inhabitants. Other damaging economic forces include clear-cutting of forested areas for cattle ranching and agribusiness, building of roads and railways through indigenous territories, illegal logging and gold mining, and land speculation along with neoliberal "transformations in the organization of capital accumulation that made processes of production and consumption more flexible and fluid."[22] Nevertheless hydroelectric dams must be singled out as being especially harmful to Amazonian environments since they destroy the ecology of aquatic, riverine life-forms and their ability to gain seasonal access to flooded forests and grasslands during annual wet seasons. Whereas indigenous economic and ritual practices are based on the annual regeneration of nonhuman environments through seasonal flooding and draining, hydroelectric dams create a permanent flood that contributes to global climate change by emitting large quantities of methane gas from the rotting of submerged vegetation.[23] Hydroelectric dams in tropical rivers and forests are weapons of mass destruction, devastating both human communities and nonhuman environments. If Amazonian rainforests act as the lungs of the world, then the rivers that annually flood and drain those forests must be regarded as the arteries and veins that nurture and rejuvenate the atmosphere's ability to breathe. Hydroelectric dams destroy this natural circulatory system and replace sustainable human communities with a form of energy production that is neither clean nor sustainable.

The current threats to indigenous Amazonian peoples and their nonhuman environments follow the historical pattern of earlier predatory modes of political and economic expansions during the colonial and early nationalist periods. Western tropes of Amazonian peoples as less than fully human beings continue to furnish ideological weapons for rationalizing contemporary violations of human rights and environmentally destructive projects of economic growth. "Given the fact that cultural and biodiversity conservation go hand in hand, and that indigenous lands often form biocultural corridors, there is a perceived risk of accelerated 'ecocide' of the Amazon along with the risk of cultural genocide brought by development and accelerated growth."[24] It is not clear what if anything will halt or slow down the massively destructive social and ecological effects of Brazil's Accelerated Growth Program.[25] However, it is important to keep in mind the adaptability and resilience of indigenous Amazonian peoples, who are not merely passive victims of progress but descendants of many generations who persisted despite several centuries of depopulation and other traumatic forms of sociocultural loss. Today's indigenous peoples living in lowland South America are speakers of approximately four hundred distinct languages and are continuing to reconstitute multilingual areal networks of trade and intermarriage as they navigate the complex and often treacherous political changes unfolding in the globalizing nation-states of twenty-first-century South America.[26]

Indigenous Amazonian Ecocosmologies

Indigenous Amazonian ecocosmologies are grounded in the idea of a primordial mythic time when there was little or no distinction between humans and nonhumans, male and female humans, animate life-forms and material elements.[27] Neither space nor time had been differentiated yet, and the world consisted of a single place that was neither here nor there but both at once. Time had not yet been divided into days and nights, wet and dry seasons, youth and adulthood. From this undifferentiated mythic primordium, specifically male and female persons as well as fully human forms of social order emerged through a series of poetic transformations. The pathways leading from primordial mythic space-time to fully differentiated human persons, families, and communities are not linear and unidirectional but can be traveled in a multitude of ways and directions through ritual and ceremonial performances in what amounts to an endlessly creative process of shaping and reshaping primordial formlessness into culturally and linguistically specific patterns of form and meaning.

The result of these poetic transformations is an ongoing construction of human social worlds that are at one and the same time clearly differentiated from nonhuman worlds of plants, animals, and material things but that are also deeply and inextricably connected to these nonhuman species and materials. Natural environments are therefore understood as living entities that are completely intertwined with human society and history rather than merely soulless exterior things that can be reduced to resources for human consumption. Indigenous Amazonian ecocosmologies are not anthropocentric, or human-centered, so much as they revolve around the relations of mutual interdependency and necessity between human beings and nonhuman environments. The focus on human to nonhuman interspecies relations and the importance of material elements and celestial phenomena in Amazonian ecocosmologies bears a strong resemblance to the ecosystem concept that developed in Western sciences beginning in the 1950s and that emphasizes the importance of interspecies relations among populations within organic communities as well as the causal pathways of abiotic matter and energy—water, sunlight, nutrients—necessary for the survival and reproduction of organic life-forms.[28] Like indigenous ecocosmologies, the ecosystem concept challenges rational-empiricist traditions of Western science to move beyond or away from an underlying metaphor of humans as detached, distant observers to a more holistic appreciation of humans as only one part of complex systems of mutual interdependencies and energy exchanges that are regulated by informationally dense cultural processes.[29]

The ecocosmology of the Arawak-speaking Wakuénai of southernmost Venezuela provides a striking example of how an indigenous Amazonian people has developed a complex cycle of mythic narratives that explains how human and nonhuman beings are integrally connected as well as how abiotic elements of the world gave rise to organic life-forms through the poetic mediation of musicalized speech.[30] Immediately after birth, the prototypic human being (Kuwái) is taken away from his mother (Ámaru) by his father, who is a powerful Trickster-Creator (Iñapirríkuli, "Made-from-Bone"). After a period of confinement in a celestial place known as "the corner of the sky," Kuwái breaks free and flies above the forests and rivers, humming and singing into being the various animal and plant species.[31] In the cycle of mythic narratives, the musical naming power of Kuwái is described as a life-giving, world-making process that combined musical sounds and verbal forms into word-sounds that came out of different parts of the body of Kuwái and that made a powerful sound that opened up the miniature world of the mythic primordium into an expanding social world filled with various kinds of nonhuman beings. "He [Kuwái] began to 'speak' the word-sounds that could be heard in the entire world. The world was still very small. He began to speak, 'Heee.' The sound of his voice ran away and opened up the world."[32]

Before this musical and verbal creation of nonhuman species could give rise to fully human socialized persons and groups, this musical naming power had to be controlled by internalizing it into human consciousness in the form of ritually powerful sung and chanted discourses (*malikái*). The process of bringing the musical naming power of Kuwái down to the ground where it could be transmitted to proto-human beings unfolded through a series of movements between celestial and terrestrial worlds. In the first instance of contact with people on the ground, Kuwái meets three nephews of the Trickster-Creator, but their relations are not yet mediated, and two of the three boys are killed and eaten by Kuwái. The third boy goes on to become the first male person to be initiated into adulthood after Kuwái has taught the sacred songs, chants, and prayers known as malikái to Dzúli, the first chant-owner and younger brother of the Trickster-Creator. By the end of this first part of the cycle of mythic narratives, the naming power of Kuwái has created a world of socialized human beings defined as poetic transformations of nonhuman animals and plants. In this proto-human social world, time has become differentiated into generations of adults, children, and grandchildren and into clearly demarcated stages of the human life cycle. The social world has itself become divided between, on the one hand, ritual specialists whose performances of malikái give them control over life-cycle transitions as well as human-to-nonhuman transformations and, on the other hand, nonspecialists who do not have access to such ritual powers.

A key moment in this first part of the cycle of narratives arises when Kuwái reveals in a soliloquy that his body is all things—water, wood, vines, stones, poison, soil, metal tools—in one being. The one thing his body is not is fire, and it is with fire that the Trickster-Creator and his family can "kill," or transform, Kuwái along with the world he has created through musical naming power. This episode occurs during the transition between the earlier episode in which the unmediated contact between Kuwái and the three nephews of the Trickster-Creator results in two of the boys' deaths and the later episodes in which Kuwái teaches the Trickster-Creator's younger brother, Dzúli, how to perform malikái songs and chants. The soliloquy is addressed to Kaalimátu, a wasp-person messenger who was sent by the Trickster-Creator to lure Kuwái down to the ground: "'Tell him exactly this way: Logs and trees, they are my body; vines, too; my body is poison, my hair; water, I live in water; soil, it is my body; steel, my body; the only thing my body is not is fire,' he said. 'With this, yes, he can kill me. That is how you will speak to him,' said Kuwai."[33]

After hearing this speech, Kaalimátu flies back to earth and tells the Trickster-Creator what he has learned. At the end of a night of drinking and dancing during which Dzúli has memorized all the malikái songs and chants for male initiation rituals, the Trickster and his brothers push Kuwái into a bonfire, and the world shrinks back to its original miniature size. Soon, however, enormous trees and vines sprout from the ashes where the body of Kuwái had burned, and these plant materials are in turn used to make the sacred flutes and trumpets of Kuwái. Groups of men and women compete for control over these musical instruments, and the world opens up for a second time as instruments' sounds are heard in various parts of the world.

The cycle of mythic narratives about Kuwái lends itself to a multitude of interpretations and serves as a metapragmatic device for explaining the hierarchical relations between the secretive, monological, opaque, musicalized, and ritually powerful discourses known as malikái and the more transparent, prosaic, and dialogical discourses of everyday social interaction. When interpreted as a key component of Wakuénai ecocosmology, the mythic narratives explain that the life-giving, world-opening musical naming power that created organic life-forms was itself the product of a poetic transformation of water and other substances. In the soliloquy quoted earlier, Kuwái reveals that his body is made of everything in the world—water, trees, logs, vines, steel tools, poison—and asserts that none of these material things or substances can be used to kill him. Fire is the only thing that his body is *not* made of, and with fire he *can* be killed. Thus an initial statement of individual being-in-the-world ("I am," or "My body is") is poetically transformed into its opposite, or negation ("My body is *not*"). But this same transformation of individual being into nothingness is also the basis for a poetic transformation of relative powerlessness ("He *cannot* kill me") into the collective agency ("He *can* kill me") of groups of men who have internalized the musical naming power of Kuwái and can use it to kill, or transform, individual beings into an expanding world of peoples, rivers, and forests by making and playing sacred musical instruments in different places.

The proto-human being of mythic narratives holds the key to understanding Wakuénai ecocosmology as a process of poetically transforming nonhuman beings—animals and plants, but also water and steel tools—into socialized individual persons and social groups who define themselves as the living and deceased descendants of mythic ancestors. Another highly significant dimension of Wakuénai ecocosmology that originated in mythic Beginning times and that continues to play an important role in the way people from different communities interact with one another is the more secular genre of collective singing, instrumental music-making, dancing, drinking, and feasting that was traditionally performed in exchange ceremonies known as pudáli. By the time of my first fieldwork with the Wakuénai in 1980 and 1981, the cycle of pudáli ceremonies was no longer being practiced on a regular basis due mainly to the influence of Protestant missionaries who saw these events as a setting for consumption of alcoholic beverages. Nevertheless the music and dances associated with pudáli were still very important as ways of giving public expression to specifically Wakuénai social identities, especially in situations of indigenous resistance to exploitative economic practices of nonindigenous merchants in local towns.[34] Also, in some villages people performed the music and dances of pudáli for entertainment and to socialize with visitors from nearby villages. In one such village I carried out an exhaustive study of the ceremonial music and dances of pudáli, learned how to perform and compose improvisatory flute duets, recorded a wide variety of vocal and instrumental musical pieces, and worked with senior men and women who remembered pudáli exchange ceremonies from their young adult years prior to the New Tribes Mission's campaign to abolish these events.

Each year at the beginning of the long April-to-July wet season, when the river levels rise over seven meters, massive schools of Leporinus fish migrate up the Rio Negro to the headwaters, where they enter into newly flooded forests to spawn before returning to the river's main channel. Pudáli exchange ceremonies were celebrations of the moment when people blocked off the mouths of streams to capture large quantities of the fish.[35] In an opening pudáli ceremony, a male "owner" (*pudalímnali*) led groups of male flute and trumpet players together with their female dancing partners in collective performances that celebrated the brief period of superabundance before the scarcity of fish during the long wet season by carrying a large gift of smoked fish into the village of their affinal relatives. As the guests arrived, they danced in circles around the baskets full of smoked fish, and each dance formation was named after a different species of Leporinus fish. The same verb (*-irrápaka*) was used to describe the movements of spawning, migrating Leporinus fish and the dancing of men and women in pudáli. The close ties between migrating, spawning fish and human ceremonial dancers were also evoked through the deep, rumbling bass sounds made with trumpets, called *kulirrína*, named after a species of large catfish (*kulírri*) with thick black stripes running down the sides of their bodies. The sound of these catfish trumpets replicated the sound of streams and rivers filled to the brim with massive schools of Leporinus fish as they migrated into flooded forests. Pairs of men playing *máwi* (or *yapurutú*, Yeral) flutes added a shrill, repetitive melody composed by alternating notes on a longer "male" flute and a shorter "female" flute.[36]

During these opening performances, men and women from the host village watched from a respectful distance. The male owner of the ceremony and the hosts' headman made formal speeches offering and accepting the gift of smoked fish, which was then put away in the host's house during a night when men and women drank manioc beer and other fermented beverages and sang personal songs (*pakamarántakan*). The singing and drinking went on all night long, allowing hosts and guests to interact in a shared, musicalized social space that began to break down the initial distance between members of the two groups. Just before dawn, a series of songs and kulirrína trumpet-playing around a bonfire signaled the beginning of a day of feasting, drinking, and dancing when young men and women from the two groups were allowed to dance together and young men could demonstrate their musical skills and creativity by composing and naming new improvisatory flute duets.[37] At the end of the day of feasting, guest men presented their hosts with the kulirrína trumpets, completing the act of generosity between the two communities but also reminding their hosts of the obligatory reciprocal gift of processed manioc pulp that was expected to be given in a ceremony that would be held several weeks later under the leadership of a female owner (*pudalímnarru*).

The cycle of male- and female-owned pudáli ceremonies gave clear expression to the complementarity of men's labor in hunting and fishing and women's labor in harvesting and processing bitter manioc into a pulp that can be cooked into flour and breads. Pudáli ceremonial cycles served as a social blueprint for constructing relations of respect and balanced reciprocity between communities of kin and affines. Music, dance, drinking, and feasting provided ways of materially and symbolically creating a direct connection between the reproduction of human social worlds through exchange and intermarriage and the natural processes through which fish and aquatic animal species regenerate themselves through seasonal migration, spawning, and foraging in flooded forests.[38] Pudáli ceremonial cycles can be understood as a symbolically overdetermined process of reintroducing nature into society, or the creation of a naturalized social space in which musical sounds are used to create a social space that acknowledges the otherness of the other through public acts of respectful, balanced reciprocity.

Concluding Thoughts

Righting the history of environmental injustice and inequity in lowland South America will require profound sociocultural changes on the part of contemporary nation-states and the global community of nations. Older stereotypes of indigenous Amazonian peoples as less than fully human beings who are frozen in time as lawless or noble savages living in states of anarchic or innocent nature must be replaced by understandings of their cultural practices as ongoing processes of poetically transforming selected species and substances of the nonhuman environment into fully human beings and social relations. Moreover these dynamic ecocosmologies have provided indigenous Amazonian peoples with cultural and linguistic tools for resiliently transforming their social worlds across five centuries of colonial and national state expansions that have unleashed pandemics of Old World diseases, military conquest, assimilationist ideologies, forced labor campaigns, and in some cases even genocide.

With the increasing evidence that global climate change is happening at an accelerated pace, it is time for nation-states to let go of programs for accelerated growth that are based on an economic model of perpetual growth and replace them with plans for constructing political economies based on sustainable communities. Indigenous ecocosmologies and small-scale, technologically simple economies cannot be amplified or extrapolated directly to the large-scale economies of densely populated nation-states with large urban centers. However, these indigenous worldviews can provide contemporary humanity with general principles or guidelines for developing policies that steer away from perpetual growth and toward sustainable communities.

Indigenous ecocosmologies start from an understanding of human beings as integral parts of a multispecies web of living beings, a network of organic life-forms that are interdependent with nonorganic elements like water and sunlight as well as natural phenomena such as the seasonal rising and falling of river levels and weather. The cultural separateness of humanity is never absolute but always defined in relation to or constructed out of the mythic primordium of animal-humans. Although such indigenous ecocosmologies clearly depart from an objective, scientific worldview by attributing communicative and other agentive, subjective powers to nonhuman species and objects, there are strong parallels between ecocosmologies and the ecosystem concept developed in Western science. The economic practices supported by indigenous ecocosmologies are survival-oriented rather than yield- or profit-maximizing. Small-scale fishing and horticultural economies like that of the Wakuénai are capable of persisting for as long as the rivers continue to flood their banks and forests are allowed to replenish soil nutrients and nourish aquatic life-forms. In theory this kind of subsistence economy can last as long as the solar system.

National governments, NGOs, and international agencies need to formulate policies and implement programs that allow indigenous communities to participate in commercial economies in ways that are ecologically sustainable, such as agroforestry, and that do not damage the delicate interdependencies between aquatic life-forms and areas of seasonally flooded forests and grasslands. Indigenous peoples who live in remotely situated forests must be allowed to continue living in voluntary isolation as part of their universal right to self-determination. The relentless assimilationist policies of the past, which relied on missionaries and other culture-bound agents as the state's representatives to indigenous peoples, need to be replaced with educational programs that support indigenous ecocosmologies and languages at the same time as they introduce knowledge and skills from Western sciences and humanities that will provide tools necessary for navigating the contemporary world. National systems of formal education must valorize the sociocultural and linguistic diversity of indigenous communities in lowland South America as a positive force in the struggle to protect Amazonian rivers and forests rather than ignore or even denigrate them as obstacles to economic growth. Above all, nationalist histories rooted in centuries-old, racist narratives of the West taming the Rest need to be replaced with accurate understandings of the tremendous losses that indigenous peoples have endured during the expansion of colonial and national states as well as the cultural creativities that have allowed indigenous peoples, their worldviews, and their languages to persist and flourish in the rapidly changing global order of the twenty-first century.

NOTES

1. Brabec de Mori, "What Makes," 46.
2. Musicalized spirit-naming also embodies the ritual power to control transformations between humans and spirits of nonhuman beings and between indigenous and non-indigenous human beings.
3. For a discussion of perpetual growth, see Bodley, *Anthropology*, 117.
4. Hill, "Welcome"; Carneiro da Cunha, "Indigenous Peoples."
5. Denevan, *Native Population*.
6. Whitehead, "Tribes."
7. Certeau, *L'Ecriture*; Whitehead, "Historical Anthropology."
8. Hill, *Rethinking*.
9. Hill, "Alienated Targets"; Karakasidou, "Sacred Scholars"; Kiberd, "Elephant"; Whitten, "Ecuadorian Levantamiento."
10. In Venezuela, for example, the administration of Rafael Herrera in the 1960s and 1970s launched CODESUR, an acronym that originally stood for La Comisión para el Desarollo del Sur (The Commission for Development of the South), referring to the region south of the Orinoco River, or the Amazon and Guiana territories in southern and southeastern

regions of the country. By the time I began conducting fieldwork in the Venezuelan Amazon in 1980, however, CODESUR was universally understood to mean La Conquista del Sur (The Conquest of the South). The semantic shift from "development" to "conquest" is significant, since it points to an underlying conceptual link between economic projects and the overpowering of enemy forces by superior military force.

11. Brown and Fernandez, *War*; Hill, "Alienated Targets"; Santos-Granero and Barclay, "Bundles."
12. Carneiro da Cunha, "Indigenous Peoples," 1.
13. "Los pueblos indígenas, como culturas de raíces ancestrales, forman parte de la nación, del estado y del pueblo venezolano como único, soberano y indivisible" (Constitution of the Bolivarian Republic of Venezuela, article 126). That this assertion needed to be made as part of the drafting of the section titled "Duties, Human Rights, and Guarantees" speaks to the fact that indigenous peoples had been radically excluded from the Venezuelan nation-state since its founding in 1830.
14. Whitten, "Long Durée."
15. Gow and Rappaport, "Indigenous Public Voice."
16. Carneiro da Cunha, "Indigenous Peoples," 1.
17. "Dilma Rousseff Inauguration."
18. "Dilma Rousseff Inauguration"; "Dilma Rousseff."
19. "Dilma Rousseff Inauguration."
20. Films on the Belo Monte dam are available at http://www.belomonteofilme.org/portal/br. A comprehensive list of websites for Brazilian, pan–South American, and international organizations that formed part of the resistance to the Belo Monte hydroelectric dam is available on the Public Issues and Action Committee (PIAC) page of the Society for the Anthropology of Lowland South America website, http://www.salsa-tipiti.org/. The list also includes links to numerous pro-indigenous and environmental activist organizations in Chile, Peru, Ecuador, Paraguay, Venezuela, Colombia, Guyana, Suriname, and French Guiana.
21. Athayde, "Introduction," 81–82. A special issue of *Tipití: The Journal of the Society for the Anthropology of Lowland South America* containing eight research articles and the guest editor's introductory essay appeared in 2014 (vol. 12, no. 2).
22. Glick Schiller, "Old Baggage," 413.
23. Deemer et al., "Greenhouse Gas Emissions."
24. Athayde, "Introduction," 83.
25. At the time of this writing, the Brazilian government has unofficially announced that it will end the policy of building Amazon megadams and attributed this decision to resistance and opposition from environmentalists and indigenous activists as well as to the privatization of the Brazilian Development Bank by President Temer, corruption surrounding construction industry (e.g., the Lava Jatos or "Car Wash" scandal), and general economic decline in the prior two to three years (Branford, "Brazil Announces").

26. Hill and Rodriguez, "Languages."
27. Sullivan, *Icanchu's Drum.*
28. Odum, *Fundamentals*; Moran, *Ecosystem Concept.*
29. The comparison between indigenous Amazonian cosmologies and the Western scientific concept of ecosystems is not new; it was made by at least two anthropologists in the 1970s. See Reichel-Dolmatoff, "Cosmology," and Whitten, "Ecological Imagery."
30. The Wakuénai are also known as Curripaco in Venezuela and Colombia and as Baniwa-Coripaco in Brazil. They inhabit the Guainía (Upper Negro) and Isana rivers and their tributaries in Venezuela, Colombia, and Brazil. I conducted fieldwork with the Wakuénai-Curripaco of Venezuela in 1980–81, 1984, 1985, 1993, and 1998.
31. For a complete translation of these mythic narratives, see Hill, *Made-from-Bone*. An earlier interpretation of these narratives and associated ritual discourses is available in Hill, *Keepers*.
32. Hill, *Made-from-Bone*, 119.
33. Hill, *Made-from-Bone*, 124.
34. Hill, "Musicalizing"; Hill, "'Músicalisando' o Outro."
35. Pudáli could also be held in other seasons whenever there was a large surplus of fish or game meat available.
36. This style of playing, having two or more instruments play notes in a sequence, is called *hocket* and is widespread throughout lowland South America as well as in the Andean highlands (Hill and Chaumeil, *Burst of Breath*). The opening duet of pudáli ceremonies is standardized and very recognizable. Nowadays this duet is almost a musical logo for the Wakuénai social identities.
37. Elsewhere I have written in detail about one of the more well-known of these improvisatory flute duets (Hill, "Soundscaping"). The duet and dance named after a species of fish, called *dzawírra*, adds to the metaphorical comparison between human dancing and fish-spawning behaviors by enacting the movements of a male dzawírra asking the female for permission to dance (*-irrápaka*) in the nest where she is protecting her eggs.
38. Access to flooded forests is especially urgent in blackwater ecosystems like the Rio Negro because there is no deposition of alluvial silts and other nutrients into the rivers (Goulding, *Fishes*). In the long wet seasons, approximately 65 percent of the forests in the region are flooded, and it is in these semiterrestrial habitats that fish and aquatic animal species are able to fatten up.

BIBLIOGRAPHY

Athayde, Simone. "Introduction: Indigenous Peoples, Dams and Resistance." *Tipití: Journal of the Society for the Anthropology of Lowland South America* 12, no. 2 (2014): 80–92. http://digitalcommons.trinity.edu/tipiti/vol12/iss2/1.

Bodley, John. *Anthropology and Contemporary Human Problems.* 5th edition. Lanham MD: Altamira Press, 2008.

Brabec de Mori, Bernd. 2016. "What Makes Natives Unique? Overview of Knowledge Systems among the World's Indigenous Peoples." Academia, January 12, 2018. https://www.academia.edu/32100605/What_Makes_Natives_Unique_Overview_of_Knowledge_Systems_among_the_World_s_Indigenous_People_2016_.

Branford, Sue. 2018. "Brazil Announces End to Amazon Mega-dam Building Policy." *Mongabay*, January 3, 2018. https://news.mongabay.com/2018/01/brazil-announces-end-to-amazon-mega-dam-building-policy/.

Brown, Michael, and Eduardo Fernandez. *War of Shadows*. Berkeley: University of California Press, 1991.

Carneiro da Cunha, Manuela. "Indigenous Peoples Boxed In by Brazil's Political Crisis." *Hau: Journal of Ethnographic Theory* 7, no. 2 (2017). https://www.haujournal.org/index.php/hau/article/view/hau7.2.033.

Certeau, Michel de. *L'Ecriture de l'Histoire*. Paris: Gallimard, 1975.

Deemer, Bridget, John A. Harrison, Siyue Li, Jake J. Beaulieu, Tonya DelSontro, Nathan Barros, José F. Bezerra-Neto, Stephen M. Powers, Marco A. Dos Santos, and J. Arie Vonk. "Greenhouse Gas Emissions from Reservoir Water Surfaces: A New Global Synthesis." *BioScience* 66 (2016): 949–64.

Denevan, William. *The Native Population of the Americas in 1492*. Madison: University of Wisconsin Press, 1976.

"Dilma Rousseff: Brazil's First Female President." *CBC News*, October 31, 2010. http://www.cbc.ca/news/world/story/2010/10/29/f-brazil-dilma-rousseff.html.

"Dilma Rousseff Inauguration: Brazil's First Female President Sworn In." *Huffington Post*, January 1, 2011.

Glick Schiller, Nina. "Old Baggage and Missing Luggage: A Commentary on Beck and Sznaider's 'Unpacking Cosmopolitanism for the Social Sciences: A Research Agenda.'" *British Journal of Sociology* 61, no. s1 (2010): 413–20.

Goulding, M. *The Fishes and the Forest*. Berkeley: University of California Press, 1981.

Gow, David, and Joanne Rappaport. "The Indigenous Public Voice: The Multiple Idioms of Modernity in Native Cauca." In *Indigenous Movements, Self-Representation, and the State in Latin America*, edited by Kay Warren and Jean Jackson. Austin: University of Texas Press, 2002.

Hill, Jonathan D. "Alienated Targets: Military Discourses and the Disempowerment of Indigenous Amazonian Peoples in Venezuela." *Identities: Global Studies in Culture and Power* 1, no. 1 (1994): 7–34.

———. *Keepers of the Sacred Chants: The Poetics of Ritual Power in an Amazonian Society*. Tucson: University of Arizona Press, 1993.

———. *Made-from-Bone: Trickster Myths, Music, and History in an Amazonian Community*. Urbana: University of Illinois Press, 2009.

———. "'Músicalisando' o Outro: Ironia Ritual e Resistencia Etnica entre os Wakuénai (Venezuela)." In *Imagens do Branco na Historia Indigena*, edited by Bruce Albert and Alcida Ramos. Saõ Paulo: Editora UNESP, 2002.

———. "Musicalizing the Other: Shamanistic Approaches to Ethnic-Class Competition in the Upper Rio Negro Region." In *Religiosidad y Resistencia Indígenas hacia el Fin del Milenio*, edited by Alicia Barabas. Quito: Abya-Yala, 1994.

———, ed. *Rethinking History and Myth: Indigenous South American Perspectives on the Past*. Urbana: University of Illinois Press, 1988.

———. "Ritual Production of Environmental History among the Arawakan Wakuénai of Venezuela." *Human Ecology* 17, no. 1 (1989.): 1–25.

———. "Soundscaping the World: The Cultural Poetics of Power and Meaning in Wakuénai Flute Music." In *Burst of Breath: Indigenous Ritual Wind Instruments in Lowland South America*, edited by Jonathan Hill and Jean-Pierre Chaumeil. Lincoln: University of Nebraska Press, 2011.

———. "Welcome (and Farewell) of Tears: The Erasure of Indigenous Rights in 21st Century Brazil." *Anthropology News, Media Notes*, May 2012.

Hill, Jonathan D., and Jean-Pierre Chaumeil, eds. *Burst of Breath: Indigenous Ritual Wind Instruments in Lowland South America*. Lincoln: University of Nebraska Press, 2011.

Hill, Jonathan D., and Juan L. Rodriguez. "Languages in Change: Discourse, Power, and History in Lowland South America." In *The Ashgate Research Companion to Anthropology*, edited by Pamela Stewart and Andrew Strathern. Farnham, U.K.: Ashgate, 2015.

Karakasidou, Anastasia. "Sacred Scholars, Profane Advocates: Intellectuals Molding National Consciousness in Greece." *Identities: Global Studies in Culture and Power* 1, no. 1 (1994): 35–62.

Kiberd, Declan. "The Elephant of Revolutionary Forgetfulness." In *Revising the Rising*, edited by M. Dhonnchadha and T. Dorgan. Derry, Ireland: Field Day Press, 1991.

Moran, Emilio F., ed. *The Ecosystem Concept in Anthropology*. Boulder CO: Westview Press, 1984.

Odum, Eugene. *Fundamentals of Ecology*. Philadelphia: Saunders, 1953.

Reichel-Dolmatoff. "Cosmology as Ecosystem Analysis: A View from the Rain Forest." *Man*, new series, 11, no. 3 (September, 1976): 307–18.

Santos-Granero, Fernando, and Frederica Barclay. "Bundles, Stampers, and Flying Gringos: Native Perceptions of Capitalist Violence in Peruvian Amazonia." *Journal of Latin American and Caribbean Anthropology* 16, no. 1 (2011): 143–67.

Sullivan, Lawrence. *Icanchu's Drum: An Orientation to Meaning in South American Religions*. New York: Macmillan, 1988.

Whitehead, Neil L. "The Historical Anthropology of Text: The Interpretation of Raleigh's Discoverie of Guiana." *Current Anthropology* 36, no. 1 (1995): 53–74.

———. "Tribes Make States and States Make Tribes: Warfare and the Creation of Colonial Tribe and State in Northeastern South America." In *War in the Tribal Zone: Expanding States and Indigenous Warfare*, edited by R. Brian Ferguson and Neil L. Whitehead. Santa Fe NM: SAR Press, /University of Washington Press, 1992.

Whitten, Norman E., Jr. "Ecological Imagery and Cultural Adaptability: The Canelos Quichua of Eastern Ecuador." *American Anthropologist* 80, no. 4 (1978): 836–59.

———. "The Ecuadorian Levantamiento Indigena of 1990 and the Epitomizing Symbol of 1992: Reflections on Nationalism, Ethnic-Bloc Formation, and Racialist Ideologies." In *History, Power, and Identity: Ethnogenesis in the Americas, 1492–1992*, edited by Jonathan D. Hill. Iowa City: University of Iowa Press, 1996.

———. "The Long Durée of Racial Fixity and the Transformative Conjunctures of Racial Blending." *Journal of Latin American and Caribbean Anthropology* 12, no. 2 (2007): 356–83.

Epilogue

Is Environmental Justice in the Andes-Amazon Region Illusive, Elusive, or within Reach?

BARBARA J. FRASER AND NICHOLAS A. ROBINS

The preceding chapters paint a foreboding landscape of the challenges facing indigenous and marginalized peoples in the Andes-Amazon region as they increasingly demand environmental justice. Their integration into both the global economy and national polities has, for centuries now, been characterized by violence, disease, death, exploitation, exclusion, and unremitting encroachment on their environments and the natural resources they contain. The unprecedented demographic collapse brought on by the Iberian conquest not only led to the transformation, and even disappearance, of many indigenous groups, but also often robbed them of the collective memory and wisdom that perished with the spiritual leaders and elders of those societies. Disease, death, and despair were commonly accompanied by forced relocation, labor, and religious conversion, all within the context of social, political, and economic marginalization.[1]

Today the quest for environmental justice in the Andes-Amazon region is fraught with unprecedented challenges. The ability of the state to selectively impose its will is greater than ever. While railroads, repeating rifles, and the telegraph opened many frontiers, governments today are endowed with even more tools. Some of these—bribes and providing desired material goods such as outboard motors and generators—are co-optive, while others are coercive, such as the criminalization of protest, electronic surveillance, and extrajudicial executions. In frontier areas, where natural resources and human vulnerability abound, government policies to protect the environment, indigenous communities, seminomadic isolated groups, and marginalized people often are inadequately implemented.

Contributors to this volume argue that injustices persist throughout the Andes-Amazon region, but we also highlight the gains that have been made along with policies and relationships that could turn these landscapes of inequity into places of just and equitable relationships. These relationships go beyond states that truly see indigenous people and other inhabitants of remote rural areas as full citizens, and include relationships between humans and other beings that make up the varied mosaic of ecosystems that is the Andes-Amazon continuum. The abyss between the Eurocentric worldview, which divides nature from culture, and the indigenous cosmovision, in which other beings are also *gente*, as described by Jonathan Hill, is a fault line running through the Andes-Amazon landscape. Fully recognizing and internalizing this understanding of ecosystem interdependence is crucial to the development of truly intercultural, multicultural, and plurinational societies. The future of the forest and riverine ecosystems, and the people and spirits who dwell in them, depend on the will to do so.

The various cases described here underscore that the concept of environmental justice in the Andes-Amazon region applies not only to individuals but also to communities, especially for indigenous peoples. As Schlosberg and Carruthers note, "By fighting for autonomy, land, respect, or political voice, indigenous activists are fighting for the capabilities necessary for their communities to function fully."[2] In doing so activists choose from an array of tools that was much more limited just a generation ago. Some work within the state system, running for local or national offices. For example, in Brazil in 2019 Joênia Wapixana became the first indigenous woman and only the second indigenous representative to take a seat in the national legislature. Others work through the legal system. Bolivia, Brazil, Ecuador, and Peru have enshrined varying types and degrees of indigenous rights in their constitutions, and Bolivia and Ecuador also recognize rights of Mother Earth. When recourse to national courts is exhausted, groups turn to the supranational Inter-American Court system or the United Nations.

Activists have an increasing number of other tools at their disposal as well. Some are time-tested, such as protests and strikes, although these may have only local and transitory impact and brutal consequences. Others, such as social media, the internet and related communications services, and geographic information systems, offer unprecedented potential to connect people globally. Not only can activist groups form transnational alliances and share lessons they have learned in their quest for environmental, social, political, and economic equity, but they can also connect with and receive financial and other support from organizations and individuals elsewhere in the world. For multinational corporations nowadays, a hazardous discharge at a remote mining site can easily translate into a public protest at a shareholders' meeting with a toxic public relations spillover.

The same is true of government and multilateral funding agencies that support the very projects that perpetuate environmental inequities. Global interconnectivity has enabled the "invisibilized" to get their voices heard, voices magnified by supporters in the North and which penetrate and must be taken into account in previously soundproof technocratic decisions. The global reinforces the local, allowing supranational constituencies to emerge in defense of local communities and their environments.

That these mechanisms sometimes fall short—or that they are necessary at all—underscores the need for better accountability systems, the enforcement of legislation, and the recognition of indigenous peoples and other marginalized groups as full citizens. As Sze and London point out, the definition of environmental justice should include the recognition of a diversity of cultural identities.[3] This implies recognizing, valuing, and incorporating into legislation and policies the relationship that indigenous people have with the natural world, including what the Peruvian anthropologist Marisol de la Cadena calls "Earth-beings," which may be physical features, such as mountains, trees, plants, or rivers, or other beings with which humans in traditional communities maintain relationships.[4]

Full citizenship also implies the right of indigenous people to participate in decision-making processes that affect them and their territories. As MacLennan and Perch point out, that implies providing adequate education as well as reliable information and helping communities develop their ability to negotiate and gain access to political and legal venues for addressing environmental injustices.[5]

The chapters in this book point to steps that the region's governments should take to reduce inequities and increase the effective participation of local communities. Decisions about land use, infrastructure, and other types of development must meaningfully involve the local communities that will be most affected, to ensure that the benefits are equitable and the costs are not disproportionate. That implies evaluation of environmental, social, and cultural impacts of projects at a larger scale than is currently common.

Environmental impact studies still tend to give short shrift to indirect consequences—the deforestation that follows road building and paving, for example, or the migration and new long-term settlements that spring up around dam construction sites. Scientists and environmentalists advocate strategic environmental assessments, which examine the interconnected impacts of multiple projects across multiple watersheds, taking into account not only the project under study but existing infrastructure and projects that are still on the drawing board. The practice is limited, partly because of a lack of political will, but there are indications of the benefits of such a broader, more sustainable perspective in development planning.[6] In assessing impacts of contamination, whether recent or legacy pollution, states must ensure that legislation, regulations, and policies give priority to the health of humans and ecosystems, stressing the precautionary principle and ensuring adequate evaluation of health and environmental risks and impacts.

Equitable inclusion in development planning also implies the need for more widespread implementation of processes for obtaining the free, prior, and informed consent of local communities affected by plans, projects, and land-use decisions. So far, few countries have enacted legislation requiring such consent, and in those that do have laws or policies, the consistency of the processes and the degree to which communities' input is taken into account varies. Questions increasingly arise about who should be consulted, as nonindigenous communities in areas of direct influence of development projects or extractive activities seek the same rights to consultation that indigenous groups have under International Labor Organization Convention 169 and the UN Declaration on the Rights of Indigenous Peoples.

In some countries, nonindigenous local communities are part of a public participation process that includes open hearings, in which government or company officials provide information about projects and those in attendance ask questions, but the prior consultation is reserved for indigenous communities whose collective rights would be affected. Although this arrangement attempts to redress a historic injustice inflicted on the region's indigenous peoples, it disenfranchises nonindigenous inhabitants of forest and riverine communities by keeping them on the margins of decisions that also affect them. Depending on local circumstances, those communities may or may not have greater political, social, and economic power than their indigenous neighbors. A challenge for Andes-Amazon countries in the short term, therefore, is to implement equitable consultation processes that truly protect the indigenous communities' right to free, prior, and informed consent while also considering the views of local nonindigenous residents who also depend on natural resources.

Meanwhile tensions are likely to persist as indigenous groups seek to expand recognition of areas used by their people for generations. Indigenous peoples like the Wampis in Peru are declaring autonomous territory and demanding recognition, but governments have generally brushed aside their arguments. Territorial claims have both geographical breadth and depth, as groups seek control over ancestral territories but often cannot gain rights to resources in those territories, especially subsoil resources like oil and minerals, but also, in some cases, forests. At the same time, they must contend with threats that are common to all countries in the region, including encroachment by small-scale, unregulated gold mining and land grabs by speculators who resell the land, sometimes multiple times, to smallholders who seek better or more accessible areas. Land speculation as a driver of deforestation is reported in most countries in the basin, but comprehensive studies of the phenomenon are few.

Forests fragmented by the expansion of transportation and energy infrastructure throughout the Amazon basin and along the eastern base of the Andes Mountains will be more susceptible to extreme climatic events. The combination of deforestation, periodic droughts, and an upward trend in temperatures threatens to push the Amazon toward a point at which it will no longer be a buffer against climate change.[7] As scientists develop a deeper understanding of the atmospheric connections between the world's tropical rainforests and other regions of the planet, there is growing awareness that implementing policies to move the Andean-Amazon region back from the brink is crucial not just for the people who depend on the forest for their livelihoods and their spiritual and cultural survival but also for those in distant places, especially the ones most vulnerable to extremes of drought, heat, and flooding.

Those extremes threaten to undermine food security and force people to leave their traditional communities and migrate in search of more stable conditions. Along with its impacts on indigenous and marginalized groups, climate change has global implications. People in the North *should* care about what happens in the Andes-Amazon region, for the environmental changes there inevitably affect what they consume, the air they breathe, and the myriad risks posed by global warming and increasing sea levels. Just as the North drives consumerism and consumption, it can also drive policies that favor renewable energy as well as equitable and sustainable development practices.

Climate change is likely to drive a global shift toward the use of renewable energy, with possible positive and negative environmental repercussions for Amazonian peoples. An increased emphasis on hydropower could further jeopardize both ecosystems and people living in areas affected by dams, as described by Philip Fearnside. A shift toward greater use of solar and wind power on a small scale could give indigenous and nonindigenous Amazonian communities access to affordable renewable energy. Large-scale commitments to renewable energy on the part of countries, however, could leave communities open to efforts by governments or companies to appropriate land for siting of production facilities or transmission lines without consultation of those affected and without compensation or reparations for damage.

Environmental upheavals wrought by climate change also drive migration to cities, which is exacerbated by a lack of good education and decent job opportunities in rural areas. This trend creates another invisibilized group, the urban indigenous people who settle, especially, on the peripheries of cities throughout the Amazon. Often they maintain ties with their home communities, moving back and forth between urban and rural spaces. In some cases, the discrimination they face in urban areas prompts a greater effort to assimilate into the dominant society to escape prejudice; in others, it leads to new ways of asserting indigenous identity in urban spaces. This phenomenon also merits further study, as well as a revision of government policies, as indigenous groups demand territorial rights within the urban areas where they have settled, often in cities that have been built over their ancestral territories.

These migrations are the reverse of those that occurred during the rubber boom, when families fled to the more inaccessible headwaters of rivers to escape the violence of the rubber barons and overseers. Despite some official recognition of the atrocities of that era, there has been little attention to their far-reaching consequences over multiple generations, and virtually no acknowledgment of the need for reparations or proposals for making restitution.[8] A first step is recovery of the historical memory of events and processes that profoundly disturbed or eliminated the lives of tens or hundreds of thousands of people, with multigenerational effects that have yet to be studied. In the Peruvian and Colombian Amazon, some young indigenous artists are turning to their own family stories to weave history and history-as-myth into their work, as a way of discovering their own identity, making their invisibilized communities visible, and to preserve the history for future generations. A more concerted effort, with support from both governments and the academic world, would enable the indigenous communities themselves to gather and preserve these stories before the last of the generation that has heard them from the original tellers are gone.

These and other issues raised by authors in this volume point to challenges for the years ahead; those, in turn, will be shaped by political, economic, social, and environmental trends and changes in the Andean and Amazonian countries and in the world beyond. The fate of tropical forests around the planet lies not just in the hands of those who inhabit them but also with the politicians and policymakers in their countries, who often are urban dwellers living far from the ecosystems and villagers affected by their decisions, or consumers in distant lands whose choices have far-reaching social and environmental effects that reverberate through forests and along rivers in the Andes-Amazon region. The problems are multifaceted, and the response must be multidisciplinary, with key roles played not only by the local communities but also by scientists, other academics, and policymakers. We hope that the issues addressed in this book have contributed to a deeper understanding and to the quest for landscapes of equity in the Andes-Amazon region.

NOTES

1. Robins, *Mercury*, 14–15, 24–25, 33–34.
2. Schlosberg and Carruthers, "Indigenous Struggles," 18.
3. Sze and London, "Environmental Justice," 1335.
4. de la Cadena, *Earth-beings*.
5. MacLennan and Perch, "Environmental Justice," 295.
6. Rega and Baldizzone, "Public Participation."
7. Marengo et al., "Changes."
8. See, for example, "Colombia Apology."

BIBLIOGRAPHY

Bebbington, A.J., D. Humphreys Bebbington, L. A. Saulse, J. Roganf, S. Sumali Agrawalg, C. Gamboa, A. Imhofi, K. Johnson, H. Rosak, A. Royol, T. Toumbouroum, and R. Verdum. "Resource Extraction and Infrastructure Threaten Forest Cover and Community Rights." *Proceedings of the National Academy of Science* 115, no. 52 (December 26, 2018): 13164–73.

"Colombia Apology for Devastation in Amazon Rubber Boom." BBC *News*, October 13, 2012. https://www.bbc.com/news/world-latin-america-19931443.

De la Cadena, Marisol. *Earth-Beings: Ecologies of Practice across Andean Worlds*. Durham NC: Duke University Press, 2015.

Lambin, E. F., H. K. Gibbs, R. Heilmayr, K. M. Carlson, L. C. Fleck, R. D. Garrett, Y. Le Polain de Waroux, C. L. McDermott, D. McLaughlin, P. Newton, C. Nolte, P. Pacheco, L. L. Rausch, C. Streck, T. Thorlakson, and N. F. Walker. "The Role of Supply-Chain Initiatives in Reducing Deforestation." *Nature Climate Change* 8, no. 2 (2018): 109–16.

MacLennan, M., and L. Perch. "Environmental Justice in Latin America and the Caribbean." *Climate Law* 3 (2012): 283–309.

Marengo, J. A., C. A. Souza, K. Thonicke, C. Burton, K. Halladay, R. Betts, L. M. Alves, and W. R. Soares. "Changes in Climate and Land Use over the Amazon Region: Current and Future Variability and Trends." *Frontiers in Earth Science* 6 (2018): 1–21. doi: 10.3389/feart.2018.00228.

Rega, C., and G. Baldizzone. "Public Participation in Strategic Environmental Assessment: A Practitioners' Perspective." *Environmental Impact Assessment Review* 50 (January 2015): 105–15.

Robins, Nicholas A. *Mercury, Mining and Empire: The Human and Ecological Cost of Colonial Silver Mining in the Andes*. Bloomington: University of Indiana Press, 2011.

Schlosberg, D., and D. Carruthers. "Indigenous Struggles, Environmental Justice, and Community Capabilities." *Global Environmental Politics* 10, no. 4 (November 2010): 12–35.

Sze, J., and J. K. London. "Environmental Justice at the Crossroads." *Sociology Compass* 2, no. 4 (2008): 1331–54.

Walker, W., A. Baccini, S. Schwartzman, S. Ríos, M. A. Oliveira-Miranda, C. Augusto, M. Romero Ruiz, C. Soria Arrasco, B. Ricardo, R, Smith, C. Meyer, J. C. Jintiach, and E. Vasquez Campos. "Forest Carbon in Amazonia: The Unrecognized Contribution of Indigenous Territories and Protected Natural Areas." *Carbon Management* 5, nos. 5–6 (2014): 479–85.

CONTRIBUTORS

Barbara Arisi holds a Ph.D. in social anthropology and is a lecturer in the Interdisciplinary Postgraduate Program in Latin American Studies at the Federal University of Latin American Integration and a researcher at Vrije Universiteit in Amsterdam.

Juan Pablo Sarmiento Barletti is a Peruvian social anthropologist at the Center for International Forestry Research. He has worked with indigenous Amazonian societies for over a decade, carrying out collaborative research examining their experiences of natural resource extraction and related issues of rights and socioenvironmental injustice. He currently coordinates a multicountry research project that focuses on the participation of indigenous peoples and local communities in multistakeholder forums.

Carwil Bjork-James, assistant professor of anthropology at Vanderbilt University, researches and teaches about grassroots autonomy, disruptive protest, and indigenous collective rights. He holds a Ph.D. in anthropology from the City University of New York and an M.P.P. in environmental and human rights policy from the University of Chicago. His forthcoming book, *The Sovereign Street*, describes how Bolivians revolutionized their political life by taking over public space.

Philip M. Fearnside is a research professor in the Department of Environmental Dynamics at the National Institute for Research in Amazonia in Manaus, Brazil. A permanent resident of Brazil, he has lived in Amazonia for forty-three years doing ecological research. He holds a Ph.D. from the University of Michigan in Ann Arbor. Fearnside's work since 1985 has been organized around the objective of converting the environmental services of Amazonian forests into a basis for sustainable development for the rural population, replacing the current pattern of forest destruction. He has authored over 650 professional publications and over 450 popular and outreach publications on issues related to the environment and development. His papers are available through http://philip.inpa.gov.br.

Barbara J. Fraser is a journalist specializing in the environment, natural sciences, public health, and indigenous issues in Latin America, especially the Andes-Amazon region. Two decades of reporting have taken her from stops on the migrant route in Mexico to an ethnobotanical park at the southern tip of South America, and from glacial lakes high in the Andes Mountains to indigenous communities suffering from oil spills in the Amazonian lowlands. Her work has appeared in *Nature, Science, Discover, Mongabay, The Lancet, Sapiens, National Geographic Online, EcoAmericas*, and other media. She holds a master's degree in environmental studies from Green Mountain College in Vermont and lives in Lima, Peru.

Jonathan D. Hill is a professor and former chair of anthropology at Southern Illinois University and a visiting professor at Vytautus Magnus University in Kaunas, Lithuania. He is the author of *Keepers of the Sacred Chants: The Poetics of Ritual Power in an Amazonian Society* (1993) and *Made-from-Bone: Trickster Myths, Music, and History from the Amazon* (2009). His research interests include ethnohistory, ethnomusicology, and verbal art as performance, with a focus on indigenous Amazonia. He did fieldwork with the Arawak-speaking Wakuénai (Curripaco) of southernmost Venezuela in the 1980s and 1990s and served a three-year term (2014–17) as president of the Society for the Anthropology of Lowland South America.

Anne M. Larson is a principal scientist and the leader of the Equity, Gender and Tenure team at the Center for International Forestry Research based in Lima, Peru. She has lived in Latin America for thirty years and conducts research on multiple aspects of forest and landscape governance policy and institutions, from local to international scales. Current research priorities include opportunities and challenges for forest tenure reforms; women's rights to land in communal forests; and multilevel governance, REDD+, and low emissions development.

Roger Merino is a lecturer in public policy and legal theory at the Universidad del Pacífico in Lima, Peru. He received a master's degree in globalization and international policy and a Ph.D. in social and policy sciences at the University of Bath. His research agenda includes critical approaches to comparative law and policy, law and development, political ecology, indigenous rights, and social and economic rights. Recently he was a visiting scholar at the Institute for Global Law and Policy at Harvard Law School (2016–17).

Felipe Milanez holds a Ph.D. in sociology (political ecology) and serves as a lecturer in the Multidisciplinary Postgraduate Program of Culture and Society at the Institute for Humanities, Arts and Sciences at the Federal University of Bahia.

Nicholas A. Robins holds a Ph.D. in Latin American studies from Tulane University and is a teaching professor in the Department of History at North Carolina State University. He is the author or editor of fourteen books and numerous articles whose research concerns Andean environmental and social history, legacy contamination, and indigenous rebellions. The recipient of three Fulbright awards, a former ORISE Fellow with the EPA, and the founding president of the Bolivian Studies Association, Robins serves as president of the Environmental Health Council (www.ehcouncil.org), a nonprofit 501(c)(3) organization focusing on heavy metal contamination issues in the Andean region. In recognition of his work in this regard in Huancavelica, in 2016 he was awarded a Doctor Honoris Causa degree from the National University of Huancavelica.

Richard Chase Smith is a founder and current director of the Instituto del Bien Común (The Commons Institute) in Lima, Peru (www.ibcperu.org). In 2011 he helped to launch the Coalition for Territorial Security of Peru's Communities (twenty-seven institutions), an effort to document the deteriorating situation of land tenure of Peru's indigenous communities and engage in advocacy to improve it (www.comunidadesdelperu.ibcperu.org). For the past decade he has worked with the RAISG coalition on regional mapping in the Amazon Basin (www.raisg.socioambiental.org). For the past forty-five years he has worked with the Yánesha people in central Peru to preserve their oral history and to map their cultural-historical landscape (www.yanesha.com). He holds a Ph.D. in anthropology and linguistics from Cornell University and was a postdoctoral fellow at Harvard University and a visiting senior scientist at the Woods Hole Research Center.

INDEX

www.ingramcontent.com/pod-product-compliance
Lightning Source LLC
LaVergne TN
LVHW091523200926
848483LV00002B/291